Prot

B & R m E.M.Ey

Henry Oldenburg

Henry Oldenburg
First Secretary to the Royal Society
Copied from the Original Portrait belonging to the
Royal Society in Somerset House
by John Bradley
MDCCCXXXVIII.

From the portrait attributed to John Van Cleefe 1668. Reproduced by permission of the President and Council of the Royal Society.

Henry Oldenburg

SHAPING THE ROYAL SOCIETY

Marie Boas Hall

UNIVERSITY PRESS

OXFORD
UNIVERSITY PRESS

Great Clarendon Street, Oxford OX2 6DP

Oxford University Press is a department of the University of Oxford.
It furthers the University's objective of excellence in research, scholarship,
and education by publishing worldwide in

Oxford New York

Auckland Bangkok Buenos Aires Cape Town Chennai
Dar es Salaam Delhi Hong Kong Istanbul Karachi Kolkata Kuala Lumpur
Madrid Melbourne Mexico City Mumbai Nairobi
São Paulo Shanghai Taipei Tokyo Toronto

and an associated company in
Berlin

Oxford is a registered trade mark of Oxford University Press
in the UK and in certain other countries

Published in the United States
by Oxford University Press Inc., New York

First published 2002

A catalogue record for this title is available from the British Library

Library of Congress Cataloging in Publication Data

Hall, Marie Boas, 1919-
Henry Oldenburg: shaping the Royal Society / Marie Boas Hall.
Includes bibliographical references and index.
1. Oldenburg, Henry, 1615?-1677. 2. Royal Society (Great Britain)–History–
17th century. 3. Science–England–History–17th century. 4. Scientists–
England–Biography.
I.Title
Q143.O4H352002 509.2–dc21 2001053128

ISBN 0 19 851053 5

10 9 8 7 6 5 4 3 2 1

Typeset by Footnote Graphics Ltd,
Printed in Great Britain
on acid-free paper by Biddles Ltd, Guildford & King's Lynn

Preface

It will quickly be obvious that without the publication of *The Correspondence of Henry Oldenburg*, ed. A. Rupert Hall and Marie Boas Hall, this biography could never have been written. To publish Oldenburg's correspondence and thereby to illuminate the business of the Royal Society between 1662 and 1677 was the idea of the senior editor, a task which he had begun a dozen years before I joined him in it in the early 1960s. We worked at first with the aid of a microfilm courteously supplied by the Royal Society except when circumstances permitted us to work directly with the manuscripts, intermittently until 1963, then more regularly, but always in the intervals of other employment. (We also of course used photographs and photocopies where necessary.) When my husband took on the editing of the Newton Correspondence a larger share of the work necessarily fell on me, which perhaps accounts for the fact that I gradually became interested in the personal side of Oldenburg's life. It had been our joint decision as editors to give only factual information in footnotes, leaving interpretation of the material contained in the letters to others. As it took us twenty years to complete our task as it was, this proved a wise decision. We have been gratified over the years that many scholars have made good use of our efforts, as I have myself in numerous articles centred around the history of the Royal Society; these have employed and extended the information in the *Correspondence* and enlarged my understanding of the intellectual world in which Oldenburg moved. I have not attempted to cite in the notes any of the many possible references to Oldenburg in modern articles. For this there are a number of reasons. It would be invidious to cite some and not others, and I could not hope to be exhaustive. If by chance I cited some articles with whose conclusions I agreed and not others with whose conclusions I did not agree it would only have confused my narrative without benefiting the reader or clarifying my conclusions. In any case, my notes are primarily intended to refer to the *Correspondence* as the primary source for quotations, as the letters alone can illustrate *Oldenburg*'s point of view. I have not attempted to refer to the many sources from which over a long life of scholarship I have amassed my knowledge of the background to Oldenburg's life and work except in a few cases, notably in the earlier chapters, where I have become aware of changing interpretations. In many cases my knowledge and interpretation have changed from those displayed in our edited volumes; in other cases inevitably I must have failed to avail myself of information which I could theoretically have acquired but have failed to do so. For this I apologise.

No one has yet chosen to use his correspondence to illustrate Oldenburg's own life, to analyse his personal role as communicator and disseminator of scientific intelligence, to see how it was done and of what value it was at the time, nor to evaluate his importance as a scientific administrator at a time when administration as a profession was in its infancy. I hope that this, which is necessarily partly a biography of the English scientific scene between 1656 and 1677, especially as it touched upon the concerns of the Royal Society, will reveal more clearly than has been done the importance of Oldenburg's work. There is of course a difficulty in writing the biography of someone who was a manipulator of his world (in this case the world of natural philosophy) rather than being a key player in that world: this is that his story easily becomes obscured by that of the greater figures who surrounded him. I have tried to tell Oldenburg's story rather than the stories of Boyle or Newton or Huygens or Hooke, to name but a few of the distinguished natural philosophers with whom he worked closely, tempting though it was to pursue the work of these brilliant exemplars of the seventeenth century scientific revolution. I have, I hope, provided enough references to works where the achievements of Oldenburg's correspondents are detailed to permit the reader to satisfy any curiosity about them which I may have aroused but failed to satisfy.

My thanks go, first, foremost and always, to my husband who has generously allowed me to treat our joint work as if it were my own to quarry at will. (But I must emphasize that the opinions here expressed are my own). He has also kindly read over my text, although most of the material is obviously all too familiar, and been long sufferingly supportive. Next I wish to express my gratitude to successive Presidents, Councils and Fellows of the Royal Society for allowing me access to their archives over five decades and giving permission for the transcription and printing of so much material in the Society's archives; I also have to thank other institutions whose archives we plundered. I am immensely indebted to successive librarians of the Royal Society who have, one and all, made me welcome and cheerfully assisted me in every way possible, especially in the final stages Sandra Cummin and Mary Sampson. I particularly thank Dr P. D. Buchanan whose diligent and patient delvings into London archives made it possible to trace more fully than could have been expected the later life of Oldenburg's children. I thank Professor Richard Popkin for sharing his knowledge of the Dury archives in Zurich and John Thackray for assistance with Grew's bibliography. I am most grateful to all those who, over the years, have sent me offprints of their writings dealing with relevant matters; I hope they will forgive me for not naming them individually, since I despair of compiling a complete list, but many are named in the bibliography. I particularly thank Michael Hunter for sending me a most useful copy of his *Guide to the Robert Boyle Letters and Papers*, Sarah Hutton for a preliminary copy of her article on Lady Ranelagh in the *New DNB* and Lynette Hunter for information about the relationship between Lady Ranelagh and Dorothy Moore. Finally my grateful thanks to Oxford University Press's efficient editors.

Here I should mention one aspect of Oldenburg's existence which has eluded me. There exists in the Royal Society a large and handsome oil portrait dated 1668, said on its label to be of Oldenburg. It has no provenance, records only noting that it was purchased (date unspecified but almost certainly in the later eighteenth century. It was certainly in the Society's possession by 1807). (There is also the charming if not exact copy by one John Bradley, dated 1838.) But why is it said to be a portrait of Oldenburg? True, if Oldenburg had had himself painted it might easily have been his choice to be painted in clerical dress—he was after all, a university graduate with a master's degree in theology, and so entitled to be considered a cleric. The sitter, as one would expect of Oldenburg, is of a serious aspect and shows the characteristic slight frown of the near-sighted which we know he was. That he is wearing his own long hair is perfectly in conformity with the customs of the time, but on the other hand it shows hardly a hint of grey, and in 1668 he was about fifty years old. He was married for the second time in this year, but would he have chosen so serious an aspect for a wedding portrait? Vanity was not one of his obvious characteristics. Then there is the watch which he holds; this is a usual symbol of the time suggesting the brevity of life, again not very suitable for a man about to marry a young wife, and he had not had any connection with the design or making of watches in 1668. The final puzzle is that it was purchased by the Royal Society at some unknown time long after his death. It is a handsome portrait but, sadly, uncertainty about its adscription does not permit one to deduce anything about the man from the painter's interpretation of the sitter.

Finally a note on style. In our original transcription of the correspondence we followed the conventions current, producing a text as near to the original as possible, although not trying to reproduce spacing in letters. Nor, at the request of the publishers, did we use raised letters for abbreviations although we did not spell them out except when not to do so would have been confusing. Here I have seen no reason for excessive fidelity to the text and have expanded all abbreviations and occasionally modernized spelling while retaining the exact reading of the original.

All seventeenth-century historians have a problem with dates in this period when England retained the Julian Calendar (Old Style) while most of the Continent had changed to the Gregorian Calendar (New Style). The English, however, were slowly beginning to start the year in January rather than in March, often employing the system of double years, whereby 15 January 1665/6 signifies what on the Continent was known as 25 January 1666. I have used Old Style dates except where indicated, with both years shown between 1 January and 25 March.

M.B.H

Tackley

1998, 2000

Contents

Introduction

Had Heinrich Oldenburg, born about 1619 in Bremen, died, as so many then did, before his thirtieth birthday, virtually no trace of his life would have survived except for a few fairly obscure academic records. Had Henry Oldenburg, as he had become by 1653, died before his fortieth birthday he would be recorded in English history as a minor German diplomat who came on a mission from the free City of Bremen to Oliver Cromwell, stayed for some time in England and was associated with the still young scientist Robert Boyle, to whom he wrote a series of letters of some interest for the history of science. Because he lived to the end of 1677, dying at the age of about fifty-eight, he achieved a European-wide reputation. During the last fifteen years of his life he was famous in the world of learning as one of the Secretaries of the Royal Society and as the founder of the *Philosophical Transactions*, the first and oldest truly scientific journal, and thus became the first professional scientific administrator. When he died the Royal Society was shaken, the continuity of the *Philosophical Transactions* was for some years put at risk, and the European scientific community mourned his loss. For fifteen years his enormous correspondence had kept the world of learning informed of what occurred in English science and kept English natural philosophers informed of what was noteworthy in European science and learning. By establishing an astonishing network of scientific communication through both his journal and his correspondence (over three thousand letters to and from him have survived in whole or in part, now edited and published in thirteen volumes as *The Correspondence of Henry Oldenburg*) he not only helped to keep the Royal Society functioning smoothly and efficiently but left an enduring monument to his skill and industry, for as Secretary he preserved the Royal Society's papers and letters and by so doing laid the foundation of its now superb archive, a unique source for the history of science since 1661.

Oldenburg's correspondence is also the most important source for the life of this able man who, like his acquaintance Samuel Pepys, forged a unique place in English administrative life in the reign of Charles II on the basis of a university degree, patronage, and exceptional personal ability. Unlike Pepys, he had no useful English family connections (except in the last nine years of his life, after his second marriage) but he was to make personal service and a gift for friendship serve instead, for it must be noted that without the ties he forged between himself and the Boyle family his life would certainly have been obscure and unremarkable. In the event, it was neither. That this was so sheds much light on the development of scientific societies in the seventeenth century, a development commonly linked to the growth of the professionalism of science itself. This it certainly was, but more than that, in the case of the Royal Society it marked the beginnings of professionalisation of its administrative officers. That Oldenburg, initially the tutor of rich men's sons, became not only an efficient and widely acclaimed Secretary of the developing Royal Society but also a scientific administrator with a professional outlook, successfully dedicated to trying to advance its aims and the work of its members, illuminates the process of such professionalization at an early stage of its development. His achievement was recognized throughout the learned European world, whose citizens esteemed him greatly for his indefatigable industry, his enthusiasm for natural philosophy, and his ever cheerful willingness to promote the concerns of his correspondents. Relaying the news of his death to the philosopher Leibniz, a French friend described Oldenburg as alert, hardworking, benevolent, and (highest praise of all) an enlightened citizen of the republic of letters. This he certainly was. His life sheds much light on the intellectual world of the mid-seventeenth century and particularly on that corner of it which encompassed the activities of the promoters and followers of the 'New Learning', more especially of the Fellows, English and foreign, of the Royal Society of London. After all the declared aim of the Society, an aim to which Oldenburg whole-heartedly subscribed, and which he ever strove to encourage, was promotion of natural knowledge.

Finally, it should be borne in mind that he had begun his adult life with a keen interest in theology, and in theological controversy, which kept him in touch with the leading religious controversialists of the age, while at the same time his personal sense of religion typifies much of what was common to educated Englishmen of his times.

PART I

The rise to prominence

I

The slow development of a diplomat 1619–54

Henry Oldenburg was born in the ancient port and commercial city of Bremen, long a member of the Hanseatic League of trading cities and an even older Bishopric. The town had been founded on the right bank of the River Weser about fifty miles from the North Sea, with a New Town gradually developing on the left bank, the two becoming formally joined in the seventeenth century. Both towns possessed fine churches, the Cathedral in the Old Town and the Ansgarii-kirche in the New, the latter, built in the thirteenth century, having a tower over three hundred feet high. As befited a self-governing Hanse town, there was (and still is) a splendid Rathaus where the town council met, in front of which was erected in 1404 a colossal statue; tradition said that while the statue stood the city would remain free. This prophecy was not strictly true, except in the sense that it referred to the free spirit of the inhabitants; it was fulfilled as long as the city resisted its powerful neighbours, the successive Counts of Oldenburg of Lower Saxony, but when it was forced to submit to the Swedish king in 1653 and, two hundred and fifty years later, briefly to Napoleon, the statue remained upright. In spite of Bremen's long history as a bishop's see, in the sixteenth century its inhabitants were much influenced by the teachings of Luther, only to be even more stongly influenced by those of Calvin later in the century. Once predominantly Calvinist, the city developed strong religious links with the Northern Netherlands, with which it had for centuries had strong commercial links.

The Oldenburg family of Bremen seems to have had no connection with the territory ruled by the Counts of Oldenburg being, on the contrary, associated with two Saxon cities which had escaped incorporation into the County of Oldenburg, namely Münster and Bremen. The family appears to have originated in the former, also an independent bishopric, situated some hundred miles southwest of Bremen. Münster remained Catholic in the sixteenth century which perhaps explains why a Johann Oldenburg, possibly Henry Oldenburg's great grandfather, migrated to teach in Bremen, already strongly Protestant, apparently in the mid-sixteenth century. From this time onwards the Oldenburgs were all teachers. There is little remarkable about them except that a Heinrich Oldenburg, grandfather of our Henry Oldenburg, spent the sum of 2000 reichsthaler in repairing the buildings belonging to the Vicaria of Saint Liborius at the Cathedral of Bremen, by then Protestant. In recompense, the Assembly of Canons promised that the income of the Vicaria, nominally one hundred reichsthaler per annum, should belong to this Heinrich Oldenburg and his descendants, as it somewhat precariously did for at least three generations. This Heinrich Oldenburg's son, also Heinrich, in 1608 received the degree of Master of Philosophy from the University of Rostock, another Hanse town on the Baltic, having studied medicine during the preceding four years. He was appointed in 1610 to teach at the Bremen Paedagogium, an evangelical school, moving in 1633 to a Professorship at the new University of Dorpat (now Tartu in Estonia), founded in 1632 by the Swedish King Gustavus Adolphus, where he remained until his death in 1634. He had two children, a son and a daughter, both probably, the son certainly, remaining behind in Bremen.[1]

Although the date of birth of this son, the third Heinrich Oldenburg, is unrecorded, it can be presumed to have been most probably in 1619. For in 1663 he was to record on his first marriage licence that he was 'aged about 43 years' and in 1668 on the licence for his second marriage that he was 'about 50'. What is certain is that in May 1633 he transferred from the Paedegogium to the Gymnasium Illustre, founded in 1528. The Gymnasium was not quite a university, for it taught some fairly elementary subjects such as in England would have been handled by a grammar school, but its later curriculum was not unlike that of a university and it conferred degrees. It had a good number of distinguished graduates like the great Oriental scholar Levinus Warnerus (1619–65) whom Oldenburg was later to recall as a schoolmate.[2] According to the records, the Gymnasium's students were given first 'a massive knowledge of godly learning', then 'a complete mastery of Latin' together with some knowledge of Greek and Hebrew, both necessary for theological students. After this came the seven liberal arts, specifically rhetoric, logic, practical arithmetic, the elements of astronomy and music (presumably the mathematical theory rather than the practice, but that would certainly have been important for a seventeenth-century Protestant minister) and finally they were to have wide reading in theology and moral philosophy.

The whole of Bremen, young Heinrich Oldenburg included, must have been greatly alarmed in 1638 by the collapse of the south tower of the Cathedral, naturally regarded as a portent. But it had no effect on the young man personally who was to receive the degree of Master of Theology on 2 November 1639 for a thesis on the relations between Church and State. This was a highly topical subject during this period when Bremen was continuing the struggle to avoid domination by the Counts of Oldenburg and equally so in succeeding years when as part of the later stages of the Thirty Years War Bremen was less successful in resisting the Swedes. It was a topic too that was to continue to interest the young Oldenburg in later years. The new graduate was all his life to demonstrate that he was indeed a complete master of Latin, which he wrote fluently, was deeply versed in theology and ecclesiastical polity and was competent in the seven liberal arts, showing considerable ability to cope with astronomy and far from elementary mathematics. However, his degree, however well earned, was no passport to a livelihood unless, presumably, he was quietly to follow family tradition into teaching or to enter the ministry. A few months after he received his degree his father died after transferring the income from the Vicaria of St Liborius in trust for 'the learned and promising young man Henry Oldenburg the younger'. The fact that it was specified that the legacy was conferred 'in trust' confirms that the youngest Heinrich, however promising, was not yet twenty-one when the document was signed. Unfortunately, the income from the legacy was never great and in the troubled years to come even after the ending of the Thirty Years War in 1648 proved difficult to collect. But the legacy was perhaps, at least temporarily, sufficient to allow the young Heinrich Oldenburg once he had attained his majority to embark on a journey to see something more of the world following a pattern familiar before and after his time. In 1641 he decided to combine the delights and educative value of foreign travel with the academic advantages of a foreign university. He chose to go to the northern Netherlands, now virtually independent of Spain after their almost eighty-year-long revolt, independence soon to be recognized internationally by the Treaty of Westphalia in 1648 but already tacitly recognized nearly everywhere. Oldenburg must have been influenced by a number of considerations: partly by the long-established commercial and religious ties between Bremen and the Protestant Low Countries, partly by personal friendship between various professors at Bremen and many distinguished scholars at Dutch universities, and possibly partly by family ties. (His sister had married or more probably was about to marry—her son was born in 1644—Heinrich Koch, a relation of a well-known professor of theology at the University of Leiden, Johannes Koch or Cocceius.) Oldenburg set out from Bremen in May 1641 armed, as was customary, with letters of introduction, in his case letters from two of his teachers to G. J. Vossius, a notable classicist and one of the most distinguished scholars of the European world of learning, then teaching in Leiden, although the young man first met him in Amsterdam.[3] In spite of several connections with Leiden and its famous university, Oldenburg tried to settle not there but in Utrecht, whose university had been established only five years earlier on the basis of the former Latin school.

Utrecht is an old medieval city and ancient bishopric. Its citizens had been overwhelmingly converted to Calvinism in the sixteenth century and most of its many fine churches converted to Protestant use. The new university was novel in more ways than one, being the first Dutch university in which lectures were given on the then radical philosophical doctrines of Descartes. These first lectures had been given by Henricus Reneri (Henri Renier) a Belgian who, on conversion to Calvinism, had migrated north to Calvinist Holland. By 1630 he was an early and ardent disciple of Descartes and as the first professor of philosophy at Utrecht after its foundation he promulgated Cartesianism in all his teaching, obviously aided by the near presence of Descartes himself. When Renier died in 1639 his role had been taken over by a pupil, Regius (Henri le Roy) who like his master was in close personal touch with Descartes. It is not fanciful to suppose that the young man from Bremen attended at least some of these lectures, for he was certainly later to show himself thoroughly familiar with Cartesian philosophy and it is difficult to believe that an inquiring young man would not have wished to familiarize himself with the latest trends in philosophy. But not all the academics of Utrecht viewed the new philosophy with favour and in 1641 certain theses presented by a pupil of Regius were to be vehemently attacked by the Rector, Gisbert Voet (Voëtius) which prompted a reply by Descartes himself two years later. Obviously the intellectual atmosphere of the University of Utrecht was lively and stimulating and like most Dutch universities was host to many theological controversies of general interest to serious students. It is not unreasonable to suppose that it was this 'modern' intellectual tone of Utrecht University which drew Oldenburg to try to settle there rather than at the older and presumably staider University of Leiden, although that had long been famous for its learning and many scholars in Bremen had at this time close connections with its scholars. Perhaps he had been particularly attracted by the presence at Utrecht of Antonius Aemilius, Professor of History and Rhetoric and so firm an adherent to Cartesianism that Descartes was to call him the principal ornament of Utrecht. Perhaps he was also attracted because the discussion of Cartesianism led to theological controversy of the sort in which he was to interest himself for many years to come.

At this point Heinrich Oldenburg took a number of irrevocable steps away from the career traditional in his family. First, he abandoned the German form of his name, following Dutch academic fashion to become Henricus Oldenburg and later, presumably after familiarity with English, Henry Oldenburg. But this was as far as he went in adapting to formal academic life, in spite of the expectations of his teachers. Circumstances and perhaps a taste for variety soon led him to relinquish his intention of pursuing his studies in the quiet of the university world, although without extinguishing his love of learning or his inherited gift for teaching. He professed himself happy in Utrecht where he lodged with a pastor, but he found the city expensive, far too much so for his purse. It is indeed difficult to see what source of income he could now have had after his father's death except the uncertain revenue from the Vicaria of St Liborius and this was apparently difficult to collect. In any event, after only three months in Utrecht he determined that he needed some more secure source of livelihood. This, together with a keen desire to travel and perhaps new knowledge of the world gained from his Utrecht acquaintances led him to seek 'some position instructing either the son of a nobleman or the son of some honest merchant—one or the other—with whom it would be possible to set out in turn for foreign parts in order to know the condition of church and state affairs in England, France and Italy' (the subject of his Master's thesis). For this purpose he consulted by letter both Aemilius in Utrecht and Vossius in Amsterdam, obviously hoping for both approval and useful recommendations.[4]

Precisely what happened next is impossible to determine, for here firm evidence stops and for the next twelve years of his life there are few traces from which to infer what happened. His letter book[5] does contain the drafts of many letters, all to young Englishmen, whose tone strongly suggests that Oldenburg had been their tutor. There is no record in his manuscripts of any non-English pupils, so presumably his first choice of nationality had been successful, though the reason for this is not apparent, and he had from the beginning been engaged by English families and succeeded in his task so well that he had then been recommended to their compatriots. It seems that with his successive pupils he travelled extensively in France, Italy, Switzerland and Holland and probably visited England as well, while over the next few years he acquired fluency in Dutch, French, Italian and English, in all of which he later wrote fluently. For his having visited England in the 1640s there is only indirect evidence: he sometimes wrote to former English pupils of having been familiar with their families whom, one imagines, he must have met before being entrusted with the care of their sons, and it is difficult to believe that he could have acquired near perfect English without a visit to England. That he had done so by 1654 we have the word of the poet Milton, Cromwell's Latin Secretary, who was to tell Oldenburg that 'You have learned to speak our language more accurately and fluently than any other foreigner I have ever known', and Milton was familiar with several Germans from the Palatinate and the Baltic regions who had lived long in England.[6] Certainly in years to come when Oldenburg was resident in England no one ever referred to his having a German or any other foreign accent.

That he should have been tutor to English boys is not at all surprising, for many English families had resided in the Netherlands even before the English Civil Wars sent both Parliamentarians and Royalists in turn into exile there. And there were other reasons for residence: for example the court of Elizabeth, sister to Charles I and married to the Elector Palatine, the ill-fated 'Winter King and Queen of Bohemia', so called because of the short time their reign had lasted, mother of Prince Rupert who was to be Charles I's general, was a magnet for many English, naturally especially Royalists. So too was the circle which gathered around her exiled nephew Charles II. Oldenburg may even possibly have there met Prince Rupert, with whom he was to have some connection when in England in the 1660s and 1670s. There is some trace in drafts of letters preserved in his Letter Book, begun in 1653, of connections with leading Parliamentarians, among them Henry Lawrence, later Lord President of Cromwell's Council of State, as demonstrated by several letters addressed to his son Edward (1633–57),[7] who was certainly educated abroad and was to be regarded as a paragon in a famous sonnet by Milton; Oldenburg wrote to him in various languages in 1654 and 1656 and recorded that in the latter year they had called together on several notable people in and around London. While abroad Oldenburg had met many Parliamentarians, notably John Pell and John Dury. Pell and Oldenburg probably met in the 1640s when Pell was successively Professor of Mathematics at Amsterdam (1643) and Breda (1646), going to England to be appointed a mathematical lecturer in London by Cromwell in 1652. In 1654 Cromwell sent Pell and Dury on a diplomatic mission to Switzerland to try to persuade the Swiss not to form an alliance with France; he was to remain in Switzerland for some years, returning to England just before Cromwell's death in 1658.[8] (He must then have made his peace with the Royalists for, after he had taken orders, he was appointed to a good living in Essex by Charles II, but remained chronically poor in spite of much patronage.) He was recognized by contemporaries as an exceptionally able mathematician although he published little. He was to be very friendly with Oldenburg after the Restoration brought him back to England and was to try to protect the interests of his family after 1677. John Dury, with whom he shared the mission to Switzerland, was the son of a Scottish minster, had been brought up in Holland, had visited Elbing in Prussia, Cologne, and Sweden; he devoted his life to an attempt to achieve Protestant unity. In the early 1640s when Oldenburg first met him he was chaplain to Princess Mary of Orange, sister to Charles II, then living in The Hague. He was Oldenburg's 'very dear friend' ten years later and, significantly for Oldenburg's future, married in 1644 a connection of the Boyle family. Oldenburg thus had connections with both sides in the closing years of the English Civil Wars and the beginning of the Commonwealth.

That this was so is supported by the letters to various young men who had probably been his pupils, to judge from the tone in which he addressed them. The only surviving draft of a letter certainly dating from the 1640s of this kind is to Henry Liddell, who had received his Cambridge B.A. in 1639–40 and who died before 1650. He was a member of a Royalist family and had either been sent to travel abroad or, since Oldenburg spoke in his letter of 'the affection by which I was long ago drawn to your parents and and all their children', the acquaintance was initially with the family, either in England or abroad, and Oldenburg thought that a tone of guidance was proper in addressing any very young man, whether his pupil or an acquaintance.[9] A more certain example of Oldenburg's professional activities is to be found in his relations with Robert Honywood,[10] a member of a large Kentish family many of them Royalists, including Michael Honywood, dean of Lincoln Cathedral after the Restoration who in the 1650s lived in Utrecht, where Oldenburg could have met him. (Some members of the family were Parliamentarians, and John Dury's stepson married into this branch.) Robert Honywood travelled with Oldenburg on the Continent, visited Zurich in Oldenburg's company and there met Jacob Ulrich, a friend and correspondent of Dury and like him him interested in the Protestant cause.[11] This young man settled in Holland, joining the Dutch army in the 1650s, visiting England in 1653 and 1654 when

Oldenburg corresponded with him, returning to Holland in 1655 when Oldenburg stayed at the Honeywood's seat in Kent. There were other young men whom Oldenburg taught and conducted around Europe between 1642 and 1653 as there were to be yet more in the later 1650s. Clearly Oldenburg had quickly learned to be adept at pleasing both pupils and their parents, had developed a wide international acquaintanceship with men important in theological and scholarly circles, and become adaptable, linguistically accomplished, diplomatic, agreeable, socially adept and, above all, cosmopolitan.

More definite evidence for Oldenburg's life suddenly reappears in 1653. Then, for what reason is not clear, he returned to Bremen, perhaps temporarily without employment. He found Bremen embroiled in the northern war which continued in spite of the Treaty of Westphalia of 1648 which had officially terminated the Thirty Years War, with Sweden continuing its campaigns against Saxony and ultimately succeeding in taking possession of Bremen in defiance of all international treaties. This put Oldenburg's title to and possession of the income from the Vicaria of St Liborius at risk, so much so that he was forced to petition Queen Christina for confirmation of his title in a letter of which he very naturally kept a careful copy in his Letter Book. This petition was apparently formally successful but in June 1653 he was forced to petition again to request that some (unknown) persons be prevented from 'cutting off a third part' of its income.[12] Again he seemed successful but two years later, no longer in Bremen, he was to petition Charles X Gustavus, Queen Christina's successor, who had recently negotiated a peace settlement with Bremen, for remission of taxes that reduced his income.[13] He put his petition and all subsequent negotiations in the hands of his brother-in-law Heinrich Koch.

Meanwhile Oldenburg had more important business to attend to. By 1651 the Civil Wars in England had been brought to a close by the execution of King Charles I on 30 January 1649. The result was the establishment of the Commonwealth, technically a republic, with some power still residing in the so-called Rump Parliament and a Council of State. Real power in the last instance lay in the hands of Oliver Cromwell, the army's commander. The next year saw the defeat of the Scottish royalists and a violent attack on the rebellious Irish, mainly Catholics, leading to the ousting of virtually all Catholic landowners in favour of Protestants, almost all English, of whom there were already many (like the Boyle family), and the beginnings of the plans to settle Cromwell's Protestant soldiers in the south of Ireland. With internal affairs now fairly settled, Parliament passed the first Navigation Act of 1651; this forbade the importation into England of any goods except in English vessels or those of the country of their origin or production. (This was partly a matter of prestige, partly the staunch belief that there was only a limited amount of wealth to be made by world trade so that when foreigners carried goods to England, the English were being deprived of their fair profit.) The Navigation Act greatly injured the Dutch as the second great mercantile country of the time although as Protestants they were potentially natural allies of England. All mercantile navies suffered, for the enforcement of the act led to the stopping and searching of ships throughout English and adjacent waters. Hence the First Anglo-Dutch War which began in 1652. One of the victims of the resultant conflict and the rules of search embodied in the Navigation Act was the city of Bremen, whose essential overseas trade was now disrupted. Naturally enough, the city very much wished to open negotiations with the English government and so sought an envoy.

There is no means of knowing whether Oldenburg was aware of this when he returned to Bremen early in 1653 or whether he merely swiftly seized the opportunity of entering into the service of his native city. In any case, he sought the appointment of envoy to England and was successful, although not without some opposition. There exists in the Bremen archives an anonymous letter whose author was fiercely opposed to the possibility of Oldenburg's appointment, on the grounds that he was totally unsuited to it, having had no formal diplomatic experience (true) and that he was totally unsuited to a diplomatic life having 'a peculiar temper which prevented him from agreeing well with others' and besides all this had no knowledge of the commerce upon which the prosperity of Bremen depended. It was true that neither he nor his immediate family were merchants, but presumably no one could have grown up in Bremen without some awareness of its commercial life and in fact the point on which the envoy would be required to negociate needed no detailed knowledge of mercantile practice. As for the slur upon his character, his 'peculiar temper', whatever that may have meant, had never prevented him from persuading men of standing, wealth, and some influence in the world to entrust their sons to his charge, while his treatment of these boys must have been satisfactory since others in turn accepted him as tutor to their sons. Moreover he had made many friends in his travels and was to keep the friendship of many of them for years. It is true that in the future he was to make a few enemies, but very few, certainly not so many as to justify the claim that he was unsuited to the role of diplomat, as in fact turned out to be not the case, nor was he temperamentally incapable of 'agreeing with others' as his whole life was to show. Presumably the anonymous author either had some personal grudge against Oldenburg (plausibly in connection with his claims for the income from the Vicaria) or wanted the post for himself. Certainly another of the accusations in this paper was to prove to be quite untrue: this was that Oldenburg would be unacceptable to the English republican government because he had supported the Royalist cause during residence in England. This accusation may have been based upon the fact noted above, that some of the boys whom he had looked after were the sons of known Royalists, that he was acquainted with scholars like Vossius who had benefited from the patronage of Queen Christina of Sweden, and that he had in the Netherlands associated with some English emigres and exiles. But against this it must be remembered that he had been equally friendly with many who supported the Parliamentary side, like Pell and Dury, and that a number of his patrons were Parliamentarians. That he had been in England in the previous dozen years is not at all improbable, but that he would prove unacceptable to the government in power was certainly not the case, as events were to prove. These accusations, fortunately for Oldenburg, did not apparently prejudice the Senate of Bremen against him, and he was soon officially to be appointed 'representative of the City of Bremen' by its Burgomeister and Rat (Council), his letter of instruction being dated 30 June 1653.[14] Although Cromwell was not to be proclaimed Lord Protector until December, he so completely dominated Parliament that even the outside world regarded him as head of state as well as head of the army. Hence the Senate of Bremen instructed Oldenburg to treat directly with 'His Excellency, General Oliver Cromwell', as he was to do.

The immediate reason for this mission from Bremen to England was the capture by English men-of-war of a ship laden with French brandy and Nantes wine belonging to a citizen of Bremen. This was contraband in the eyes of the English as specified in the Navigation Act but normal free trade in the eyes of the owner and the Senate such as all citizens of Bremen were entitled to pursue. According to the formal instructions issued to Oldenburg he was to seek assurance that Bremen's neutrality in the Anglo-Dutch War would be respected in future for 'the commerce of our city depends upon our ships'; he was to endeavour to have the German Hanseatic factory or counter (trading post) at Bergen in Norway included in the neutrality agreement; and he was to assure the English Parliament that the Senate of Bremen would license no ships except those belonging to citizens of Bremen. Since, as noted above, the city of Bremen was suffering from the refusal of the Swedish government to accept the terms of the Treaty of Westphalia which had technically guaranteed Bremen independence, with the result that Sweden was levying contributions on the city and erecting fortifications on the Weser, free trade on the high seas was of peculiar importance to Bremen. But for the moment the Senate did not expect the English to intervene on their behalf. Oldenburg arrived in England in late July 1653 as he dutifully notified the Senate, writing for their benefit in German, a language which he wrote, as he admitted, with some difficulty. (Once he even wrote to a friend in French because, he said, he did not have time to write in German and indeed German of the 1650s was not well suited to diplomatic language and many necessary words had to be written in Latin and in Latin script.) On 29 July he was able to present his credentials to the House of Commons which in turn referred the whole affair to the Council of State. Oldenburg was granted an audience with the Council on 4 August at which time he submitted a written account of Bremen's grievances in Latin. A month later the affair was further referred to the Committee for Foreign Affairs. Discouragingly this chain of events was repeated in the late autumn, beginning on 21 November. It must be realized that at this moment affairs of state in England were in considerable disarray and on 12 December 1653 Cromwell was to dissolve Parliament, to be proclaimed Lord Protector of the Commonwealth of England, Scotland, and Ireland four days later. This in some ways made Oldenburg's mission easier, for less than a fortnight later he was able to appeal directly to Cromwell himself, as the Senate had intended him to do, with a firm statement of Bremen's desire for neutrality and for reassurance. His letter is written in clear and tactful English accompanied by a Latin version; it is not certain whether this was his own English or whether he had received some assistance for this and for the English translation of his credentials. He was certainly capable of making such translations, but possibly Milton as Cromwell's Latin Secretary may have given some assistance. There is no record of any response by Cromwell to Bremen's plea but fortunately for Bremen and its inhabitants war between the English and the Dutch ceased in April 1654, as Oldenburg duly and cheerfully reported to the Bremen Senate. He therefore returned his credentials, at the same time reporting Cromwell's despatch of John Dury and John Pell to Switzerland as a matter of interest to the devout Calvinists of Bremen, who might be expected warmly to support the cause of unity among the Protestant states of Europe for which Pell and Dury were instructed to negotiate. Their mission was also of private interest to Oldenburg who used the occasion to send a letter to Jacob Ulrich, the Zurich divine whom he had met when escorting Robert Honywood some years earlier. He was still in touch with the Honywood family and in the summer of 1654, after the end of his diplomatic activities, he stayed at the Honywoods' house in Kent whence he wrote to Bremen a fairly full account of what was occurring in English politics. He also corresponded with Milton, who addressed him as 'Agent for Bremen to the English Parliament', although, properly speaking, he no longer held that appointment.[15] Their correspondence concerned theological controversy, then rife, and especially Milton's written defence of his writings against violent Royalist attacks, a work of which he sent a copy to Oldenburg for comment.

Oldenburg was soon to be summoned to London by a letter from the Bremen Senate asking whether he would be prepared to act once again as its envoy to Cromwell.[16] The Senators hoped that the Protector might be persuaded to serve as an intermediary between the Swedes and Bremen, for the Swedes were now attacking the city directly, endeavouring to absorb it into the Duchy of Bremen (which they already held), once again ignoring the relevant terms of the Treaty of Westphalia to which they had agreed six years previously: these, although endorsing the Swedish conquest of the Duchy, had been intended to secure its ancient rights to the city of Bremen as a Free City within the Holy Roman Empire. Although Bremen was receiving some help from other German states it would clearly be unable to resist the Swedish army much longer. It was hoped that Cromwell, on behalf of a fellow Protestant state (but Sweden was of course also Protestant) might intercede in some way. Oldenburg tentatively agreed to act as Bremen's envoy once again, only inquiring how long the negotiations might be expected to take and exactly what would be required of him. (It is possible that he was then seeking a post as tutor to some English youth.) Without waiting for a reply he wrote a fortnight later to report that Parliament had met on 3 September and to give a lengthy summary of Cromwell's peremptory speech demanding total loyalty from Parliament now that he was officially Lord Protector. Oldenburg took advantage of the occasion to seek assistance for his private affairs, sending under cover of his letter to the Senate letters to his brother-in-law about the delinquency of his tenant who refused to pay the rent on his house (presumably something to do with the Vicaria but possibly a reference to a house bequeathed to him by his dead father); a week later he wrote again to ask the Senate to intervene to secure the rent when paid to his sister and brother-in-law.

Tentative although Oldenburg's acceptance of the request of the Bremen

Senate was, the latter jumped at the chance of employing him and promptly sent him new credentials: a formal letter to Cromwell as Lord Protector, copies of relevant documents, and, most welcome, the sum of one hundred reichs-thaler for Oldenburg's necessary expenses as envoy. He was instructed to go to London (where in fact he already was) in order to present the Senate's letter and his own credentials to Cromwell personally and to ask not only for the Lord Protector's good offices in arranging a peace between Bremen and Sweden but also promise of active help if a recently agreed armistice should come to an end in November as well as 'a sizeable, indefinite loan'—a great deal to ask for. Oldenburg found it easier to deal with Cromwell direct than with Parliament: he sought and promptly obtained an audience on 20 October 1654, an audience which, as he reported, would have been granted earlier in the month had Cromwell not been thrown from his coach in Hyde Park on 29 September and been forced to rest for a fortnight. Even then, three weeks after the accident, as Oldenburg duly told the Senate, Cromwell was still weak and attending to little business. Oldenburg's petition was favourably received but before taking any action Cromwell wished to be sure that the Dutch would not object and Oldenburg therefore had to seek acceptance of the proposal from the Dutch Ambassador in London, a matter of diplomatic negotiation. In reporting his work Oldenburg was fairly gloomy about the prospects of success, telling the Senate that there was no question of England's offering anything other than moral support at best, military and financial aid being quite out of the question. As he wrote, he rather suspected that nothing at all would happen, especially as the Senate of Bremen had left him to act on their behalf without providing him with any proper funds. The money which had been sent was, he found, quite insufficient 'to represent Bremen with dignity (and dignity does carry weight here)'. As he ruefully wrote, he had been forced to appear before Cromwell, and no doubt before the Dutch Ambassador also without due ceremony and without even a servant to accompany him. As he informed the Senate he should, in consideration of the extreme kindness of both Secretary Thurloe and Sir Oliver Fleming, Master of Ceremonies, 'have presented a barrel of good Rhine wine to each of them, and a couple of good sable furs to their wives' if he had been able to afford it and, as he carefully noted, if the Senate had given him permission to do so. (In seventeenth-century terms, this was not bribery but normal polite practice and even grand officials expected such gifts.) Moreover, as he stated, he had had to give a silver florin (two shillings) to each of the under-secretaries who had given him copies of documents he needed. He did not mention it, no doubt assuming that the Senate would take it for granted, but probably he had had to 'fee' several minor officials in order to gain admission to Cromwell and also to Thurloe and Fleming, for this was normal practice.

Remarkably, and a tribute to his diplomatic skill and personal charm, Oldenburg had some success in his discussions with Cromwell in spite of his lack of money and 'dignity': only a week after Oldenburg's audience with him Cromwell dispatched a strong letter to Sweden (probably, as it was naturally written in Latin, composed by Milton) deploring the strife between two Protestant states, so dangerous for the welfare of the Protestant faith generally, and offering his good offices towards effecting a peace. Oldenburg triumphantly sent a copy of Cromwell's letter to the Senate, mentioning in his own covering letter the cost of securing the copy and the fact that all the money originally sent to him had been expended on 'food, clothing, travel, whether by carriage or by water in such a large city as London' and hinting strongly that he needed more funds. (These he probably never received.) There are no more letters in existence to indicate any further exchange between Bremen and Oldenburg nor any trace of letters from the Senate to Cromwell, Thurloe, or Fleming. Cromwell's letter had no real effect, no doubt nullified by his decision to make an ally of Sweden, and Bremen was soon compelled to accept Sweden's overlordship. Perhaps the Senate felt that, this being the case, Cromwell deserved no thanks, but neglect of formal gratitude was unusual.

There, with mixed success, Oldenburg's diplomatic career ended, defeated by the international situation. Personally he could rest fairly satisfied with his own work. He had proved himself to be a skilful envoy, obeying instructions, getting a favourable reception from Cromwell, showing tact towards the important men with whom he came in contact, getting on good terms with men close to the Lord Protector, not only men like Thurloe and Fleming but also, as already noted, becoming friendly with Henry Lawrence, Lord President of Cromwell's Council of State but, more lastingly, developing a real friendship with Milton, both in his official position and, subsequently, in his guise as a Puritan theologian. (It is worth remarking that there is never any reference in Oldenburg's surviving correspondence to Milton as a poet.) All this was to have important consequences for Oldenburg's subsequent career and his personal development. First, having developed real diplomatic skills he was to use them throughout his life in his relationships with others, generally with the same success he had showed as Bremen's envoy. Secondly, and initially even more directly important for his career, he had gained admission into Milton's personal circle, which included many of the most distinguished and best connected members of the Parliamentary party as well as humbler members, all of whom pursued aims with which Oldenburg sympathized. Among the more important of the former group was Lady Katherine Ranelagh (1614–91), daughter of the first Earl of Cork, to be a lifelong acquaintance, a woman of learning and great charm, personally extremely devout. She had considerable influence in Parliamentary circles and a close friendship with Milton. She was almost universally esteemed for her personal qualities and her piety. As contemporaries noted, the characteristics of her husband, Arthur Jones, second Viscount Ranelagh, were the exact opposite of those of his wife, his character a little redeemed by the fact that he, belonging like the Earl of Cork to a family of early seventeenth-century English settlers in Ireland, was often there overseeing his estates; he died a quarter of a century before his wife, not much lamented in the circles in which she moved. It was through his acquaintance with Lady Ranelagh with whom he soon became friendly that Oldenburg was to advance his career, for it must have been through her that he met her favourite, much younger, brother, Robert Boyle (1627–91).[17] In 1653 he was already passionately devoted to natural philosophy and especially chemistry, highly esteemed by the natural philosophers in Oxford where he was then resident, although not yet known outside this small circle. He was soon to become Oldenburg's patron and, with the passing of years, his very good friend.

About this time Oldenburg met Samuel Hartlib (d.1662), possibly also through Lady Ranelagh or Milton, possibly through Dury or Mrs Dury, the latter of whom, like Lady Ranelagh herself, was a constant correspondent of Hartlib's at this time.[18] He was highly thought of in Parliamentary circles and was enormously active in trying to pursue an immense number of schemes for the benefit of mankind, as he saw it. Like Oldenburg he was a German from a Hanseatic town, in his case Elbing in Prussia, but unlike Oldenburg he had many English relations, his mother having been English, the daughter of an Englishman resident in Elbing, while his father had been a merchant to the King of Poland, the original overlord of Prussia. In 1628 Hartlib settled in London and devoted himself to English affairs. As is often the case with the lives of people in the past it is difficult to understand how he made a living, except that it is known that in the 1640s and 1650s he was the recipient of assistance from the English government in the form of a grant intended to help him to promote his many schemes. These ranged from plans for the improvement of agriculture, the development of new technology, the improvement of alchemy, especially for medical use, to plans for the reform of education in line with the work of Continental reformers like the Czech Comenius and novel means for the dissemination of knowledge. The activities for which he was esteemed by contemporaries and for which he is now remembered brought in little or no income and he spent much money for, as he saw it, the public good, either in publishing books advocating his schemes or in encouraging others, particularly young inventors, to make their ideas public. In association with Dury with whom he was particularly close he planned and in 1648 published a prospectus for his 'Office of Publick Addresse for Acommodations', intended to be an international clearing house for knowledge in all the many subjects in which he was interested. The prospectus includes a list of those subjects which were to be registered so that information about them could be readily obtained: they ranged from 'religion' and 'learning' to 'ingenuities' and many more. Dury's mission abroad fitted the first topic. The second corresponded closely, as Hartlib understood them, to the aims of Francis Bacon for 'The Advancement of Learning'. The third, of very particular interest to Hartlib, was intended to include every kind of invention then practised as well as alchemical discoveries useful for medicine. (Hartlib's son-in-law Clodius or Clod was an energetic alchemist.) There were also plans for a College, Office, or Agency of 'Universal Learning' of which Hartlib was to be the agent, presumably in modern terms the manager.

By 1647 the young Robert Boyle had met and been influenced by Hartlib who reinforced his existing interest in knowledge that'hath a tendency to use', as Boyle himself put it, and was convinced of the necessity of pursuing Baconian natural history (that is, the collection of facts about nature which, it was hoped, would ultimately lead to a true and profound understanding of nature itself). By the time that Oldenburg met him, Boyle was well advanced on his life's work, which demanded a far more sophisticated approach to an understanding of the natural world, while still being interested in Hartlib's projects, all of which required cooperation from the like-minded, just as Bacon had foreseen. In 1655 Hartlib was to publish *Chymical, Medicinal, and Chyrurgical Addresses Made to Samuel Hartlib*, which he must have been working to collect when Oldenburg first knew him and which, incidentally, includes Boyle's first published essay, 'an invitation to Communicativeness'. This, in furtherance of Hartlib's beliefs, argues for openness in scientific work which, he argued, would lead to the public good in contrast to the secrecy practised by alchemists and, often, medical practitioners, secrecy intended to gain profit for the discoverer. Two years later, in 1657, Boyle, Dury, William

Petty, Benjamin Worsley and others of Hartlib's proteges and sponsors (all, except Dury, associated with Ireland) petioned the Council of State for money 'for the Advancement of Universal Learning' in accordance with Hartlib's 1648 scheme.[19] Money from forfeited estates in Ireland (forfeited in the wake of the Irish rebellion) was set aside and a committee set up to administer the funds by the Council of Ireland, but Worsley and Petty, already at odds over their respective share in the survey of these and other Irish estates, quarrelled bitterly and Hartlib was left, as he sadly noted, without the promised money.

Oldenburg soon became very friendly indeed with Hartlib and was, initially, much influenced by him. It is very probable that Hartlib provided Oldenburg's first introduction to Baconianism in its simplest form, that which stressed the importance of understanding the work of practical men and the necessity for collecting both histories of trades (accounts of manufacturing and craft processes) and, even more important for Oldenburg's later interests, of accounts of the natural history of various parts of the world leading to 'a universal history of nature' such as he was to strive constantly to collect in later life. He was initially to show himself particularly receptive to Hartlib's desire to collect facts, facts about nature, facts about what went on in theological, political, and social circles throughout Europe, and later more remote parts of the world, facts about inventions and advances in agriculture and medicine, as well as advances in the invention of mechanical contrivances. In 1656 he undertook to translate some work (nature and language unspecified) into English on Hartlib's behalf, of which however no more was heard.[20] From 1657 to 1660 he and Hartlib were to be indefatigable correspondents, and Hartlib then, and until his death in 1662, acted as Oldenburg's postal agent, his address being King Street, Westminster or Charing Cross, half-way between the City of London and the seat of government in Westminster.[21] Altogether Oldenburg's meeting with Hartlib was to be of great influence on his future life and work even though, as a result of his close association with Boyle after 1656 he became more intellectually sophisticated than he in many areas. But although his became a more complex view of Baconianism and the proper study of the world of nature than any Hartlib ever held, he never neglected that life-long devotion to the collecting of facts which might permit the writing of that universal natural history which was to become his goal.

2

Learning the art of scientific communication 1655–61

With his diplomatic mission finally concluded, Oldenburg was clearly at a loose end. He temporarily retired to the country, probably directly to the Honywoods' house in Kent where he had certainly arrived by May 1655. From there he wrote, or at least drafted, a letter that month addressed 'To the Right Honble ye Lady' [*sic*] plausibly but not necessarily to Lady Ranelagh.[1] This letter reveals that at Petts he was in company with others and that there was much religious discussion amongst them. For almost the only time in his life he may then have experienced religious doubts, but it is more likely that the doubts he expressed in his letter about the nature of that 'Spirit' from which the human spirit is born related rather to doubts about religious practice and the custom in some sects of expecting a clear, personal revelation of religious conviction by a descent of the Holy Spirit than to any real religious scepticism, for he was to remain deeply religious throughout the rest of his life.

At Petts, where he seems to have spent nearly a year, Oldenburg had time to think about his future. First, as already recounted, he tried to clear up the remaining problems connected with the income from the Vicaria in Bremen as he was advised to do by a Bremen correspondent. He wrote frequently to his former pupil Robert Honywood, now in Holland, during the succeeding year, as he had done for some time.[2] At first these letters contained mainly moral and educational injunctions but gradually, no doubt as he realized that the young man was maturing, he added political news. He also maintained contact with various learned men whom he had met in previous years, his letters to them revealing some tantalizing glimpses of both his past and his future life. He wrote a very revealing letter in June 1655 to the philosopher Thomas Hobbes (1598–1679)[3] which suggests that the two men might have met in Paris as early as 1646. Hobbes was then living there (as he did from 1641 to 1652 during the period when he published his most famous and influential works on political philosophy, including *Leviathan*, 1651), and also, it is said, acted as mathematical tutor to the future King Charles I, and Oldenburg could well have been there also with one of his pupils). Oldenburg did not approach Hobbes about political or philosophical issues, but rather with a question about mathematics, a subject in which Hobbes was passionately interested although he never really understood its fundamentals. (This was to be revealed when, shortly after this, he began to engage in a seemingly endless dispute with John Wallis, Savilian Professor of Geometry at Oxford from 1649, a dispute in which Wallis scathingly and fiercely exposed Hobbes's many errors, although Hobbes could never admit them.) Since Oldenburg shows no qualms about approaching Hobbes, it is reasonable to suppose that they must have been on good and probably friendly terms. What is significant about this letter is not so much Oldenburg's acquaintance with Hobbes as the fact that it contains his first recorded attempt at scientific correspondence and foreshadows his remarkable ability to extract information from one correspondent for the benefit of another. Here he asked Hobbes for information about books which contained discussions of the possible usefulness of mathematics, saying that he wanted the information for a friend. Though he did not name this friend, it is very likely that he referred to Robert Boyle (whose name was probably quite unknown to Hobbes) for at this very time Boyle was engaged in writing a series of essays on the usefulness of various disciplines, later to be published in two volumes as *Some Considerations Touching the Usefulnesse of Experimental Naturall Philosophy* (1663 and 1671). (An essay on the usefulness of mathematics appeared in the second volume, although composed much earlier.) Oldenburg, here approaching a correspondent privately, already used the methods of eliciting information he was to employ in future years when acting officially as the by then world-famous Secretary of the prestigious Royal Society of London.

About this same year Oldenburg wrote, or at least drafted, a decidedly curious letter[4] to a young man already in his mid-twenties who cannot by any stretch of the imagination be thought to have been his former pupil. This was addressed to Philip Howard (1619–94), ultimately a Cardinal, already a notable member of England's leading Catholic family and a prominent member of the Dominican Order on behalf of which he had come to England to raise money. Oldenburg had apparently met him recently in London but must have known him earlier for he spoke of 'the affection I formerly bore you'. The tenor of the letter (about the problems involved in reconciling the Protestant and Catholic Churches) seems curiously simple-minded as addressed to a mature and responsible Dominican, but perhaps Oldenburg was relying on old acquaintance and had forgotten how many years had passed since that time. Not surprisingly, there is no trace of any reply. A few months later Oldenburg wrote to William Cavendish (1641–1707), ultimately the first Duke of Devonshire, to whom he probably had acted as tutor in the past.[5] (Curiously, Hobbes had been tutor a generation earlier to this young man's father.) This letter resembles several others written about this time, including some to Edward Lawrence, now no longer Milton's pupil but acting as amanuensis to his father, the Lord President of Cromwell's Council of State with whom Oldenburg had recently had dealings and here he spoke of having seen much of the young man in London.[6] He also wrote to Thomas Sherley (1638–78),[7] the son of a royalist and later a successful physician, probably also one of Oldenburg's former pupils. The tone of all of these letters is that of an older man (Oldenburg was now in his late thirties) to young friends, stressing the importance of keeping up their learning and leading pure and moral lives. This was natural enough, and moreover the tone makes it impossible not to believe that in some way Oldenburg hoped to remind the recipients and their families of his existence in case his services as a tutor might be welcomed in looking after other young gentlemen of their acquaintance. That he was a little reluctant to return to his former life, as seems to have been the case, is hardly surprising for he had now experienced two years of freedom from supervising the young, presumably, had certainly made many important friends, and greatly enjoyed England and varied English society. But at the same time he obviously badly needed paid employment, and could hardly expect to spend more years with the Honywoods at Petts, delightful although his experience there had been.

But although, as he put it, he 'had once both thoughts and occasions for another course of life'[8] (what this might have been he did not specify) obviously no opportunity had presented itself for the desired change. So, needing employment and, as he discreetly put it to Lady Cork, sister-in-law to Lady Ranelagh and Robert Boyle, persuaded by 'the exemplary piety of Lady Ranelagh and the singular probity and worth' of her brother, together with some unspecified cause for gratitude to them both (had they perhaps assisted him financially in this difficult period?) to be prepared to undertake service under the Boyle family. Indeed he was soon to show himself eager to do so, entering into complex negotiations with the Earl of Cork (1612–97/8), the eldest surviving member of the Boyle family (as well as the richest) with a view to undertaking the guidance of his two sons on a foreign tour. These plans fell through, possibly because the boys were not then in good health, but by April 1656 he had had better luck with Lord and Lady Ranelagh.[9] It was agreed that he should act as tutor to the Ranelagh's son Richard Jones (1640/41–1712) whom he was to shepherd around England and the Continent for the next four years. Jones was regarded as a promising youth and he had even received some instruction from his mother's great admirer, Milton. As it turned out he and Oldenburg got on well and enjoyed their travels, so that it was fortunate for Oldenburg that the prospective post with the Cork boys had fallen through. More important for Oldenburg's future, this seemingly routine post was to bring him into intimate contact with men he might never otherwise have met. Even more fruitfully, he then began service for Robert Boyle who was thenceforward to be his life-long patron, his principal task initially being to purvey to him scientific news from the European scene. In order to do this successfully, Oldenburg had to learn to seek information about men, books, and ideas of which otherwise he would probably have remained ignorant. Moreover, in the course of his subsequent travels with Jones Oldenburg was to learn those arts of correspondence which were later to serve him so well and to be of inestimable benefit to the intellectual society in which he was later destined to move.

The first step in his guidance of Jones was to leave London for Oxford, arriving there by the end of April 1656, where he was to enjoy a rich intellectual life, educational for himself as well as his pupil. He found pleasant lodgings near Wadham College which made it easy for him to renew his acquaintanceship with its Warden, John Wilkins, whom he had met previously in London.[10] (Wilkins was high in Parliamentary circles, having married Cromwell's sister). Oldenburg enrolled himself in the University in June 1656 or, as Anthony À Wood was to put it, 'entred as a Student', his matriculation certificate designating him 'nobilis Saxo'. (This was an unwarranted piece of self-aggrandizement as far as the nobility was concerned but a citizen of Bremen might justifiably, if he wished to do so, call himself a Saxon).[11] Here in Oxford he awaited the conclusion of his negotiations with the Boyle and Jones families, possibly with some trepidation; meanwhile, though he did nothing more about his enrollment in the university, he much enjoyed the intellectual delights available to him. For Oxford in the spring of 1656 had recovered pretty completely from the traumatic events of the 1640s: fire in 1644 followed by plague, siege, and devastation by the armies of both the Royalists who for some years held the city and the Parliamentarians. It was by now an exciting place for someone of Oldenburg's ability who had access, as he did, to the intellectual circle gathered around Wilkins, and he was to take full advantage of it.

Settling in Oxford, even with a new pupil in tow, gave Oldenburg new friends, new interests, a new way of life and much intellectual novelty in the course of the next year. He was of course by any standard now a mature man, had travelled widely and mixed with a great variety of social and intellectual circles. He had retained, as he was to do throughout his life, his early interest in theology and was on friendly terms with many of the most up to date controversialists and their writings, as Milton with whom he continued to correspond had ample opportunity of learning, while it must be remembered that academics were almost all clergyman and very many of them wrote on theological topics. Oldenburg had a notable talent for entering into the concerns of men of very varied interests and had shown the ability to use his international contacts to obtain information about men and books, an ability which he was to employ to the full in years to come. Even in England, where he had so far been only a visitor, he had useful connections: a recommendation from Cromwell's Latin Secretary could not fail to be helpful in Parliamentary circles, while, as already noted, the Boyle family had influence with both Parliamentarians and Royalists. His entry into the intellectual academic circles of Oxford was not in any way adversely affected by his status as tutor to a young nobleman entered as an undergraduate. Only a few years earlier one of the Oxford natural philosophers[12] had written to a well-to-do friend about a similar case, speaking of 'a speciall friend' of his who was 'an excellent scholar, philosopher, Mathematician' and who needed employment because he had just finished seeing a young gentleman through the university. The writer evidently viewed this as a perfectly honourable way for a learned man to earn his living. Like this man, Oldenburg had a respectable academic past, being a Master of Theology from Bremen and having studied at the University of Utrecht, even if that had been for so short a time. Moreover it must be remembered that a university degree automatically conferred the title and the status of a gentleman upon a man of any social origin, as had been the case in England since as early as 1583 when it had been noted[13] 'as for gentlemen, they be made good cheape in England. For whoever studieth the lawes of the realm, who studieth in the universities, who professeth liberall sciences . . . and will bear the port, charge and countenance of a gentleman, he shall be called master [that is, Mister] for that is the title which men give to esquires and other gentlemen, and shall be taken for a gentleman.' And although in the sixteenth century this may have been a complaint against those of humble birth who rose in the social scale by means of learning (as had occurred ever since the Middle Ages), by the mid-seventeenth century it was mere established fact that professional men were, *ipso facto* gentlemen, whatever their social origin.

Oldenburg's position in university society was strengthened by his having such unimpeachable connections as he had with some of the best minds of Parliamentary circles in London, and perhaps also by the fact that he was intimate with many foreign theologians of note. Certainly he was well received, to his great pleasure. As he wrote proudly to his close Dutch friend Adam Boreel of Amsterdam[14] himself a former Oxford student, 'I was not only summoned to Oxford in a creditable way, but am pressed to remain here.' And this welcome continued to be true as he settled in with his pupil, to become speedily well integrated into Oxford life. He can have had nothing to do with ordinary undergraduates preparing for the requirements of the university's Bachelor of Arts degree, for this was all done by College Fellows. His instruction of Jones, much the same age as that of most undergraduates, may have been similar although Jones was very possibly better prepared than they were. Whatever his precise duties, clearly Oldenburg found time to join the company of those Oxford residents, mostly connected in some way with the university, who pursued research in natural philosophy and Biblical languages at the highest level. While still greatly interested, as he was to continue to be throughout his life, in theological controversies like those he discussed in his letters to Milton and Boreel at this time, Oldenburg found himself enormously impressed by the novel approach to learning of his new friends, more novel to him even than that of the Cartesians whom he had met fifteen years earlier in Utrecht, though he must have noted that they were all familiar with the ideas of Descartes. To Boreel he boasted of his campanionship with 'some few men who bend their minds to the more solid studies' and 'are disgusted with Scholastic Theology and Nominalist Philosophy' being 'followers of nature itself and of truth, and moreover they judge that the world has not grown so old, nor our age so feeble, that nothing memorable can again be brought forth.' This last comment would seem to mark Oldenburg's introduction to the strain of Baconianism then fashionable in the Oxford circle, which was far more concerned with the latest forms of experimental and theoretical natural philosophy than was Hartlib's simpler, more practical Baconianism.[15] This reinforced the generally optimistic tone of the natural philosophy current at Oxford; it was an ideology which Oldenburg accepted eagerly and which was to inspire him for the rest of his life.

The Oxford 'companions' of whom he wrote were the now well known group, all of them proponents of what they called the 'New Philosophy', who had met regularly for some years at Wadham College under the aegis of its Warden John Wilkins, appointed in 1648. Wilkins, like most of the other men, was an ardent follower of Copernicus, Kepler, and Galileo, as shown in his numerous popular works. From Wilkins and the others Oldenburg learned what acceptance of the 'New Philosophy' entailed. In astronomy it meant accepting Copernicanism, Tycho Brahe's denial of the existence of material celestial spheres, some knowledge of Kepler's work on planetary motion (and Ward went beyond mere acquaintance) and familiarity with Galileo's telescopic discoveries (the terrestrial nature of the moon, the starry composition of the Milky Way, the existence of Jupiter's satellites, the phases of Venus, the existence of spots on the sun). In mechanics it meant knowledge of Galileo's work on falling bodies and Descartes' mechanical conception of the universe as being based upon matter and motion together with his laws of impact motion and inertia, whether the details of his system of philosophy were accepted, as they were by some, or not. At the same time it meant an understanding of Bacon's ideas on the importance of experiment and of compiling natural histories in order to advance the understanding of nature generally. Many of those then in Oxford had participated in similar discussion meetings begun in London as early as 1645, meetings now generally agreed to have been the first origin of the Royal Society and which continued throughout the 1650s.[16] At that stage Oldenburg was apparently not aware of them which is not surprising since his particular mentor Boyle was not yet involved. By 1649, those now in Oxford had formed a 'Clubb' which by 1652 consisted of about thirty persons each of whom, as Seth Ward, a prominent member, was to put it,[17] had 'gone over' some aspect of 'naturall philosophy & mixt [applied] mathematics' collecting 'onely a history of the phenomena' from books available from 'our library'—that is, surveying the current state of natural philosophy in individual topics, and, what is also significant, sometimes 'trieing experiments'. Ward noted that there were also smaller groups concerned with special topics: some eight persons created a chemical laboratory and were making experiments.

Others had been building 'a slight observatory' at Wadham and equiping it with astronomical instruments. Yet others were concerned with anatomical and physiological investigations, deriving often from Harvey's earlier discovery of the circulation of the blood. This Club, which was to continue in only slightly altered form for almost the rest of the century, had formal rules for admission by ballot, the exaction of admission money to pay for instruments, and specific duties in fulfilment of which all members were required to read a paper or perform an experiment when called upon to do so. The topics dealt with were widely conceived, ranging from the highly theoretical to the practical: in 1655 a visitor writing to Hartlib noted with satisfaction that although these men were all what would now be called high-powered scientists they saw natural philosophy as being directly relevant to 'Husbandry' in all its aspects in a manner very close to Hartlib's interests.[18] This visitor, like other curious visitors to Oxford in the 1650s including Oldenburg, was much impressed by the glass beehives kept by Wilkins in the garden of Wadham College which permitted a careful study of the habits of bees. On the other hand his own tastes and those of Hartlib led this visitor not to expatiate on any more complex subjects discussed by the group nor on the diversity of their tenets of scientific method. Probably all the members agreed with Wallis both about the supreme importance of the broad Baconianism that overrode any other ideological differences between individuals and about the immense importance of Galileo's work in directing natural philosophy in the way it should go. Not unnaturally there were important doctrinal differences of opinion on various subjects, notably on the nature of matter and the causes of the physical properties of bodies, some of the Club being Cartesians and believing in the infinite divisibility of matter while others were atomists, although all agreed that the properties of matter arose from the motion of the parts of bodies.

As it happened, most of the members of what was to become the Oxford Philosophical Club were to be Oldenburg's associates after the Restoration as founders of the Royal Society. Although only Wilkins and Boyle (who joined the group a year before Oldenburg's arrival in Oxford) were known to him beforehand, friendship with these two important members sufficed to introduce Oldenburg favourably and quickly into the society of the club. The best known of the original group were fewer than ten in all. The leaders were Wilkins, the eldest (1614–1672) Warden of Wadham since 1648, John Wallis (1616–1703), appointed Savilian Professor of Geometry in 1649, an excellent mathematician who had served Cromwell by decoding enemy documents, who was to become the patron of younger promising mathematicians, Seth Ward (1617–1689), made Savilian Professor of Astronomy in 1649, a highly original theoretical astronomer, later successively Bishop of Exeter and of

Salisbury (Sarum), unusual in being a staunch royalist, also a patron of younger scholars who was later to befriend Oldenburg in times of trouble, Dr Jonathan Goddard (1617–75) Warden of Merton College since 1651 and by the time that Oldenburg met him Gresham Professor of Physic in London, physician and chemist. All these were Cambridge men and, except for Ward, had impeccable Parliamentarian connections and all, except for Ward and Boyle, who had no direct university affiliations, 'intruded' into their Oxford posts in place of dismissed Royalists. Among the younger men were Thomas Willis (1620–75) soon to be noted for his work in anatomy and physiology and later chemistry, and Robert Boyle (1617–91) who had been invited to join the Oxford circle by Wilkins and who settled in Oxford in 1654 highly esteemed by his colleagues although they found his theoretical, physical approach to chemistry puzzling when he tried to explain chemical phenomena by means of a corpuscular theory of matter and not yet much known outside this group since he had published as yet virtually nothing on natural philosophy. Younger still were Christopher Wren (1632–1723), at this time a Fellow of All Souls College regarded as a 'wonder youth' for his work in mathematics but also active in anatomy and physiology although not yet in architecture, and Richard Lower (1631–91) of Christ Church who was working with Wren and Boyle on experimental physiology, and Robert Hooke (1635–1703), also of Christ Church, laboratory assistant to, successively, Willis in chemistry and Boyle in pneumatics before becoming a distinguished experimental natural philosopher in his own right. There were of course many lesser men who were drawn into this circle and with whom Oldenburg was to work in the future, as he probably also met at least some of the many distinguished orientalists whom he certainly knew later. It is worth remarking here that while all the older men were clergymen, this was by no means the case with the younger men.

It is no wonder that Oldenburg found it exciting to be included in this circle, to be allowed to listen to their reports and discussions, to witness their experiments, to be thereby introduced to what was for him a whole new world, involving the investigation of nature by non-traditional means, the pursuit of the 'new philosophy' in contrast to the old academic philosophy of his earlier education, probably not much corrected even during his short stay in Utrecht. To his credit, he saw its interest and importance at once. And although he himself never became an original natural philosopher, he entered as fully as possible into this new world, becoming a virtuoso, one keenly interested in the activities of those who performed the work of the investigation of nature. He came to understand thoroughly the aims and importance of what his contemporaries were doing in natural philosophy and even to know and understand enough about it to assess the worth of what a wide variety of men were trying to do. Moreover he soon found that he could serve their needs to some extent and devoted himself to doing so. It speaks much for his engaging personality that he was so quickly and whole-heartedly accepted into this new world of natural philosophy and for his intellectual flexibility that he could adapt to it. For although his London connections were good and Boyle was already his patron, still his previous interests had been, as noted, primarily theological even in his relations with Hartlib's circle. How quickly and easily he broadened his interests is demonstrated by the fact that it was not because they were learned clergymen that Oldenburg esteemed his new colleagues but because they were dedicated natural philosophers. And it was to their Philosophical Club that he was proud to be admitted during the next twelve months while there is no evidence that he took any part in the theological controversies then raging in contemporary Oxford, although he could still concern himself a little later in such matters when writing to Milton. He was never to lose the strong, even passionate conviction which he then acquired of the importance of investigating nature nor of the belief that this should be done in a broadly Baconian manner, combining experiment, observation, and theory, as well as collecting facts.

It was soon to be apparent that Oldenburg's closest connection was to be with Boyle as his patron, who now began to employ him as literary assistant, collector and purveyor of information, and publishing agent. But it was not only because it was to his advantage to be so that he became interested in what engaged Boyle: on the contrary he quickly came to appreciate Boyle's aims and soon accepted them as central to the pursuit of natural philosophy in which he now took the genuine interest which was to remain with him in years to come. In response Boyle overcame his determined reticence and made Oldenburg even more familiar with his writings and ideas than he had done the members of the Philosophical Club. He even treated him as a sufficiently close associate to be allowed to see what he had written on various aspects of natural philosophy, work about which Boyle was at this time still inordinately shy, even secretive. It was during this year at Oxford that Oldenburg read some of Boyle's unpublished work. For example he must then have read Boyle's essay on 'Poisons as Medicines', never published in its early form but to be utilized by Boyle in the first volume of his *Usefulnesse of Experimental Naturall Philosophy* of 1663, for he both referred to it in a letter of 1657[19] and then or later copied it into his *Liber Epistolaris*[20] together with a short essay which was to be expanded into Boyle's famous *Sceptical Chymist* only published in 1661. (These copies are to be found at two points in the *Liber Epistolaris* on pages which lie between copies of letters dated 1660, but since nearby is entered a summary of a book published in 1653 and in any case Oldenburg used both ends of the book at the same time, the dating is clearly subject to uncertainty. However, it seems most likely that he copied these works during his Oxford year because by 1660, the next date at which he could have seen the originals, Boyle was already preparing the full version of *The Sceptical Chymist* for publication). All this was excellent preparation for the work Oldenburg was to undertake for Boyle during his subsequent travels on the Continent when, among other things, he was to search out chemists and endeavour with some success to extract information from them for Boyle's benefit.

In his year at Oxford Oldenburg was by no means entirely free to savour the intellectual delights of the university and the Philosophical Club, for he had his living to earn. Evidently his supervision of Richard Jones was satisfactory for there was no suggestion that it might stop. For some months there was still the further possibility that the Cork boys, Charles and Richard Boyle, might also become his pupils once again. But the negotiations dragged on inconclusively until ultimately their parents chose Peter du Moulin, also a protege of Robert Boyle, a man much older than Oldenburg and a Cambridge and Oxford Doctor of Divinity; he was to look after them while they attended the university. There were still vague plans for all three young men to be taken together on a Continental tour, the standard form of finishing the education of upper-class young men at this time, one which had been experienced by their uncles. However, this plan fell through—Oldenburg was not the only person hoping to act as tutor to the boys on the Grand Tour and indeed it is clear that someone, presumably Walter Pope who was eventually to be chosen for the position with the Boyle boys, resented Oldenburg's competition for the post, no doubt regarding him as a pushing foreign upstart.[21] (Pope, Wilkins's half brother, was some years younger than Oldenburg; he was then at Wadham as a disciple of Seth Ward and after 1660 was to become Gresham Professor of Astronomy and thus clearly more deeply versed in natural philosophy than Oldenburg could claim to be.) Partly apparently because of this clash of personalities which he clearly found distressing[22] Oldenburg determined to leave Oxford for the Continent in early May 1657, having obtained parental permission to take Jones there alone. And so they left, Oldenburg laden with instructions and requests from Boyle, after arranging with Hartlib to act as postal agent for incoming and outgoing letters.[23]

By the end of June Oldenburg and Jones had crossed to France, paid a brief visit to Paris, without which no journey to France would have seemed complete to an Englishman, and settled in Saumur on the left bank of the Loire, downstream from Tours.[24] Saumur was then a very large city for the time, containing perhaps as many as 25000 inhabitants; it was an important Protestant centre as it had been since the beginning of the previous century.[25] The bitter religious wars of the sixteenth century had ended with the conversion of Henry of Navarre, formerly head of the Protestant party as well as next in line to the French throne, to the Catholic faith with the famous declaration that 'Paris was worth a Mass'. His aim was, first, to be accepted by the Catholics as Henri IV, King of France, and second, the establishment of peace and security, both to be consequences of his conversion and succession to the throne. As King he faithfully observed the Catholic faith but he never forgot his erstwhile Protestant colleagues and subjects and worked diligently to protect them. Foremost of these attempts was his promulgation of the Edict of Nantes in 1598: by this Edict the Catholic faith was to be re-established in all areas where it had been suppressed by the then dominant Protestants in previous years, but Protestants were given the freedom to live everywhere in France, to exercise their religion wherever it had existed before 1598, to maintain churches (*temples*) in the suburbs of all towns, cities, and villages in which they had existed in 1598, and to retain a number of fortified places and cities which they had held at the same date. This was not quite full religious freedom, for Protestants were not allowed to build new churches nor to worship openly except in designated areas, but it went a long way towards it and seemed likely to prove acceptable. And so it was to Protestants, but not to the Catholic majority and after the assassination of Henri IV in 1610 they began actively eroding Protestant rights wherever and however they could. Saumur felt the effects of this change, for its handsome chateau overlooking the Loire had been newly fortified by 1598 and by Oldenburg's time it was no longer a Protestant stronghold, having been lost to Catholic forces after a siege by the army of Louis XIII lasting twenty-six days. Indeed the Protestants had lost all of their fortified places by 1620. But Saumur remained predominantly Protestant and retained a number of privileges granted in 1598, one of which was the right to maintain its Protestant Academy, the most famous of those founded by the national Synod of Protestant clergy held at Montpellier (also long a Protestant stronghold) in 1598. These Synods, of which that of 1611 was held at Saumur and that of 1659 at Loudon only twenty miles away, were the supreme governing bodies of the combined Protestant churches to which the regional synods all sent delegates. Originally it had been stated that they could be held at the discretion of the members of the Protestant congregations but this was one of the privileges which had been eroded by mid-century and national synods were held only by express permission of the government and were correspondingly valued.

Of particular concern to Oldenburg was the educational structure which the Montpellier synod had been at great pains to establish. At the base were schools to provide elementary instruction, so essential to Protestantism with its emphasis upon the importance of reading and studying the Bible. Above these were colleges, not unlike English grammar schools of the time, consecrated chiefly to the study of Latin and Greek but of course also including much religious instruction. Most of these colleges were attached to the Academies to which their attendance was intended to lead; hence the pupils came from a wide geographical area and lived as part of the families of their teachers. Having finished the college course of studies the pious student proceeded to the Academy, in effect a Protestant university which possessed a Faculty of Arts and a Faculty of Theology. The two-year course in Arts led to a Master of Arts degree and comprised study of logic and moral philosophy in the first year and physics and metaphysics in the second. The Theology Faculty, intended for the training of future ministers, possessed Professors of Hebrew and Theology. It must have reminded Oldenburg of his own education at the Bremen Gymnasium. It is not difficult to see why Saumur was considered an ideal city for the education of a young Englishman of good family like Richard Jones and other such young men; as Oldenburg described it, 'the air is good, the countryside beautiful and the people kind and cheerful'. And then, as now, it was noted that the purest French in the whole country was spoken in those parts. The Protestant Academy together with its introductory college provided instruction for the young without any danger of contamination by Catholic doctrine, an important consideration for a member of a devoutly Protestant family, and it appears that an English youth like Jones could attend courses at both institutions. Saumur was well accustomed to foreign Protestant visitors and warmly welcomed them. There was besides intellectual stimulation for both tutor and pupil so that while Jones attended the college and the Academy, Oldenburg was soon on friendly terms with at least one of the staff of the Academy. This was the elderly Moïse Amyrault,[26] one of the most celebrated Protestant theologians of France and a Professor of Theology at the Saumur Academy, through whom he must have met other members of the Academy faculty. The religious and intellectual life of the city was calm and fruitful, for at this time the Protestants there had known some thirty years of peace and relative toleration. Not until 1661, over a year after Oldenburg and Jones had returned to England, did the slow but forcible movement towards complete destruction of all religious freedom in France, culminating in the Revocation of the Edict of Nantes in 1685, begin to make life for Protestants throughout the country increasingly uncomfortable, oppressed, and finally all but impossible except in very remote regions like the Cevennes. When they were there, however, Oldenburg and Jones found life in Saumur enjoyable and remained there until the early spring of 1658. Jones learned French thoroughly, pursued classical studies in the college (these included two-way translation between French and Latin and some Greek) and also studied history and theology, these latter presumably at the Academy. He was allowed some 'lighter' studies in the afternoons, like practical geometry and fortification and probably riding and fencing, all necessary for a gentlemen.[27] Oldenburg presumably saw to it that he took his studies seriously by overseeing much of his work, but he was not directly responsible for all of this instruction and so must have had some leisure for his own pursuits. But he certainly took his position seriously, as no doubt he had always done, and hence professed himself to be much occupied. He planned to reward Jones's diligence with a tour of Anjou at vintage time, for which he secured permission from the young man's parents. No doubt both men would have enjoyed the tour but bad weather at the time prevented this small indulgence.

In the later summer and autumn of 1657 Oldenburg found time for a more extensive correspondence than he had been able to undertake in preceding months, from which the only surviving letters are dutiful reports to Lord and Lady Ranelagh. He now wrote frequently to Boyle, not only sending news but urging him, vainly, to come to Anjou for the sake of his health. He also wrote to older acquaintances and correspondents about his own interests as well as news from the intellectual world of Saumur. Thus he wrote to Milton about the religious scene with special reference to Milton's controversies with various French theologians. He wrote to Manasseh ben Israel (d.1657), rabbinic scholar and milleniarist, whom he had, it seems, met in London in company with Edward Lawrence; with him he mainly discussed Messianic prophecies, a subject which always interested him. He wrote (apologizing for not having done so sooner) to Boreel with whom he had intended to correspond on a regular basis, his excuse being his 'responsibility for developing the studies and character' of his pupil; now he revealed that he had become especially interested in the genealogy of Christ. He wrote twice to Dury in Latin, describing the warm aproval given by Amyrault to Dury's mission for Protestant unity and outlining Amyrault's published views on the subject; it was perhaps because of the topics covered that he wrote in Latin.[28] Of more significance for Oldenburg's later career is a surviving letter to Boyle in which he sent a receipt for an invisible ink which he said he had tried himself;[29] this is a very early example of Oldenburg's talent for eliciting information which he knew would be of interest to his correspondents and in which he therefore interested himself, a talent which he was to exercise freely and importantly in years to come. (There was great interest in both invisible inks and ciphers during the years of the English Civil Wars and their aftermath.) Above all, Oldenburg wrote frequently to Hartlib,[30] mainly now about private matters relating to Hartlib's role as postal agent for the Ranelaghs and Boyle. No doubt (but some of Oldenburg's letters to Hartlib are lost) he wrote to congratulate Hartlib on the apparent success of his latest scheme for the Advancement of Universal Learning of which Boyle, a patron of the scheme, had informed him.[31] In return Hartlib sent news and information for Oldenburg to keep in mind in his future travels, to use as bargaining counters in order to extract information from those he met, this consisting mainly of news of new inventions made or projected by Hartlib's proteges. This correspondence with Hartlib demonstrates yet again Oldenburg's increasing ability to respond to the varied interests of his different correspondents.

Clearly this year in Saumur was a profitable one as far as Oldenburg's tutorial reponsibilities were concerned although he necessarily found it somewhat dull in regard to his own interests. No doubt too Jones was becoming restless in the quiet of Saumur, especially once he had mastered French, and longed for the promised beginning of his real grand tour. It is not at all surprising that in the early spring of 1658 Oldenburg sought (and obtained) permission for further travel. The obvious place to visit in the summer of 1658 was Germany and in particular the city of Frankfurt-am-Main where the election of the next Holy Roman Emperor, then no mere formality, was due to take place. So in March the travellers set off for Frankfurt, going first to Paris, not merely because it was a natural starting point but because 'some important occasions' (unspecified) made the visit vitally necessary.[32] No doubt young Jones was delighted to see more of Paris than had been possible the previous year and no doubt Oldenburg too was pleased to be there and took the opportunity to renew old acquaintance, for he must have known Paris well from previous visits. They stayed in Paris for about six weeks, long enough to sample the pleasures of what was reputed, then as now, to be the most enjoyable city in Europe; Jones must have particularly enjoyed the fact that a six-weeks stay was considered too short a period to necessitate the resumption of his studies. It is tempting to believe that Oldenburg also devoted much of his time to amusement, for he seems to have written no letters during this period nor in subsequent correspondence did he ever refer to anything of note that occurred during this visit.

Oldenburg and Jones left Paris in early May, going via Geneva where they stayed a few days and picked up the letters forwarded by Hartlib. They then proceeded to Basle where they stayed long enough to make a few calls. Once arrived in Frankfurt they promptly immersed themselves in the excitements of the Imperial election. Frankfurt was in any case an exciting city in which to stay; it was filled with medieval churches both Lutheran (the majority) and Catholic and possessed many fine houses built for medieval merchants. A famous book fair was held there annually and commerce throve. At this time all interest centred on the Imperial election, successive elections having been held in the city since the ninth century. Since the fifteenth century the business had taken place in the appropriately named Kaisersaal which was also the room where coronation banquets were held. Judging by his surviving letters to Hartlib Oldenburg seems to have spent much of his time during the next month listening to gossip and rumours about the expected outcome of the forthcoming election (but clearly he knew nothing of Cromwell's vital but vain concern to secure a Protestant Emperor).[33] The contest was finally decided in early July in favour of Leopold I, already King of Hungary and Bohemia. Oldenburg also undertook a small personal favour for Hartlib, attempting to collect from the young Graf ('Count') Christian von Hohenlohe a sum of money he owed to Hartlib, either borrowed during his visit to England during the previous year or promised by him in the guise of Hartlib's patron. In the attempt (only partially successful) Oldenburg was to spend much effort during this year and the next.[34]

Almost as soon as they learned the results of the election Oldenburg and Jones set out on a five-day journey through the Palatinate. This was friendly territory to any Englishman, since it was ruled by a grandson of the English King James I, a brother of Prince Rupert who, like his brother, had been in England during the Civil Wars and, although more briefly than Prince Rupert, had served in the Royalist army. They do not seem to have visited the court at Heidelberg, or if they did were not impressed enough by it to comment in letters; perhaps there was still too much devastation left in the aftermath of the Thirty Years War. They were to be greatly impressed, however, by Mainz, an ancient city still containing many traces of Roman settlement as well as much medieval and later building. Mainz, unlike Frankfurt, was still largely Catholic and its Archbishop was one of the Imperial Electors. Although it had suffered in the Thirty Years War from occupation by, successively, the Swedes and the French, it still contained numerous fine churches and the electoral palace was at this time being restored and extended. In addition to sightseeing, Oldenburg in his usual manner sought out interesting individuals. He was particularly struck by J.J. Becher (1635–82) whom, in what was to become his characteristic fashion, he summed up as a young man of promise and interest—and rightly so, for Becher was to become an important chemist and later the deviser of what was to become the phlogiston theory of combustion, so central to much eighteenth-century chemistry. Becher, who had only recently settled in Mainz after extensive travels, told Oldenburg about his perpetual motion machine, in fact a clock which required no winding; although he showed it to Oldenburg he did not ever reveal any details of its mechanism. Naturally enough, Oldenburg reported on this device and Becher's achievements at some length to Hartlib, always eager to learn about every sort of mechanical device. Oldenburg was to take pains to call on Becher again before leaving Mainz and indeed to keep in touch with him in the future, following his progress in both invention and chemistry.

When the two travellers returned to Frankfurt to find what he called 'the solemnities' over, Oldenburg determined to leave for Dresden by a route partly determined by the possibility of visiting interesting cities, partly by the need to avoid political disturbances like those which prevented travel via Ratisbon to Prague, a route they would have preferred. Their first stop was at Schillingfürst where Hohenlohe's elderly mother (d.1659/60) and several of her sons were then living, all of whom they met.[36] She was a redoubtably pious Protestant as were her elder sons, in marked contrast to her younger son Christian, Hartlib's debtor, whose Protestantism was even then shaky and who was to be converted to Catholicism in 1667. After this they went to Nuremberg, detouring as travellers often did then to see the small walled town of Altdorf, to hich city Nuremberg had moved its university in 1575, fifty years after its foundation in the larger city. It was thus one of the older German universities, destined to remain for centuries the smallest of them, although it possessed a high reputation in the seventeenth century. From Nuremberg Oldenburg and Jones turned north-east to visit Dresden; this old city had then few buildings surviving from its medieval past, having been burned to the ground in 1491 and largely rebuilt in the sixteenth century, so that even its fortifications were relatively modern. After Dresden they returned to Frankfurt through Leipzig, Jena, Weimar, Gotha and Fulda, a rapid six weeks' journey which yet gave ample time for meeting notables in each of the cities through which they passed, although it left Oldenburg with no time to write letters.

During this tour Oldenburg combined his old interest in theology with the newly acquired interests which derived from both Hartlib and Boyle, to whom he reported very differently, suiting his news to the very different interests of the two men. So to Boyle he wrote of his encounters with chemists while to Hartlib he wrote of the inventions of which he had heard reports and asked for accounts of similar inventions in England, about which many of those German inventors he met were much interested.[37] Gotha and Weimar were the seats of, respectively, the two very pious Dukes of Saxe-Gotha and Saxe-Weimar, both interested in what Oldenburg could tell them of Dury's missions to endeavour to unite the Protestant sects of Europe, especially those in Germany. At Gotha Oldenburg also met Job Ludolf, then in the Duke's diplomatic service; he was one of the most learned Orientalists of the day and Oldenburg was to correspond with him for some time to come. The Duke was anxious to learn the construction of the 'fire-machine' devised by J. S. Küffler of which he had heard. (In fact it was probably only demonstrated by Küffler, for he was the son-in-law of the Dutch inventor Cornelis van Drebbel, the probable true inventor of this machine intended as a means of blowing up ships and under-water fortifications. Both men had worked in England where Drebbel had famously demonstrated a submarine on the Thames and both had been then known to Hartlib, from whom Oldenburg now promised to try to extract details of the 'fire-machine').[38] After leaving Gotha and Weimar Oldenburg wrote carefully formal letters of thanks; rather curiously, that to the Duke of Saxe-Gotha, of which a draft survives, was written in German, while to the Duke of Saxe-Weimar Oldenburg seems to have written only a short note in French leaving the more formal thanks (also in French) to be written by Jones.[39]

In Dresden the travellers found, as Oldenburg reported, 'an excellent physician' named Gansland, a keen believer in the importance of antimony and its compounds as drugs together with other chemical and iatrochemical remedies.[40] He also believed in the existence of a 'vivifying' or 'celestial' spirit by means of which, he claimed, he could do great things if only he could render it material (that is liquid or solid). Oldenburg struck up a friendship with him which led to a brief correspondence, as so often. Oldenburg had already met Francis Mercury Van Helmont, son of the more famous chemist J. B. Van Helmont,[41] Hermeticist and a religious mystic, a great traveller who was to come to England in the 1670s where he associated with Lady Anne Conway and with Henry More, Cambridge Platonist; he finally settled in Berlin as a friend of Leibniz. Now they met him again when he had just been been made a Baron by the Emperor. Oldenburg had many conversations with him but was forced to confess that he could not understand Van Helmont's theory of generation; he could only say that he believed that his view and Gansland's were similar as far as he could understand either of them. At Leipzig Oldenburg met Dr Michaelis, Professor of Medicine,[42] another spagyrical or chemical physician, also a great believer in the use of antimony in both medicine and chemistry, on whose behalf he begged Boyle for information about such mysterious subjects as the elder Van Helmont's *ens veneris*, information which Boyle politely supplied. It was also apparently in Leipzig that Oldenburg met Freiherr von Freisen, a great linguist and an influential Protestant, also interested in medical chemistry and in mechanical contrivances.[43] With most of these men whom he then met, as with Becher, Oldenburg became sufficiently friendly to engage in correspondence during his subsequent travels, extracting information from them for the enlightenment of both Hartlib and Boyle, and in some cases continuing to keep in touch in years to come. To assist him to continue the contact he begged his English correspondents to send him information about a great variety of subjects so that he could transmit it to Germany, requests to which both men responded cheerfully.

The German tour amply displays a number of important characteristics which Oldenburg had acquired in the preceding years, characteristics which were to stand him in good stead in years to come. He was entering middle age now, almost forty years old, well schooled in the proper manner of approaching all ranks and conditions of men. He was at ease in both court and university circles. His entree into the former, it is true, was facilitated by the presence of young Jones who usually travelled under his father's title as Lord Ranelagh, as the custom was on the Continent, for English lords were always welcome in minor German courts. Clearly, however, once admitted into these court circles as a young lord's tutor and companion, Oldenburg was rapidly valued for himself, for his conversation and for his extraordinarily wide knowledge of men and learning. He could discuss theology in a well-informed, even learned, manner and was keenly interested in the politics of religion, subjects of absorbing interest at this time. Equally he could talk about medicine, medical chemistry, mechanical contrivances and knew the latest books on all these subjects. He had learned how to extract information even from the normally secretive, how to 'insinuate oneself into the goodwill of such men', as he put it,[44] using the promise of exchange of information to learn the secrets of those he met. He must have had a very attractive personality to suceed so well in the task that he had set himself, namely to gather information on the subjects which interested his patron Boyle and his friend Hartlib. Unfortunately there is no independent record of his personal appearance then or later, the only certain information, derived from a casual reference by himself, is that by this time he was slightly short-sighted but he needed no spectacles either then or, as far as is known later, although he obviously in years to come used his eyes intensively. He was evidently sufficiently personable to win friends quickly and easily as he was to continue to do throughout his life. Possibly to modern taste he would have seemed somewhat pompous but a grave mien was than regarded as suitable to a man of his age and position, when the great expected deference from their social inferiors.

As soon as the travellers had returned to Frankfurt Oldenburg was careful to write letters of thanks to all those who had entertained them or conversed with them at length. They then set off without haste for France. First they stopped in Mainz so that Oldenburg could learn more of Becher's contrivances from the inventor, for Hartlib's benefit. They then stopped, but only for a day, in Augsberg, where, as Oldenburg told Hartlib,[45] 'We did not neglect to see Wiselius'. This was Johann Wiesel, an optician and instrument maker, one of whose telescopes had been acquired by Boyle through Holland in 1651. Now Oldenburg bought for himself a perspective glass (short telescope or monocular) 'fitted for my sight, which is somewhat short, that costed a ducat'. But to his regret, they did not have time to stay long enough to learn some of Wiesel's craft secrets. The two travellers then proceeded fairly rapidly through Munich and Switzerland to Lyons, where they arrived about a month after leaving Frankfurt. Here they paused only briefly before setting out for the south of France via Marseilles, to spend the winter in various Provençal and Languedoc towns, notably Castres, Pézenas, Méze (on the coast), and Montpellier. This was a suitable region for a winter stay not only on account of the climate but because, even more than on the Loire, it was a region which was still firmly Protestant. For example, in Castres, then a city of about eight thousand inhabitants, half the population was Protestant and it not only possessed two Protestant churches but at least three Protestant ministers. It had not suffered greatly in the religious war of 1627–29, unlike other centres of Protestantism in Languedoc, so that there was an atmosphere of security there for the native Protestants. Montpellier was also fairly secure then in its Protestantism, with a quarter of the population belonging to the Reformed Faith; although it had been besieged in 1622 that attack was less savage than most of those launched in 1627 and Montpellier had never been forced to capitulate. It had lost its fortifications but had been allowed to survive as an important Protestant centre, retaining its Protestant Academy which flourished. The old medieval university flourished also but from this all Protestants had been banned by the energetic action of the Catholic bishop. The relative good fortune of these southern Protestant cities contrasts strongly with the fortunes of those in the north where even those with a strong Protestant population like La Rochelle had experienced fearsome sieges in the late 1620s and as a result had lost many of their former privileges when forced to capitulate to the royal armies. Oldenburg was always careful to take Jones only to Protestant regions of France, with the understandable exception of Paris, no doubt at the insistence of his mother, for pious mothers like Lady Ranelagh were always fearful lest travel on the Continent should lead to their sons being converted to Catholicism. This perhaps explains why Oldenburg never won permission to take his charge to Italy, although his uncles had survived the experience.

Even more to Oldenburg's taste than the religious and meteorological climate of this part of southern France was its intellectual climate. It appealed to him greatly for it was a region abounding in provincial academies, not only educational institutions suitable for Jones but also societies or clubs composed of small groups of people of similar tastes who met regularly for conversation and discussion of literary and philosophical topics. The tone was lively and informative even if informal, for these were amateurs of learning in the strict sense of the word. Montpellier had long been a centre for visiting English travellers for the sake of its famous university, noted for its medical faculty and although this was not at the time accessible to Protestants there was still both the Protestant Academy and a very active intellectual academy. Jones was probably forced to resume his studies under Oldenburg's guidance but it seems that Oldenburg also took him to meetings of the various informal academies and societies. He was also introduced to fashionable Montpellier society, where he apparently did not acquit himself in a wholly satisfactory manner, especially with the ladies; for this lapse, whatever it was, Oldenburg later felt obliged to ensure that a proper apology was sent on behalf of both men.[46] Oldenburg himself clearly revelled in the learned society he found in the academies, and especially in the opportunities these gave him to meet men with an informed interest in medicine and chemistry, these two subjects being generally combined in the concerns of one individual as they were joined in Boyle, to whose interests Oldenburg was clearly turning more and more. He was to keep in touch by letter for several years to come with men whom he then met, sending them news of the academies very like their own then flourishing in Paris and later still news of scientific gatherings in London.[47]

That he was well received by these provincial academicians is clear. He was, for example, able to give a letter of introduction to one of them (M. Pradelleis, a lawyer who acted as Moderator of the Academy of Montpellier) to a young Anglo-Irishman, Robert Southwell, recommended to Oldenburg himself by Robert Boyle; Southwell was to become a Fellow of the Royal Society of London and in the 1690s its President. Presenting this letter of introduction, as Southwell wrote to Oldenburg, ensured that he was very well received in Montpellier, so well that he could write that Pradelleis's 'civility made me understand the powerfull influence of your recommendation', no small tribute to Oldenburg's importance in the eyes of those whom he met, far above what would normally have been accorded to the mere tutor of a young English lord.[48] Oldenburg seems to have plunged so deeply into the society of the region during the time of his stay there as to write very few letters during the whole of the four months which elapsed between his leaving Lyons in the autumn and his arrival in Paris in March of 1659. But it is difficult to believe that he did not report regularly to Lady Ranelagh about the welfare of her son or share with Hartlib and Boyle at least some indication of the intellectual stimulation which he found in the society of the academies he so assiduously frequented. Certainly he informed Boyle of the difficulty he sometimes experienced in learning 'secrets' from medical men and chemists without something to offer in return, as indicated by Boyle's promise to send him some useful recipes to 'trade with among your philosophical merchants'.[49] Hartlib sent him news of inventions and alchemical affairs, the latter learned from his son-in-law Clodius or Clod, as well as accounts of new books; unfortunately Oldenburg's acknowledgements have not survived. This visit to southern France reinforced the activity developed in the latter part of the visit to Germany by which Oldenburg perfected methods of extracting information, usually what would be of use to Boyle or to Hartlib, from those whom he met, even when they seemed reluctant to divulge their knowledge. He evidently had learned how to show himself to be a person worthy of knowing such secrets, a man of discretion and of wide knowledge in a variety of subjects, and one who had learned how to secure the acquaintance of those whom he expected it would be profitable to meet. As time went on Oldenburg was to become ever more skilled in this 'trade', coming to handle its operations so discretely that it was not obvious to those from whom he extracted news and 'secrets' that he was doing so. In this 'trade' he was to develop changing interests over the years, becoming less and less concerned with the kind of inventions and improvements beloved by Hartlib and more and more interested in astronomical, physical, medical, and chemical discoveries and ideas, first according to Boyle's interests, later on a broader front. At this time although chemistry and medicine retained, as they were to do for many decades, much concern with supposedly secret remedies and processes, the chief subjects of natural philosophy were rapidly becoming ever more open. True, some natural philosophers retained a reluctance to publish for many reasons, but they could be, and by Oldenburg often were, encouraged and enticed into communication and even publication by careful and tactful handling. In this important activity Oldenburg was to play a key role in years to come.

By March of 1659 Oldenburg and Jones had left Montpellier, travelling west and north to La Rochelle. There, as was their habit by now, they managed to meet several physicians interested in chemistry during their short stay,[50] arriving in Paris by the end of the month. There they were to remain for the next year. Without any doubt Oldenburg had been greatly stimulated intellectually by his attendance at the provincial academies of southern France and acquaintance with their lively members and was thus well prepared for the intellectual riches which Paris had to offer to a man who wished to enter into the right society. He was helped in this by the fact that Jones, now eighteen years old, was of an age to share his tutor's activities and had apparently matured sufficiently to share his interests also, as his mother and uncle wished him to do—although whether voluntarily or not it is impossible to determine. He was still thought of by his tutor as in need of further education and one of Oldenburg's first duties was, as he put it, to see about 'putting ourselves into an academy', which seemingly restricted their activity and so must have been of an educational nature.[51] Hence time which otherwise might have been used to extend their acquaintance was occupied with whatever were the concerns of this academy: very possibly the polite arts of riding, fencing, and dancing to render Jones fit for truly polite society and to correct those awkwardnesses which he had displayed at Montpellier to his tutor's embarrassment. His intellectual needs were not forgotten either and he was encouraged to write to several of the chemists whom the travellers had met in previous months, no doubt with an eye to the interests of his uncle Robert Boyle.[52] He was made to accompany Oldenburg to meetings of various groups in Paris, as the latter told Boyle, noting that[53] 'We have several meetings here of philosophers and statists [men active in affairs of state] which I carry your nevew to, for to study men, as well as books.'

Academies for various purposes and various intellectual subjects had existed in Paris since the sixteenth century and were to proliferate in the seventeenth. Most were effectively private clubs or societies centred, generally, around an individual patron. The exception was the Académie Française, founded in 1629 as a private body, which acquired government patronage, support and control in 1636; it was, as it still is, dedicated to the maintenance of an official, pure French language through the compilation of a great dictionary. The government in the person of Cardinal Richelieu provided it with a meeting place, a charter, and pensions for its members, membership being then as now a limited honour; it was in all this to be copied by the Académie royale des Sciences founded in 1666. Oldenburg never had any links with the Académie Française, nor would one have expected it, given his interests and the fact that he was a foreigner, although he occasionally met some of its members in other contexts. But there were plenty of other, more readily open, academies from which to choose, into almost all of which he found it easy to gain admittance once he had displayed interest.[54] Earlier in the century the historian de Thou had employed two brothers, Pierre and Jacques Dupuy, to look after his library, housed in the great Hotel de Thou on the left bank of Paris; to this the brothers added their own books, making it an extremely important collection. The triumverate soon turned into a society composed of men of letters, scholars, lawyers and intellectuals of all sorts, after which it was referred to as, successively, the Cabinet du President de Thou and, after 1617, as the Cabinet des frères Dupuy. It remained as the Cabinet Dupuy even after the deaths of the Dupuy brothers in, respectively, 1651 and 1656, the meetings being then presided over, as they had been for some years, by Jacques de la Rivière, librarian at the Hotel de Thou since 1647. The Cabinet Dupuy was very popular and its meetings, which were fairly informal, were well attended, almost every possible subject being discussed. In 1659 the chief subject was politics or, as Oldenburg put it, in 'the circle which assembles every day at Mr de Thou's . . . they talk impartially on all topics but chiefly of what is happening in the world and of interesting books as they come fresh from the press.'[55] It was to these meetings, surely, that Oldenburg took Jones 'to study men'. Oldenburg often spoke as well of inspecting the library at the Hotel de Thou, admission to which was easy to gain and he was to remain in touch with de la Rivière by letter after leaving Paris with happy recollections of both the library and the Cabinet.

In 1659 the only academy which concerned itself primarily with natural philosophy was the Montmor Academy, begun a few years earlier. Its organizer and patron was Henri-Louis Hebert de Montmor, as a Maitre de Requetes a leading member of the legal world of Paris, an immensely wealthy man who used much of his wealth as a patron. Together with others of his family he had been an early member of the Académie Française, but his real interests lay elsewhere. He became successively a follower and then a friend of Descartes and he drew to his meetings most of the natural philosophers active in Paris in the late 1650s and 1660s. There were many skilled mathematicians. The young genius Blaise Pascal (1623–62) in 1659 was just in the process of giving up mathematics after a severe illness which turned his thoughts to religion but Oldenburg met him as a mathematician at the Montmor Academy in April of that year. He also met the much older Roberval (1602–75), with whom he was to have a desultory but continuing correspondence for many years to come. He was doubly Professor of Mathematics for he had been in 1634 appointed to the chair founded by Petrus Ramus (Pierre de la Ramée) a century earlier and then in 1655 had succeeded Pierre Gassendi in the chair of mathematics at the Collège Royale, founded in the early sixteenth century by King François I. They and other Parisian mathematicians were keenly interested in devising pneumatic experiments to prove the existence of atmospheric pressure, following earlier work by such Italians as Torricelli, whose work was made known in Paris very soon after its success through the medium of correspondence. The Montmor Academy also contained medical men like Gui Patin, dean of the Faculty of Medicine of the University of Paris whom Oldenburg does not seem ever to have met, and the innovative Jean Pecquet (d.1674) discoverer of the thoracic duct, who had publicised many of the French pneumatic experiments in his little work of 1651, *Experimenta Nova Anatomica*, well known in England, and whom Oldenburg knew. There was Adrien Auzout an able observational astronomer, and pneumaticist with whom Oldenburg was later to correspond, and Pierre Petit, engineer, astronomer and mathematician who had been an associate of Pascal in one of the first performances of the Torricellian experiment in France and with whom both Oldenburg and Jones became friendly. As he told his French correspondents in the south of France, Oldenburg became keenly interested in the pneumatic discussions at the Montmor Academy; that he said little about them to Boyle and Hartlib must have been because he was aware that Boyle already knew of the chief experiments, as he was to show when in 1660 he published his first scientific book, a work he finished in 1659, which contained descriptions of the airpump with which he worked and the experiments he made with it. Oldenburg also met Jacques Rohault, physicist and dedicated Cartesian, who in 1659 offered a series of weekly lectures on Cartesian physics which Oldenburg attended.

The Montmor Academy was unique among Parisian private societies, then and later, in having a formal constitution with regulations for the conduct of meetings.[56] The purpose, these regulations stated, was not to be 'the vain exercise of the mind on useless subjects', but, together with 'the clearer knowledge of the works of God', always to be kept in mind, those gathered together should consider 'the improvement of the conveniences of life, [together with] the Arts and Sciences which seek to establish them'. This ideal was, clearly, fully consonant with the Baconianism which Oldenburg had learned in different guises from, successively, Hartlib, Boyle, and the Oxford natural philosophers. In theory the Montmorian statutes which Oldenburg heard re-affirmed in July 1659 dictated that a subject was to be chosen for each meeting and the president was to nominate two well-informed persons to open each meeting by reading one or more papers, after which the discussion was to proceed in an orderly manner.[57] It is clear that this procedure was not always followed nor was the discussion always as orderly as was envisaged in the rules, leading Oldenburg frequently to express his doubts as to whether the French natural philosophers would ever accomplish anything 'solid'. Admission to the Academy was to be by the consent of two-thirds of those present and was limited to 'persons curious about natural things, medicine, mathematics, the liberal arts and mechanics'. This was hardly a very restrictive qualification, and Oldenburg and Jones gained admission at the end of July without apparent difficulty, no doubt helped by the fact that they had already met many of the mathematicians (who included Pascal and Roberval) as well as a number of the iatrochemists who attended the Academy, all before they met Montmor himself, as they did in mid-June.[58]

Oldenburg was to become an assiduous member of the Montmor Academy as his correspondence bears witness. He seems at first to have been doubtful whether he was fitted to belong to it fully: after the first meeting which he attended he wrote diffidently to Hartlib, 'I hope, we shall continue to meet with them, though we can make no great discourses' which suggests that he recognized the worth of the many and dedicated natural philosophers who normally frequented the Academy, as distinct from the many interested amateurs also present. On the other hand, he was also sometimes very critical, writing that 'I wish only, these discourses may not rather tend to speculation and shew of wit, than usefulness to the life of man.'[59] Perhaps this was said because he knew it was what Hartlib (to whom he wrote) would wish to hear, or perhaps it was that his opinion of the Academy improved with further acquaintance with its meetings, for he was soon to become more and more in favour of its activities. And by mid-November he had found that there was something which he himself could offer to the Academy, in the shape of a French translation of the newly published anatomical work by Lodowijk de Bils.[60] Bils had discovered a new method of preserving cadavers for dissection. The Dutch original had been translated into English by John Pell at the request of Boyle who wrote a preface addressed to Hartlib and also saw to its publication. It is not obvious whether Oldenburg translated the original Dutch into French or used the English translation since he was fluent in both languages and both versions had been published earlier in this same year.

Oldenburg's interests were not confined to those of the Montmor Academy, for he was still equally attentive to the contributions of such learned orientalists as those whom he had met recently. So he reported to Hartlib that Ludolf, still acting as tutor to the young princes of Saxe-Gotha, was working on an Ethiopian dictionary and a history of Abyssinia. Oldenburg was always to remain keenly interested in contemporary work on all the languages of the Middle East, Coptic being of especial importance at this time for its possible revelations of the nature of early Christianity, the Copts being among the most ancient surviving Christian sects.[61] During the summer of 1659 Oldenburg and Jones were to be very busy. They still attended the meetings of the Montmor Academy regularly, they listened to the weekly Cartesian lectures of Jacques Rohault, and they went to the daily meetings at the Hotel de Thou; after leaving Paris the following May Oldenburg wrote wistfully of his fond recollections of the Cabinet Dupuy and its members.[62] They also visited libraries, which led Oldenburg to report to his corrrespondents in England on the rare books and manuscripts which they found there, like the oriental writings of which he sent Boyle news for the enlightenment of 'Mr Pocock and others', that is the leading orientalists of Oxford.[63] There was still time for other intellectual activities like reading, about which he often reported to

Hartlib knowing that Hartlib did not care to hear about 'the philosophical discourse of our Clubs'.[64] He reserved his more strictly scientific news for Boyle and for his recently made friends in the south of France. But he did write to Hartlib about Christiaan Huygens's most recent astronomical discoveries and his horological inventions, clearly recognizing their importance and believing that Hartlib would do so too. Yet at the same time he told Hartlib that he doubted the ability of the French natural philosophers (it should be remembered that Huygens had not yet settled in France as he was to do six or seven years later) to 'produce any great matter in point of Tubes [telescopes] or chymistry or any mechaniques' because, so he averred, 'They have not that required steddiness'. (This very English comment comes oddly from such a truly international being as Oldenburg had become, and the French were soon to prove him quite wrong.)[65] He also remarked that they suffered from lack of patrons, a puzzling view considering that all the Parisian academies enjoyed patronage of some sort, private or public, while no English society did, in spite of Hartlib's earnest efforts.

Oldenburg's various reports of his association with members of the Montmor Academy show his increasing sophistication, greater depth of knowedge of natural philosophy and increasing self-confidence in expressing his own opinions, compared with the case when he had left England in 1657. That he was able to become so friendly as he seems quickly to have come to be with so distinguished a man as Roberval speaks for itself and he was apparently now able to converse with him and others like him if not on an equal footing at least from the position of a well read and well informed amateur. He reported that he had discussed with Roberval the possibilities inherent in the employment of non-spherical lenses in telescopes, a topic in which there was to be much interest in the next few years, and when he replied to Hartlib's request for a copy of Roberval's projected work on algebra (it was in fact never to be published) he could confirm

> there is such a work written, which I think I might get leave to transcribe, if I had leasure; without which it can hardly be done, because no Cop[y]ist will be found, that will doe it to purpose, the matter requiring one, that hath some insight and understanding of the subject itself,

showing that he himself did reckon to understand mathematics, since he believed himself competent to understand Roberval's text.[66] And when he wrote to his French friends he showed himself very knowledgeable about the pneumatic experiments made by Rohault who was repeating those made by various French natural philosophers including Roberval and Auzout, most of which Pecquet had described a few years earlier in his *Experimenta Nova Anatomica* (1651). As Oldenburg also reported, Rohault also now performed some experiments of his own devising to demonstrate, as Oldenburg put it, that it was 'the equilibrium between the exterior air and the mercury' which caused 'the uniform suspension' of the mercury in the Torricellian tube, an experiment which Oldenburg fairly called 'entirely convincing'. He also commented learnedly on Rohault's views on the nature of colours, adding that he did not yet know whether Rohault regarded light as a body, adding that he himself had 'no doubts at all about its corporeality' thinking 'it nothing else but fire (rarefied to the highest degree and purified . . . which is a body'.[67] This was a common enough tenet, but worth noting to show how much natural philosophy Oldenburg had absorbed since his stay in Oxford in 1656, and how much his confidence in his own knowledge had increased, making him an entirely worthy member of the Montmor Academy, able to converse on a serious plane with natural philosophers everywhere. Meanwhile, on Hartlib's behalf he visited craftsmen like the telescope makers of Paris, reporting especially on one Bressieux with whom he became friendly which caused him to ask Hartlib to procure and send to Paris some good English glass blanks with which Bressieux might try to grind spherical lenses.[68] (Evidently his earlier poor opinion of the abilities of Parisian natural philosophers did not extend to Parisian craftsmen.)

All this fascinating intellectual life did not lead Oldenburg and Jones to neglect the sights of Paris, on which Oldenburg also reported to Hartlib. Indeed he was soon to ask Hartlib to send a 'perspective glass' so that they might 'take a distinct view of the pompe we are like to see here in October next' on the occasion of the forthcoming marriage of Louis XIV and the Spanish Infanta Maria Theresa, a spectacle to be much postponed, unfortunately for them.[69] They did go to Fontainebleau in late July to watch the deputation setting out for Bayonne to welcome the Spanish princess to France, with or without the desired perspective glass of which there was no further mention. Oldenburg was careful to collect news of the Infanta's progress towards Paris and to report it to Hartlib from time to time. No doubt he and Jones watched other colourful events taking place in the French capital as befited foreign visitors and mingled with less serious society than that of the natural philosophers whose talk so pleased and benefited Oldenburg, if not Jones.

Towards the end of the year Oldenburg's correspondence with Hartlib touched on more serious subjects. For example he reported that he was spending much time and effort endeavouring to obtain a manuscript of Josephus's history of the Jewish Wars owned by the widow of a former Protestant minister at Nimes which he wanted to send to Johannes Koch (Coccejus), Professor of Hebrew at Bremen, a relative of his brother-in-law.[70] And in December he suddenly renewed his long neglected correspondence with Milton, sending news of the Protestant community of the Loire valley, as he had done a couple of years before. To this letter Milton replied promptly, but the renewed correspondence did not flourish, perhaps on account of the changing political situation in England.[71]

In early 1660 Oldenburg was an active and international correspondent generally, with ever wider interests. He wrote to his friends in the south of France with news of the Parisian academies; he wrote to Toll, the chemist in La Rochelle whom he and Jones had met on their journey from the south of France to Paris; and he continued to write to Becher in Mainz, trying to exchange news of inventions. He also continued his English correspondence, exchanging letters with the Ranelaghs, with Boyle and with Hartlib, as he had done regularly for the past three years and through them kept in touch with English events. Hartlib now introduced a new subject for his consideration by telling him of gardening interests in England, notably of the work of John Evelyn, on whom Oldenburg had tried to call in 1656 before leaving England with Jones, for what particular purpose is not known. Hartlib also told him about John Beale (1608–1682/3) who was to become a correspondent for the next seventeen years.[72] Beale was an enthusiast for many very different subjects but most of all for horticulture, cider making and gardening. On behalf of both Evelyn and Beale, Oldenburg gathered news of recent French books on gardening, at this time full of ideas quite novel to the English; Evelyn was to translate and publish several French books on gardening. Boyle sent Oldenburg information about his publishing plans for he was just then bringing to fruition a number of works on which he had been gathering material for several years; Oldenburg was later to assist him in their publication. Both Boyle and Hartlib sent copies of English books to be passed on to acquaintances in France, for English books, except those written or translated into Latin, were not easily procurable on the Continent, only Latin texts being printed there, usually in Amsterdam. This activity may even have added to Oldenburg's income as it was to do in later years when he frequently acted as a book agent for Continental friends and correspondents.

Clearly Oldenburg was enjoying a happy and fertile intellectual life in Paris and he, and very possibly Jones also, would have been quite content to spend the remainder of 1660 there. But political change summoned them back to England and marked the end of Oldenburg's employment as tutor in the service of the Ranelaghs, as must have happened soon in any case. Because events in England moved so fast and because Oldenburg and Jones left Paris in some haste, there is no remaining evidence of precisely when or how they were summoned; we can only be sure that in Paris English events were by no means difficult to follow so that they must have had some knowledge of the changing English scene.

When Oliver Cromwell died early in September 1658 his son Richard had succeeded him as Protector peacefully enough and the Commonwealth seemed destined to continue in being. But Richard Cromwell was quite unable to control public affairs in the expert and forceful way in which his father had done. Some indication of the resulting shift in public opinion in favour of a restoration of the monarchy was picked up by Oldenburg in the summer of 1659, at which time he told Hartlib that he had heard rumours of a possible overthrow of the Commonwealth and that many royalists in Paris were claiming that the monarchy would soon be restored. This was in June and later in the summer he reported that the number of those in favour of the monarchy was said to be increasing daily.[73] This was probably pure optimism but by the spring of 1660 optimism had become reality, and the restoration of the monarchy seemed certain. Once this was so, either the Ranelaghs urgently summoned Oldenburg and Jones back home or they asked and received permission to return to witness these exciting events, but in any case they did arrive in London towards the end of May. They were there some days before Charles II entered the city as he did on 29 May 1660, riding in triumph and reaching Whitehall in the early evening, all to the universal rejoicing of the crowds, made wilder because it was the King's birthday. People thronged the streets to watch the King ride by, houses were hung with tapestries and fountains ran with wine instead of water.[74] The rejoicings lasted for several days and the few letters surviving from this period bear witness to the fact that Oldenburg participated fully in the festivities.[75]

Parliamentarian by religion and friendship Oldenburg may have been, but like most moderates he was prepared quickly and fully to accept the restored monarchy once it was in being, as a bringer of peace, security and tranquillity, marking the end of the troubled last days of the Commonwealth, as did his patrons and his friends. This turning away from the past was made easier by the leniency wisely displayed towards former enemies (always excepting regicides) by Charles II to a most remarkable degree: he rewarded the loyal and won over former opponents by rewarding them too. This made men like Wallis and Wilkins securely part of the new regime equally with Ward, the only royalist among the natural philosophers whom Oldenburg had known during the early years of the Commonwealth. The Boyle family had, for the past fifteen years, managed to accommodate adherents of both the Parliamentary and royalist factions without, remarkably, producing internal family strife. As far as any official opinion was concerned, Oldenburg had nothing to fear on account of his former diplomatic dealings with Cromwell nor his friendship with many of Cromwell's supporters. Guilt by association was not the fashion in 1660 and in any case Oldenburg had had as many connections with staunchly royalist families like the Honywoods as he had had with families adhering to the Parliamentary side. His only problem was to find fresh employment. England seemed destined to be a peaceful country in which to live if he could find a place in it with the help of his previous patrons. What was to materialize for him Oldenburg could never have predicted.

3

Towards a settled life 1660–65

This Curious German having well improved himself by his Travels, and pursuant to the Advice of Montaigne, rubbed his Brains against those of other People, was upon his Return into England entertained as a Person of great Merit, and so made Secretary to the Royal Society[1]

So, with a suspiciously superior choice of phrase (he did not greatly like the English) did Samuel Sorbière describe the turn of events which was to give Oldenburg a settled way of life in England, a post of responsibility and a distinguished position in the world of learning both at home (now firmly England) and abroad. In fairness it must be remembered that Sorbière had known Oldenburg in Paris as tutor to Richard Jones, an auditor rather than a member of the Montmor Academy. And indeed Oldenburg did not rise above his earlier circumstances immediately upon his return to England and was in danger of dropping into obscurity. Fortunately the three years after his return had put him into the distinguished position in which Sorbière found him, although it must have seemed a slow change to Oldenburg himself.

Return to England had inevitably meant loss of employment, for Jones, now nearly twenty years old and restored to his family, no longer required a tutor; indeed very little over a year later, on attaining his majority, he was to enter the Irish Parliament. The prospects for Oldenburg must have seemed poor; not surprisingly, his personal affairs for the next couple of years are a little obscure and, as was to remain the case for virtually the whole of his life, it is difficult to determine how he obtained the requisite amount of money to permit him to live in reasonable comfort. To begin with he almost certainly lodged with Hartlib, now living in Westminster, not far from either Whitehall; that is, the Court, or Parliament.[2] However, any hope that Hartlib might have had of government support for his schemes had quickly vanished with the restoration of the monarchy and he was to suffer financial difficulties for the remainder of his life. (He died in March 1661/2.) Several of Oldenburg's friends were also adversely affected by the Restoration: Dury decided that he needed to remain on the Continent while his wife, who did return from time to time, was herself in some financial difficulty when Hartlib appealed to her for help in 1661, although her connection with the Boyle family (she was the aunt by her former marriage of Robert Boyle and of Lady Ranelagh and such connections counted for much in the seventeenth century) no doubt helped her to secure the property which she was to bequeath to her daughter. Milton was forced to go into hiding and when he gave himself up he was briefly imprisoned and fined (although he probably never paid the fine), after which he devoted himself to literature. Others among Oldenburg's acquaintance came off better and those who were former royalists were now likely to prosper. Oldenburg himself must have continued to look to Boyle and Lady Ranelagh, who throve under the monarchy as they had done under the Commonwealth, and perhaps to the Honywoods as well.

Life in London in 1660 and 1661 was full of excitement and spectacle. Like everyone else who had sufficient leisure, Oldenburg hung about Westminster Hall, watching events as they occurred and the activities of the Court.[3] He met there both courtiers and many who, like him, had leisure to look about. He now somehow met both John Evelyn and his friend John Beale, with whom he had corresponded during the previous year.[4] It seems that they discussed missionary work and optics, but over the years many other topics cropped up arising from Beale's interests of horticulture, cider, and the improvement of agriculture. He visited instrument makers, perhaps in the company of Boyle, who was also in London to witness political and courtly events. He met natural philosophers returning from exile and renewed his acquaintance with many among the Oxford philosophers. He kept in touch with his French friends, reporting the London news.[5] He wrote to his old friend Adam Boreel in Amsterdam, expressing the hope that he might himself soon visit Holland and so speak with him once again if personal affairs permitted: for, as he confessed, he found himself without financial resources, remarking a little bitterly that he had not found that having served as a tutor had led to riches, although others had, he knew, been more fortunate.[6] (He later estimated that such employment could result in the sum of perhaps £100 a year, not great riches even if the expenses of travel and subsistence were paid for, but there was, he hinted, always the possibility of an annuity at the end of the engagement, something the Ranelaghs clearly never supplied.)[7] In 1660 and 1661 he was probably being helped financially by Boyle, about whose first scientific books, then in course of publication, he wrote to his correspondents, showing intimate and sometimes advance knowledge of their contents: for example, in October 1660 he sent to Pierre Petit a copy of the engraving of the airpump from Boyle's *New Experiments Physico-Mechanical touching the Spring of the Air and its Effects*, published at Oxford that same year.[8] That he was able to secure a copy of the illustration suggests that he was, in some way, working for Boyle, possibly translating the book into Latin as he was to do for many of Boyle's later works.

Immensely important for Oldenburg's future, for it was to change his whole way of life, was an event which occurred on 28 November 1660. This was the organizational meeting of what was to become the Royal Society of London, of which he was destined to serve as Secretary from 1662 to 1677. This post was to make him a permanent resident of England, give him the opportunity to develop his skills as an administrator, ultimately to provide him with a small but useful annual salary, and to allow him to be active in and to serve the world he loved best, that of philosophers and learned men. It also established his reputation in this world, so that his name and his contributions were known to natural philosophers from the Near East to all parts of Europe, and the Americas. It will be recalled that Oldenburg had had a close connection with the Oxford Philosophical Club; very many of its members had earlier followed the practice of meeting regularly in London after 1645, seeking relief from the disturbances of the Civil Wars by discussing 'Philosophical Inquiries, and such as related thereunto', as Wallis remembered it,and banning all discussion of religion, politics and day to day affairs.[9] This not only gave them relief from the tensions of the time but permitted the inclusion in the group of both Parliamentarians (like Wallis, Wilkins, Dr Goddard, Dr Ent, Dr Glisson) and royalists (like William Balle and Dr Scarburgh) without any danger of quarrels and dissensions. These men often met regularly at the lectures of Samuel Foster, then Professor of Astronomy at Gresham College in London. (This is the first mention of a connection long to be maintained between the Royal Society and Gresham College, founded in 1597 under the will of the rich London merchant Thomas Gresham, enunciator of Gresham's Law, to provide public lectures for Londoners on the seven liberal arts together with law and medicine.) The meetings thus established continued in London after Wallis, Wilkins and Goddard had migrated to Oxford and established their own meetings, which soon included Seth Ward, although he was a staunch royalist, and several promising young men by the time that Oldenburg joined them in 1656. According to Wallis, the Oxford natural philosophers continued to attend the London meetings when they themselves were in that city.

In 1658, while Oldenburg was in France, a meeting had been called for 22 November, the attendance being expected to include Lawrence Rooke, who had been at Wadham College under Wilkins and became successively professor of astronomy and of geometry at Gresham College (but was to die young in 1662); Christopher Wren, whom Oldenburg had known as a young member of Alls Souls College when at Oxford, now Rooke's successor in the chair of astronomy at Gresham College; Lord Brouncker, a royalist who had spent much time during the Civil Wars retired in the country, studying mathematics in which he had become proficient and was to become a Commissioner of the Navy after the Restoration; Sir Paul Neile, also a royalist who, like Brouncker, had spent the Commonwealth years in the country, in his case studying astronomy; and a number of others, most of whom played little part in Oldenburg's life.[10] Presumably this practice of meeting continued during the next two years although no further record of it survives before what was to be a momentous meeting on 28 November 1660 when, as the surviving record states, those present 'according to the usuall Custome of most of them' met at Wren's astronomy lecture at Gresham College. After the lecture, again 'after the usuall Manner', they remained for discussion. Those among them whom Oldenburg certainly knew were Boyle, Wilkins, Goddard and Wren; there were also present Brouncker, Sir Paul Neile, and Rooke, who had attended the 1658 meeting; and, among others, William Petty, originally a successful medical man who had gone to Ireland in 1652 as physician general to Cromwell's army, when he had instructed Boyle in anatomy. Three years later he volunteered to survey land belonging to defeated Irish landlords which was to be given to the troops of the conquering army in lieu of pay; this was the well known Down Survey which Petty carried out together with a compilation of maps for most of the country. His achievement was rewarded by the bestowal of much land and so, becoming a man of property, he abandoned medicine. Like many others he went to London in 1659 while awaiting the Restoration and the possibility of ingratiating himself with the new government. Six months after the Restoration, according to his own account, he brazenly told the King that he had never had any 'desire or designe to do him harm'. The King, he reported, 'seeming little to mind apologies as needless' (no doubt knowing all too well how little they were worth) 'replied: "But, Doctor, why have you left off your inquiries into the mechanics of shipping?"' about which they then conversed, not long after which Petty was knighted, presumably for his work on 'the mechanics of shipping'. Nothing could better illustrate the remarkably wise clemency shown by Charles II to any who, although thriving under the Commonwealth, were prepared to accept the monarchy peaceably, provided that they had not had anything to do with the execution of Charles I nor taken any active role against the interests of the exiled Charles II. (No doubt Petty had made the King laugh and he never could resist that.) Other men present at the 1660 meeting with whom Oldenburg was later to have dealings included William Balle, astronomer, already mentioned as a royalist; Sir Robert Moray, statesman and courtier, a Scot and staunch royalist who had spent the later 1650s in exile in Paris and the Low Countries, where he had met Christiaan Huygens with whom he corresponded for many years and who might have met Oldenburg, whose very good friend he was to become; Alexander Bruce, later Earl of Kincardine, another Scottish royalist, who had spent the past few years in Bremen, whence he had corresponded with Moray and where he must at least have heard of Oldenburg's family; and Abraham Hill, son of a city merchant, who was to prove an able administrator for the Royal Society in years to come.

The discussion on this momentous evening turned, probably under Petty's lead, on 'a design of founding a college for the promoting of physico-mathematical experimental learning'. The group did not chose to undertake so ambitious an organization, but it did, collectively, decide to continue weekly meetings (on Wednesdays) in term-time in Rooke's rooms in Gresham College and in vacations in Balle's lodgings in the Temple. Further decisions were taken in respect to admission and subscription charges to pay for 'occasional expences' just as the 1645 group had done to cover the cost of experiments performed at the meetings. Furthermore this was to be an entirely formal organization with officers: a President (Wilkins to begin with), a Treasurer (Balle) and a Register (Secretary) (William Croone, physician and professor of rhetoric at Gresham College). Those present were asked to suggest the names of 'persons' whom they 'judged . . . willing and fit to be joined by them in their design'. The result was a list of forty names, including that of Oldenburg, a mixture of Londoners, Oxford men, parliamentarians, royalists, courtiers, clergymen, country dwellers and townsmen.

By no means all of those proposed or later elected were ever to be very active members, for a vague and general interest in some form of natural philosophy sufficed, provided that the potential member was of a suitable social standing (although some instrument makers had been members in 1658, and there were to be others in the eighteenth century, there were none in the early years), known to others of the group, and likely to pay regular dues. Many of those named were true 'virtuosi'—that is men keenly interested in some branch of natural philosophy and more or less knowledgeable in it, and these were most likely to contribute to the content of the meetings in years to come, either by attendance at meetings with experiments to show or books to report on or accounts of interesting events in the world of learning. Many, like Samuel Pepys, still a minor and hardworking civil servant in the Navy Office, were simply fascinated by experiments and accounts of experiments even though, by Pepys's own admission, he at least understood their importance and true meaning very little. Others again, like Oldenburg's pupil Richard Jones, elected 11 September 1661, were really only slightly interested, in his case probably only dutifully so; he paid his dues for the next two years (a very useful contribution) and then vanished from the Society, as was common with many, although by no means all, men of affairs. There were many provincial members who virtually never came to meetings but were anxious to be kept informed of what was being done at the weekly meetings and might contribute accounts of natural history, weather, 'curiosities', anatomical abnormalities and so on. All such men were welcome, especially at first, if they professed interest and, very important, paid their subscriptions, for money was always to be in short supply. Besides, men of affairs and noblemen lent prestige and might help by their influence and patronage, as they did when the king accepted the Society's desire to be a Royal Society, and subsequently granted it a charter. The solid core of the Society, however, to whom those mentioned above were, so to speak, fringe members, remained always a smallish number of dedicated and committed men determined to make the new society work and prepared themselves to bend their efforts into making sure than it did work.

When the founding group met for the second time, on 5 December 1660, formal rules for the election of members and officers were proposed. It was also decided to employ certain 'servants', namely an amanuensis to assist the Register and an operator to assist in the performance of experiments, both posts to be permanent for many years, with more or less fixed sums awarded either per annum or according to the work done. By this time the royalist courtiers had taken steps to ensure that the new organization should have real royal backing, the King being known to favour serious discussion of the kind of problems likely to be considered, especially any practical problems. As the minutes of this meeting duly recorded, 'Sir Robert Moray brought word from the Court, that the king had been acquainted with the design of the meeting, and well approved of it, and would be ready to give encouragement to it' as he was to do a year and a half later.

Although Oldenburg was not formally proposed as a candidate until the fifth weekly meeting on 26 December 1660, he had obviously before this time learned much about its affairs from one or more of his friends. On 13 December, writing to his friend Boreel in reply to Boreel's query about what had happened to those close to Cromwell after the Restoration, a query commonly being asked by foreigners at this time, Oldenburg hedged, writing cautiously that 'The things you desire to know concerning the surviving relatives and friends of Cromwell cannot be so safely sent by letter' demonstrating that, for all the king's lenient attitude, it was still unsafe to discuss some things openly and presumably many of those like Milton who had escaped the initial trials and executions were still in hiding. Oldenburg then continued,

> However Dr Wilkins [who, it must be remembered, had married Cromwell's sister, although he now accepted the Act of Uniformity swearing to uphold the Anglican Church] lingers in this city; he has been made dean of York and elected President of the new English Academy very recently founded here under the patronage of the King for the advancement of the sciences. It is composed of extremely learned men

ten of whom he named, noting that there were twenty-one in all. He then added, 'Whether foreigners will be admitted I doubt very much, though some say so'.[13] This doubt may explain the slight delay in his election, nearly a month after his name had been listed as a probable candidate, but although he himself did not, until the last year of his life, have English nationality, as the permanent resident of England which he soon became there was no question of the suitability of his election. And indeed quite soon distinguished foreigners like Huygens were being elected on much the same footing as provincial members, there being no separate category of 'foreign member' for many decades.

By the time of Oldenburg's election at the end of December 1660 the structure of the new society had been settled and meetings had assumed nearly the form in which they were to continue for over a century. The President, until 1662 elected monthly, took the chair, elections to the membership were held in formal fashion, no one being elected who was not proposed in advance except for royalty and nobility, the minutes were carefully kept, being taken by the Register and copied out by the amanuensis, committees were appointed to consider subjects for future meeetings and members were held responsible for performing experiments either as proposed by themselves or as directed at a previous meeting. Meetings therefore included discussion, the reading of papers and letters, and the performance of experiment with comments and occasionally the introduction at random of topics which interested individual members. The minutes appear to be fairly full (they were certainly not unduly brief) so that quite a good idea can be gained from them of what went on and who was active, although, unfortunately for the historian, they never, then or later, listed the names of those members present but only the names of those who spoke, formally or informally, were apppointed to a Committee, or were asked to undertake some task. So it is impossible to learn how well attended these early meeting were.

Nor is it possible to say how many of the weekly meetings Oldenburg attended before that on 6 February 1660/1 when he was appointed to 'A committee for considering of proper questions to be inquired of in the remotest parts of the world' an appointment very natural considering his experience of travel and his already extensive correspondence. Those named to the committee, sixteen in all, included, besides Oldenburg, Brouncker, Moray, Petty, Goddard, Rooke, Boyle, Wilkins, and Evelyn, a truly distinguished company. No minutes survive for this or any other very early committee, but it is probable that it met, or at least some of its members did so, and reasonable to suppose that Oldenburg would have played an active role. Other traces of Oldenburg's close involvement with the society are somewhat scattered, but he was clearly frequently at the meetings in the next few months and had much to do with the society's members. When Huygens was in London in April 1661, he mentioned meeting Oldenburg on two occasions in his diary: first on the second of the month when Boyle came to visit him in the morning followed by Oldenburg in the afternoon; and then on the eighth when he saw Oldenburg in the morning before attending the meeting of the society at Gresham College in the afternoon.[14] No doubt Oldenburg was present on at least some of the other occasions when Huygens attended meetings or was in the company of members of the society, which helps to explain why Huygens was to treat Oldenburg as a familiar acquaintance when they met at The Hague in the following summer. But it seems that Oldenburg was not one of the party (which included Moray, Neile, Wallis, and Huygens) who all went to Windsor Castle to watch the installation of the new Knights of the Garter on 15 April, which Huygens very reasonably found a most impressive ceremony, so much so that he went to Whitehall a few days later to see the creation of the new Knights of the Bath (19 April). But although Oldenburg seems not to have witnessed these ceremonies, or at least is not recorded as having done so, it is difficult to believe that he did not, as Pepys and Huygens both recorded that they did, (Huygens from a balcony at Charing Cross in company with Prince Maurice) manage to witness the splendid sight of Charles II riding in procession from the Tower of London to Whitehall on 21 April 1661, on the eve of his coronation. Then, as Pepys recorded, it was 'a brave sight . . . The King, in a most rich imbroidered suit and cloak, looked most nobly'. For this occasion triumphal arches were erected over many streets, the streets themselves being newly covered with gravel, householders hung carpets from their balconies and those in the procession wore their best clothes, often richly embroidered with gold and silver. But it is doubtful whether Oldenburg managed to squeeze himself into Westminster Abbey the next day for the coronation itself, as Pepys did.

All these festivities seem to have interrupted the meetings of the new society, not surprisingly, and none were recorded for three weeks after that of 10 April. The society met again in May when there is only one recorded example of Oldenburg's presence at any of the meetings this month. On 15 May he was appointed to a new commitee with a curious double purpose, namely 'for erecting a library and examining the generation of insects' as the minutes have it. For the first objective, namely the instituting of a library, he was well fitted, although nothing came of this proposal for several years. The second objective was in the short term more fruitful, for the minutes of 22 May record that 'The committee for the generation of insects was appointed to meet on the Monday following [27 May], at six of the clock at Mr Boyle's lodgings'. As usual, there are no surviving minutes, but presumably Oldenburg did attend and was asked (or volunteered) to read the 'treatise on insects' which had apparently come to hand but it is not obvious why he was chosen, except possibly on linguistic grounds, nor is it clear that the committee did more with the topic, except that when Oldenburg wrote to Beale on 30 May, according to his surviving memorandum he 'Gave him some account of what Gresham College [i.e. the society which met there] is doing about poisons, insects, infections'.[15] And a fortnight after this, on 13 June, Oldenburg read his promised account of the treatise assigned to him.

A few days later Oldenburg left London for what was to be his last Continental journey.[16] He went first to Bremen, where presumably he stayed with his relations; probably his visit had something to do with the Vicaria of St Liborius, the collection of whose income had always been difficult. Possibly he had now decided to remain in England for the rest of his life, as he was in fact to do, and seized the opportunity of a period of relative inactivity to take a holiday and see family and old friends once more. He stayed in Bremen not more than two or three weeks, going from thence to Amsterdam, then Leiden, then The Hague where he arrived on 22nd July and, finally, going the very next day to Rotterdam. Even on this brief visit he managed to see friends and make new ones: in Amsterdam he must surely have talked long with Boreel, with whom he had had so long and so fruitful a friendship and met Giuseppe Francesco Borri, a notoriously esoteric alchemist and iatrochemist, widely regarded as little better than a charlatan, an Italian who, after founding a mystic society, was condemned by the Church as heretical and so fled to Amsterdam where he spent several years before (probably) making contact first with Queen Christina and then with the Danish Court.[17] He ended his life in Italy again, first as a prisoner of the Inquisition and then, after performing some notably successful medical cures, under noble patronage in Rome. He was even more profoundly alchemical than Francis Mercury Van Helmont and hence widely distrusted, especially in later life. Nevertheless Oldenburg seems to have regarded him as being of interest, especially after Borri gave him a fascinating piece of 'incombustible wood' to be passed on to Kenelm Digby, promising to send Oldenburg an account of the method of making it—as he did, and Oldenburg read the account to the society. (Digby, elected to the society a fortnight before Oldenburg, was passionately interested in the occult. A Catholic who had lived much in France before the Civil Wars, and during them he was to act as 'Chancellor' to the widow of Charles I, while at the same time spending much time negotiating with Cromwell in an attempt to secure freedom of worship for Catholics. He was a serious natural philosopher, much influenced by Descartes and by Hobbes, as shown in his important work, the *Two Treatises* of 1644.) He was also well known for a little treatise on the 'powder of sympathy' which, so it was alleged, would cure wounds when it was rubbed on the weapon which had caused the wound, all of which must greatly have appealed to Borri. Oldenburg knew Digby as the author of a 'Discourse concerning the Vegetation of Plants' which he had read to the society in January 1660/1, a serious work on plant physiology.

In Leiden Oldenburg saw his relative by marriage, Johannes Koch, who here used the form Coccejus, as Oldenburg's nephew was later to do; Koch had migrated from Bremen to Leiden as professor of theology, yet another example of the intimacy between Bremen and the Low Countries. More significantly, it was at Leiden that Oldenburg made an important acquaintance, introduced either by Koch or by Boreel: this was a then young man not yet well known in philosophical circles, but soon to be regarded as very eminent, Benedict de

Spinoza. He had moved about five years earlier from Amsterdam to the village of Rijnsburg near Leiden, where he supported himself by lens-grinding, as can still be seen by those who visit his cottage, now a museum. Oldenburg was, very rightly, greatly impressed by this young man and on his return to England was to institute a correspondence which was to be particularly intense over the next four years. On the visit itself, as Oldenburg recalled in the letter written a week after his return to London on 9 August 1661,[18] they had

> conversed about God, about infinite Extension and Thought, . . . about the nature of the union of the human soul with the body; also about the principles of the Cartesian and Baconian philosophy.

It must have been a prolonged discussion since it touched on so many deeply philosophical subjects; evidently Oldenburg, although obviously not philosophically in the same class as Spinoza, was yet able to listen appreciatively and talk intelligently, as Spinoza's letters in reply clearly demonstrate. Indeed on the last of the topics listed Oldenburg was almost certainly the better informed of the two. In his letter Oldenburg requested Spinoza to clarify his thoughts on all these matters, but consciously or unconsciously he diverted Spinoza's attention from them in mid-August by sending him a copy of the recently published Latin version of Boyle's *Certain Physiological Essays* and asking for his comments on its argument.

This produced a now famous exchange, a fascinatingly clear example of the unbridgeable division between (to simplify) the Cartesian and Baconian points of view, or rather between those who believed that reason was the all-important tool for attempting to understand nature, completely overriding experimental evidence and those who believed in the essential importance of experiment in investigating and understanding nature. Certain 'Baconians' like Boyle and later Newton even believed that experiment could *prove* the truth or falsity of a proposition in natural philosophy, while logical reasoning, all important to Cartesians, could investigate and initiate but never prove such a proposition. Spinoza, like Descartes himself, believed that, without any question, logical reasoning alone permitted certainty even in the face of empirical evidence seemingly to the contrary, for he equated logical reasoning with mathematical or geometrical reasoning, both in its method of argument and its certainty of conclusion. He could therefore by no means accept what Oldenburg presented as Boyle's firm belief in the essential importance of experimental evidence, an importance which meant that it transcended normal reason. In the subsequent long drawn out exchange between Spinoza and Oldenburg the latter sought Boyle's help in countering Spinoza's arguments but he almost certainly needed this help only in technical matters of fact, being well able to hold his own even with such a mind as Spinoza's in the discussion of the purely philosophical arguments involved. This correspondence, which continued to go over the same ground for the next four years,[19] appears to be so much an exchange of ideas between Boyle and Spinoza that Oldenburg has often been regarded by modern historians and philosophers as acting only as Boyle's mouthpiece. However, it is clear that it was Oldenburg's interest in the discussion rather than Boyle's that was responsible for keeping the correspondence alive and that Boyle helped to maintain the discussion only under considerable pressure from Oldenburg, whose personal share in the exchange must be taken seriously. One can indeed detect a decidedly missionary zeal on Oldenburg's part which Boyle himself would never have displayed in private correspondence and which he normally disliked. Presumably he accepted it here partly because he did not know what precise use Oldenburg made of the information with which he was supplied and partly because it was purveyed with little effort on his part and he was willing to oblige Oldenburg when asked to help only by answering questions on specific points. It was certainly Oldenburg who kept the controversy alive. Clearly the greatest thinker in this three-cornered debate was Spinoza, but equally clearly he unconsciously proved to the unprejudiced observer that reason alone did not suffice for the profitable study of nature. To take a fairly trivial but clear example: Spinoza could not understand how chemical change could differ from a mixture, and so he was baffled when assured that two chemical substances which had been mingled together in a certain way (so that, in modern terms, a reaction took place) could not be separated by any physical means. In this now well-known and philosophically important correspondence, which Spinoza carefully preserved, showing that he too regarded it as important (it survives in printed form in Spinoza's *Opera Posthuma* published in 1677, the year of the death of both protagonists) Oldenburg displayed to the full his ability to understand arguments relating to both philosophical and scientific issues and, even if he needed help in compiling his facts, to set them out in clear and logical Latin. He had learned much about the discussion of problems inherent in approaching discussion of natural philosophy and this was to stand him in good stead for the remainder of his life.

After his stimulating and challenging visit to Leiden Oldenburg went on to The Hague, arriving on 22 July 1661, presumably merely to see Christiaan

Huygens whom he had found so congenial the previous spring when Huygens had visited London. He told Huygens about what had been discussed at the Society's meetings in the past couple of months while Huygens in return offered to show Oldenburg the moon through his best telescope. (This proved impossible since Oldenburg left for Rotterdam the next day.) In Rotterdam Oldenburg spent some time with Lodowijk de Bils, author of the anatomical tract which Oldenburg had translated into French while in Paris. Here he stayed perhaps as much as a week before setting out for London where, as already noted, he had arrived by 9 August. As was his custom, he promptly sent letters to both Spinoza and Huygens (and perhaps others) to thank them for their kindness in receiving and conversing with him. From this time onwards, his acquaintance with these and other foreign friends was to mature chiefly through correspondence.

A week after his return to London Oldenburg attended the meeting of the society held on 14 August 1661 when, as the minutes note without further explanation, he 'exhibited' a piece of camphire wood which, it may be supposed, he had brought back from Holland, for the Dutch merchants had a wide trade with the East. At the next meeting which was held on 28 August, Oldenburg, so the minutes record, took a more active part in the proceedings. First he read from a letter from Borri to himself, a letter now lost, its contents known from the substance of Oldenburg's reply dated 7 September 1661.[20] This letter contained Borri's account of his method of preparing 'incombustible wood' and, according to Oldenburg's report to the author, the account 'aroused unusual admiration [wonder] in me' but 'our members are very doubtful about your success', showing that Oldenburg had not yet entirely absorbed the critical spirit of the leading minds of the society. This is perhaps the first example in Oldenburg's correspondence in which he acted as a mouthpiece for the new society rather than writing from a purely personal point of view, and it displays to the full his talent for doing so, a talent which he was to refine and display to great advantage in years to come. It also shows that Oldenburg's correspondence was not infrequently to be more informative about the discussions at the society than the formal minutes. At this same meeting Oldenburg for the first time showed an experiment, as he was hardly ever to do after he became Secretary and was no longer a simple member. At the previous meeting Croone, possibly less as Register than as a chemically-inclined physician, was 'desired' (a very usual expression) to obtain a sample of 'salt of cabbage . . . to try whether it would destroy the taste of wine, as it was reported to do'. The minutes say nothing more about the matter, so one can only speculate that the report had been made by a member present at this or the previous meeting in the course of the often random discussion characteristic of the meetings and that whoever was taking the minutes, whether Croone or the amanuensis, had perhaps inadvertently omitted both the topic and the name of the proposer. Whether Croone obtained the requested sample is not stated; it was in fact Oldenburg who actually 'tried the experiment' at the meeting on 29 August. The trial was regarded as unsuccessful because, although the salt of cabbage 'much abated the taste' it did not make the wine tasteless but 'made it a mixture of vinous and lixivious [alkaline]' as one might expect since the salt was probably mainly potash; in the discussion it was sensibly 'supposed, that all lixiviate salts would do the same'. Oldenburg's performance clearly gave satisfaction, for a week later (4 September 1661) 'It was ordered' (again a customary expression) 'that a collection of all [!] quicksilver experiments be made, examined and brought in by Mr Oldenburg.' This request arose because Goddard had performed the experiment of the 'vacuum within a vacuum' (that is, a Torricellian tube placed inside a larger Torricellian tube, an experiment first performed in France where Oldenburg had probably seen it).[21] Boyle was then asked to 'bring in' his own similar experiment performed in the receiver of an airpump, as described in his published book; this he did at the next meeting after Croone had performed several other experiments to demonstrate the force of atmospheric pressure. Presumably Oldenburg was merely to act as recorder of all such experiments made in the society and, possibly, to add to them those he had witnessed at the Montmor Academy of which perhaps he had spoken. What he actually did is not mentioned in the minutes nor is there any trace of a paper on this subject by him in the archives.

Unfortunately there is now little surviving of what must have been an extensive correspondence between Oldenburg and his foreign friends during 1661 (presumably if these letters were kept they remained, as one would expect, in Oldenburg's own files) so it is impossible to know from whom he received the letters from which he frequently extracted excerpts to read to meetings. That such letters did exist is obvious, but little now survives intact except for the correspondence with Spinoza, which was preserved by the latter. What information does exist comes from references in letters by others. For example, in late April 1661 Thevenot, a very active member of the Montmor Academy, told Huygens that Oldenburg had sent him an account of Boyle's forthcoming *Certain Physiological Essays*; in the future Thevenot, a great traveller and compiler of travel books, was to correspond frequently with

Oldenburg.[22] So too there is evidence in a letter of 19 September 1661 from Robert Southwell, returned from his travels in France and living in Ireland, sometimes on his father's estate at Kinsale, Co. Cork, sometimes in Dublin. Southwell after expressing his pleasure at the good reports sent by Oldenburg of the doings of what Southwell somewhat prematurely called 'the Royall Society', referred to 'your quaere concerning Sounds and Ecchoes' saying that he had given a written account to Boyle of trials made to determine the velocity of sound by the Accademia del Cimento of Florence and listing 'whispering places' he had encountered (that is, rooms and buildings in which sound was amplified so that it was possible to speak in a whisper in one corner and have the sound heard over a considerable distance) and echoes.[23] Although the minutes do not record any discussion of sound during August and September 1661, it seems probable that the subject had arisen casually, as it was to do again in December, and that Oldenburg had been asked to make inquiries of known travellers, in accordance with what later became an invariable custom.

In the autumn of 1661 the minutes several times record that Oldenburg read papers and letters received by himself from correspondents both at home and abroad. Thus on 16 October he 'read a paper concerning a liquor to be had out of animals like the alkahest' of which he had heard, he said, during his journey through Germany in the summer of 1658.[24] This seems to have stimulated Goddard to bring in to the meeting of 30 October 'inquiries concerning the liquor alkahest, made of animals' which the amanuensis was to copy and give to Oldenburg for consideration. He in turn dutifully (on 6 November) 'brought in an answer to Dr Goddard's queries concerning the liquor alkahest of animals', but it is still not clear what this substance was thought to be. At this period there was a particular interest in histories (accounts) of trades: for example, on 30 October an account was read of the method of making 'China varnish' which stimulated the production of accounts of other sorts of varnish while 'Mr Oldenburg [was] to write about . . . the making of steel and lattin [brass] plates'. His qualifications for writing on this subject are not obvious and there is no record that he did write on it, although one would have expected that, having been asked to do so, he would have complied with the request. But the record of such matters is far from complete even in the minutes and less so in the archives.

What Oldenburg did at this time besides attending the Royal Society (to follow Southwell in convenient anachronism) is not at all clear. He possibly continued to live for a time with Hartlib, but this is not certain; he probably did continue to live in London; it is virtually certain that he worked for Boyle, now no longer as a mere purveyor of scientific news (although this was a role he played throughout his life) but acting as an intermediary between Boyle and his printers. It must have been at this time that he began his long career as Boyle's 'publisher', as the seventeenth century called what we might rather term 'editor'. He organized Boyle's manuscripts as far as he could, Boyle being a self-confessedly disorganized author, mislaying parts of his manuscripts, finding missing sheets at the last minute, changing his mind about the order of sections and so on. For although he had amanuenses since his sight had been poor since his early years as an author, none of these was capable of doing much more than writing down a text from dictation as scribes, not assistants. And to compound the confusion, Boyle always worked on several treatises at any one time, and after 1660 had the problem of having Latin texts of most of his English works. Oldenburg saw very many of Boyle's works through the press, writing prefaces, proofreading, mediating between printer and author generally. In future he was to translate into Latin most of Boyle's works published in the later 1660s and early 1670s. For all this work Boyle was to pay him at agreed rates, and possibly let him have the profits of the books in which he was concerned. He also acted as Boyle's unofficial book agent, as his correspondence bears witness. Only thus can one interpret the existence of a curiously flowery letter addressed by Oldenburg to Vincenzo Viviani, mathematician and mainstay of the Florentine Accademia del Cimento, Galileo's last pupil.[25] This was in October 1661, but the practice continued for many years.

Oldenburg was proud to be in Boyle's service without any doubt, regarding him as a worthy patron as well as an almost ideal natural philosopher, whose views he wholly accepted and was pleased to pass on to his correspondents. But more than this, Boyle was a friend as well as a patron. They were clearly not social equals, but Oldenburg was, by virtue of his university degree and his later attainments, as good a gentleman as most of the society's members when that mattered more than economic standing. Boyle was so very great a gentleman, for all that he was only the youngest son of the first Earl of Cork who had been a mere gentleman himself before his aggrandizement as a great landowner in Ireland, that to be in his employ offered status in the eyes of others. Oldenburg as a member of the Royal Society was on terms of perfect equality with the other members, whatever his source of income. That it largely came from Boyle was less important than the fact that it was Boyle who had, and was to continue to have, an overwhelming importance in broadening Oldenburg's interests and making him recognizably a virtuoso or lover of natural philosophy, worthy of inclusion in the new society of natural philosophers. Almost certainly it was Boyle who had suggested that he be included in the nascent society, but once elected it was his own talents, efficiency, industry, and interest that advanced his position within that society.

This was to appear significantly in the next year, 1662, the year when the informal society was transmuted officially into the Royal Society of London, with Oldenburg taking an important place in its structure. The new year began quietly enough. On the first of January 1661/2 Oldenburg 'read a paper concerning a new manner of cutting the stone out of a man's bladder' probably an extract from a foreign letter now no longer extant. (Indeed few letters survive from this year.) A fortnight later he was, naturally enough, asked to translate a German description of gunpowder manufacture sent by Prince Rupert who was familiar with the society because of his close relations with the Court and his acquaintance with very many of the society's members including probably Oldenburg himself. (Prince Rupert was to be made an honorary Fellow of the Royal Society in 1665.) In May 1662 Oldenburg 'produced a letter concerning a level of air', no doubt bubble-level, perhaps that of Thevenot, although no name is mentioned. In late May and early June (the entries in the minutes are incomplete) there was a report of an 'engine' (a kind of pump) designed by a man named Mr Towgood (otherwise unidentified) and Oldenburg and Petty were both instructed to take steps to organize a comparison of this 'engine' with an ordinary ship's pump. The trial was scheduled for 7 June but there is no further mention of the pump until 10 December 1662 when Wilkins reported that 'Mr Towgood's sucking pump' could raise water 42 feet—hence, whatever it was and assuming that the report of its lifting power was correct, it could not have been a suction pump. All these activities demonstrate what a thoroughly normal member of the society Oldenburg had become, a man whose interests and abilities were very little, if at all, seen as different from those of others. Rather more surprisingly, on 2 July 1662 'a new astronomical hypothesis of a stranger was referred to the consideration of the bishop of Exeter [Seth Ward, newly appointed], Dr Wren, Dr Pope, Mr Croone and Mr Oldenburg'. This is surprising only because Oldenburg had no particular association with astronomy, unlike Ward and Wren, both of whom had made significant contributions to the subject, but perhaps he was chosen for his knowledge of Continental astronomers and wide reading, perhaps merely because, like most educated men of the day, he was expected to be tolerably familiar with astronomy. Tantalisingly, there is nothing in the subsequent minutes about the 'stranger's' hypothesis. Oldenburg continued busy, on 9 July reporting on 'a paper of collections concerning the generation of insects' when he 'was desired to translate the whole book', a request possibly connected with his earlier appointment to the committee to consider the subject, but possibly merely to his skill as a translator. In any case he was by now clearly seen as a faithful, useful and able member of the society.

As already mentioned, there is no certainty about Oldenburg's place of residence during 1662. Hartlib had died in March, and so could no longer provide even a postal address. Few of the existing letters to or from Oldenburg in this year give any clue at all—many surviving only in printed form without address. Not until the end of 1662 are there any letters with the postal address intact, and then these were addressed to him at Lady Ranelagh's in Pall Mall, whether because he had lodgings nearby, as he was later to do, or because he had so much work to do for Boyle that it was convenient for him to collect letters from the establishment of Boyle's sister where the servants could put them on one side for him it is impossible to determine. To anticipate, it was not until the new year, 1662/3, that he was certainly an established householder, to be addressed 'at Mr Herbert's house in Pall Mall' which was variously said to be 'near Lady Ranelagh's' and 'about the middle of Pall Mall'. This was a good address in the recently developed area south of St James's Square and near St James's Palace, although his lodgings may have been small and poor; he was to change them later in 1663 to an address nearby, still in Pall Mall. (These changes may be deduced from the addresses on letters from John Beale and certain other English correspondents, although some, like Wallis now re-established in Oxford, continued to address him in care of Lady Ranelagh.)[26] However, by the end of 1662 correspondents who took the simple way out and merely wrote 'Mr Oldenburg, London' on their covers could expect to have the letters safely delivered.

This was as a result of the great change in Oldenburg's life and status which occurred in mid-July 1662 when a royal charter was granted to the society officially incorporating it 'under the title of the Royal Society' as the First Charter states, a title changed less than a year later in the Second Charter to 'The Royal Society of London for promoting natural knowledge'. With the First Charter came a more formal structure than previously, and for Oldenburg an official and lasting place within this structure.[27] The charter named officers—a President (Lord Brouncker), to be elected annually like the other officers, all at the Anniversary Meeting held on 30 November (as it still is). The other officers were a Treasurer (Balle in the first instance) and two Secretaries (Wilkins and Oldenburg)—together with a Council of twenty-one members, including the officers. The members of the Society from this time on were to be known as Fellows, as they still are. Important privileges now granted to the Society were the right to licence books for publication and, what made Oldenburg's future career possible, the right

> to enjoy mutual intelligence and knowledge with all and all manner of strangers and foreigners . . . without any molestation, interruption, or disturbance whatsoever. Provided [that this was for] the particular benefit and interest of the aforesaid Royal Society in matters philosophical, mathematical, or mechanical.

(Strictly speaking this applied only to letters written by or on behalf of the President in the Society's name and sealed with its seal. But this proviso was never rigidly adhered to and it was to become increasingly difficult in years to come to distinguish between official correspondence to or from the Secretary and private correspondence to and from Henry Oldenburg, a fact which was to cause him major difficulties in future, especially in 1667.) Other Fellows did continue their own established interchange of letters, as for example Sir Robert Moray who for the next three years corresponded regularly with Christiaan Huygens, but this sort of exchange became ever more private and was often channelled through Oldenburg.

Although the other officers were to change frequently, Oldenburg, as did Brouncker, was to retain his post until 1677 and to undertake the bulk of the Secretaries' duties. He now had position, which counted greatly in seventeenth-century society, and soon became not only a major figure in the Royal Society but well-known, even famous, throughout the learned world. No salary came with the position and so, although he was to work very hard as Secretary and to maintain an ever increasing correspondence at home and abroad, now at least semi-offical and often entirely official, he still needed employment to make a living. He had his work for Boyle, which did pay, but he needed to develop other sources of income. This he slowly succeeded in doing so that somehow he managed to survive financially, but not surprisingly he was to find himself perpetually short of money in the years to come.

His official duties as Secretary seem to have begun on 13 August 1662 when he read the letters patent for the incorporation of the Royal Society to a meeting at, as usual, Gresham College. It was then voted that, as soon as the

King returned to Whitehall from Hampton Court where he then was, which was of course a considerable ride from the City of London and Gresham College, the whole Council together with as many Fellows as possible should wait on the King to return thanks for his gift in a formal speech, which, plausibly, Oldenburg may have helped to draft. They did so wait upon him in Whitehall on 29 August in what was presumably Oldenburg's first attendance at Charles II's Court, although he had in the past certainly hung about Whitehall, as at Westminster Hall, to see the King go by as the custom was and to collect news and gossip. On 30 August the Society's delegation waited on the Lord Chancellor armed, as on the previous day, with a formal speech of thanks made by Brouncker as President.

From the summer of 1662 onwards it becomes possible to trace Oldenburg's activities as Secretary of the Royal Society. For one thing, the minutes of the meetings become generally fuller and as it is not likely that the new charter generated increased activity at meetings it must rather be the case that Oldenburg as Secretary was more thorough in his taking of them than Croone as Register had been. From now on it was always Oldenburg who took the minutes, wrote them up from his rough draft (some of these survive) and gave them to the amanuensis to copy into the Journal Book. On the rare occasions when he was absent there were, as often as not, no minutes kept. Other examples of increasing orderliness become apparent: for example there is no reference in earlier minutes to a Register Book into which papers read at meetings were to be entered when so ordered (the usual term) as happened frequently from this time onwards. On the other hand, Oldenburg's own name appears less frequently in the minutes except to record when he read all or part of a paper or letter addressed to him as Secretary. It seems that he was no longer usually asked to undertake the compiling of information and he modestly never records himself as taking part in discussions, although he must have done so. From this time onwards Fellows, especially provincial Fellows, entrusted him with letters which were in effect papers which they hoped to have read at meetings and foreigners, whether Fellows or not, increasingly did the same. It therefore becomes more and more impossible to separate the activities of Oldenburg as Secretary from the activities of Oldenburg the man.

In November 1662 another event occurred which was to have a considerable impact upon Oldenburg both professionally and personally. This was the appointment of Robert Hooke (1635–1703) to be Curator of Experiments, a new post specified in the Charter and not yet filled. This appointment was greeted with enthusiasm by the Fellows present for, busy men as most of the more active among them were, they found it difficult to make time for the design of experiments to be shown at meetings. For it was both difficult and expensive to provide the apparatus requisite for many experiments, while at the same time experiment was, as they all agreed, at the heart of their purpose as a society. Hooke was a dedicated, ingenious, and skilful experimenter, well known to many of the Fellows as an Oxford graduate and as Boyle's laboratory assistant. He had, as everyone knew (for Boyle had acknowledged it in print) designed and built Boyle's first airpump to Boyle's specifications and assisted him in the pneumatic experiments which Boyle had published in 1660 and 1662. Hooke himself had published a little book in 1661 about certain 'Phaenomena' (the rising of water in very small tubes—by capillary action as it came to be called much later) which he had come across during his work for Boyle and it had indeed been mentioned in Boyle's 1660 book. This little work Hooke dedicated to Boyle very much in the manner, allowing for the more formal language of the times, of a modern research student to his supervisor presenting work which had begun under the supervisor's auspices and probably at his suggestion. It had been discussed at a meeting of the Society soon after 10 April (there are no minutes for the relevant meeting). Now Boyle was thanked for 'dispensing with' Hooke in favour of the Society (he obviously could not work with Boyle in Oxford and with the Society at its weekly meetings in London at the same time). Boyle was not contemplating any more work on pneumatics at this time, but in any case Hooke corresponded with him and called on him whenever Boyle was in London. When there, Boyle invariably stayed with his sister Lady Ranelagh. As nothing was immediately said about paying Hooke, it is probable that Boyle continued to support him at least until late in 1663 when it was voted that he be paid a salary of £30 a year by the Society, not enough to live on and not officially added to until 1665 when Hooke became Gresham Professor of Geometry, with rooms in the College. It was intended that as Curator of Experiments Hooke should 'bring in' as the term was, one or more experiments to each meeting, as he was pretty faithfully to do for a good number of years to come, either showing the experiments to the meeting or describing and commenting on them. He was elected a Fellow in June 1663, and it should be noted that he was never a 'servant' of the Royal Society as were the amanuensis who assisted Oldenburg and the Operator who assisted Hooke, but a salaried officer and so, socially within the Society, the equal of any other Fellow, a gentleman by virtue of his university degree. He and

Oldenburg were thus on much the same footing except that Oldenburg was not to receive any salary for another five years. The two men were to work harmoniously together for the good of the Society for the next dozen years: they of course met weekly at Gresham College at the Society's meetings; they met frequently during the interval between meetings to discuss Society business; they were often together at the Coffee Houses to which Hooke was partial and where many Fellows gathered after meetings; they often encountered each other at what Hooke called 'Boyle's', although it was in fact Lady Ranelagh's house; and when that did not happen Hooke frequently called on Oldenburg.[29]

On St Andrew's Day, 30 November 1662, there should have been an election of officers and Council but, as Brouncker told the meeting on 26 November, it had been decided, presumably by the Council,that it was necessary to apply to the King for a new charter. By vote of those Fellows then present it was agreed that the officers and members of Council named in the first charter should remain in office until the new charter had been granted, of which there seemed to be no doubt. In the winter of 1662–63 the Society continued its regular meetings, elected new Fellows, witnessed experiments performed by Hooke and others, discussed scientific news and information reported by Oldenburg and other Fellows derived from their correspondence. There is no record of the discussion which must have taken place about the amendments to be enshrined in the new charter but presumably this was left to the officers and Council, in which case Oldenburg must have taken a hand in its formulation. The resultant draft was presented to the King in the early spring, he approved it quickly and it became official on 22 April 1663.

It differed from the first charter in small but significant ways. To begin with, the Society's name was changed as has already been noted although on all ordinary occasions it was (as it still is) referred to simply as 'The Royal Society'. The King now declared himself to be its founder and patron and bestowed on it a splendid silver mace which was and is carried before the President at meetings. (Charles II subsequently kept in slight touch with 'his' Society, being usually aware of some of its activities, famously teasing its Fellows 'for only weighing air', but although foreigners often assumed otherwise, he never granted the Society any money.) Those members admitted within the two months after 22 April 1662 were called Original Fellows (mostly but not all those already Fellows under the first charter); Fellows elected after this time were to be elected under strict rules as to elegibility. The most important change was that the Royal Society was now allowed to frame its own statutes which, unlike the charter, might be freely altered, and so it became in effect self-governing. The first statutes drawn up in 1663 covered the regulation of meetings, admission of Fellows, and the duties of the officers. The Secretaries, or either of them, were to attend all meetings of the Society and of the Council (which now met fairly regularly, with formal minutes which were preserved), to read the previous meeting's minutes, to take fresh minutes which the clerk (or amanuensis) was to copy into the Journal and Council Minutes Books, and to read out the names of any candidates to be considered for admission to the Fellowship, to be voted upon subsequently. They were also to 'draw up all letters to be written to any persons in the name of the Society or Council' to be approved at a subsequent meeting before being sent. Although it was originally envisaged that both Secretaries should be active, in practice from 1663 until his death Oldenburg acted as the principal Secretary, attending the meetings, writing the minutes, overseeing the clerk, and handling the bulk of the correspondence, much of it on his own initiative. Wilkins, who remained Secretary for the next five years, played some role in the Society's affairs, for example overseeing Thomas Sprat's *History of the Royal Society* (history in the Baconian sense of natural history or survey) begun soon after this although, after many delays, not published until 1667. This outlined the constitution of the Society and its purpose, gave examples of the work of its Fellows, and defended its activities in the face of some contemporary criticism. Wilkins was to be succeeded in 1668 by Thomas Henshaw, a civil servant replaced in 1672 by John Evelyn when Henshaw was appointed an envoy to Denmark, whence he was to send, addressed to Oldenburg, a series of vivid letters about the natural history of the country, returning to England in 1675 to resume his position as Secretary. Neither Henshaw nor Evelyn made much of a contribution to the routine duties of the Secretaryship. In 1662 and 1663 it was clearly not foreseen what a burden was thus placed on the Secretaries nor, of course, that in fact it was to be Oldenburg who, single-handed, conscientiously carried it out. No other man would ever attempt it. No remuneration was offered Oldenburg until after about five years he pleaded with the Council for financial assistance, made necessary not only to give him a better income than he would otherwise have had, but to cover the increasing cost of his ever-expanding correspondence, and it was to be some years before he managed to secure an arrangement which relieved him of the cost of foreign postage (paid in those days by the recipient).

Even before the passage of the Second Charter, Oldenburg had begun to initiate correspondence in the name of the Society in the manner soon to be expected of him, no longer as a private individual but as an officer of the Royal Society with which he soon became closely identified.[30] Even domestic correspondents like Beale, continuing an established practice of writing to Oldenburg's private address as to a friend, habitually inserted an express injunction that the contents should be shared with the Society (as they faithfully were). So too Wallis, to become extremely friendly over the years, from the beginning assumed that his letters were intended for the information of the Society as well as Oldenburg. There were soon new domestic correspondents and new foreign ones as well.

One of the first and most important of foreign scientists was Johann Hevelius (1611–87), the distinguished Danzig astronomer. Hevelius had travelled in England in the 1620s, probably visiting Hartlib with whom he shared a birthplace, and in France where he met many important scientists; since his return he had been occupied with the family brewing business which supported his avocation of observational astronomy, at which he was singularly skillful. He had built an observatory, stocked it with excellent instruments, and installed his own printing press with which he printed a succession of beautiful and authoritative books filled with his own observations. In 1647 he had published *Selenographia* containing the most careful study of the surface of the moon up to that time. He had sent copies of this to various astronomers in Oxford, including Wallis and Ward and in turn Ward had dedicated to him along with other distinguished astronomers his own *Astronomia Geometrica* of 1656. He was therefore well known and esteemed in England. In 1662 Hevelius published *Mercurius in Sole Visus*, an account of the transit of Mercury across the sun in the previous year, seen by few. To this account he appended the previously unpublished *Venus in Sole Visus*, an account of the transit of Venus across the sun, a rare event also little observed at the time, by the English astronomer Jeremiah Horrox (1617?–41), from a transcript of the original supplied by Wallis. Hevelius's book had been mentioned at Royal Society meetings, probably by Wallis, and it was this which stimulated Oldenburg to draft a letter to Hevelius telling him about the aims of the Society, urging him to continue his good work, and asking him to share the results of his labours with the Society by means of letters to Oldenburg (to be sent to his private address at this time).[31] The letter was apparently drafted on Oldenburg's own initiative; for some unknown reason he waited for nearly a month before, as was proper, obtaining the Society's approval of the contents, dispatching a corrected version a week later on 18 February 1662–3. To anticipate, Hevelius welcomed the approach by

Oldenburg and the Society and, replying at the end of 1663, he described the work on which he was then engaged and the books which he intended to publish.[32] Postal communication was slow between London and Danzig because letters either went via Hamburg which possessed a postal service with London or, more securely and speedily, by the hand of a ship's captain making an appropriate journey to or from Danzig, this being the only possible method for parcels. Not surprisingly letters were from time to time, and especially in wartime, lost in the post, which explains why Oldenburg commonly summarized his previous letter each time he wrote. On this first occasion Hevelius sent his reply by a young relative, probably a merchant, who was also charged with an attempt to recover a debt owed to Hevelius by Hartlib, necessarily a futile mission since Hartlib had died nearly two years earlier, but he could and did deliver the letter to Oldenburg.

From this time onwards correspondence between the two men was constant, although necessarily slow and irregular. On his side Oldenburg demonstrated that he was quite well enough versed in current astronomical issues to understand what Hevelius wrote, to summarize it to Society meetings and to explain to Hevelius what had been the reaction of the astronomical Fellows to what Hevelius wished them to know, all of which he accomplished with tact and skill. He and Hevelius developed a mutual and always friendly respect for one another and Oldenburg willingly carried out various commissions for Hevelius, undertaking the sale of copies of the astronomer's books as they arrived in his hands, purchasing for him books available in London but not in Danzig and finding safe methods of conveying them to him. Other correspondents outside England at this time continued to be found in France, especially Paris, and the Netherlands. There were also correspondents in the English speaking world: these included Southwell in Ireland, who kept Oldenburg and hence the Society supplied with the partial success of Petty's original 'double-bottomed' boat (a catamaran)[33] and, more ambitiously, John Winthrop in New England who sent a considerable amount of information to the Society through Oldenburg after he had learned of Oldenburg's position as Secretary (probably either through Lord Brereton, a patron of Pell and of Dury, or through Moray).[34] The subjects which interested Winthrop were various: agriculture, the weather, the Indians, and astronomy, among others.

Here it should be noted how many and various, at this time and later, were the technical subjects with which Oldenburg had to deal. There was always astronomy. There were the philosophical problems connected with Boyle's theory of matter about which he was then writing to Spinoza. There was agriculture of various kinds, especially in letters from John Beale ranging from the 'hereditary interest' in cider for which he was renowned to new crops, and also in letters from Beale's friend John Evelyn who however lived near enough to London (in Deptford, Kent) to attend Society meetings from time to time. Evelyn shared Beale's interest in cider and cider apples about which both men wrote extensively: Evelyn's well-known book *Pomona* (1664), a compilation, was in considerable part based on letters at least ostensibly addressed to Oldenburg. At this time his chief interest was in gardening, particularly French gardening (he had published a translation of a famous French treatise on the subject when the French style was newly popular, and was to translate others) but also in forestry and in horticulture generally. It was probably through Evelyn that Oldenburg began a correspondence with a famous French horticulturalist Jean de la Quintinye.[35] He was an expert on the cultivation of melons and was to send samples of seed which Oldenburg distributed among those Fellows who were particularly interested in horticulture, including Evelyn, Sir Paul Neile, and Charles Howard, a younger son of the fifth Duke of Norfolk.

The existence of the Royal Society and its potential importance quickly became known abroad and in 1662 was fully realised by many natural philosophers in Paris, above all by the members of the Montmor Academy, so many of whom had not only known Oldenburg when he attended their sessions but had kept in touch with him after his departure from Paris. They all seem to have regarded him with respect, naturally even more so after he became Secretary of what they regarded as a new royalist foundation. In 1663 there were no fewer than four foreigners who attended the Society's meeting on 10 June: Christiaan Huygens, once again in London, this time with his father Constantijn, diplomat and poet, who had often lived in England and who now or later began a close and amicable friendship with Oldenburg, Samuel Sorbière, an active member of the Montmor Academy, and Balthazar de Monconys who had also known Oldenburg in Paris. Both of these latter subsequently wrote accounts of their visits[36] (Sorbière was to annoy the Montmorians by seeming to present himself as an official representative of their academy, although both he and Oldenburg denied his doing so, and he annoyed the English later by unflattering and superficial descriptions of the English character in his book.) Both men admired the Royal Society unreservedly, especially the formal manner in which its meetings were conducted: as they both noted, the President sat with the Secretary at a square table with the mace before the President, while the Fellows who were seated on benches did not chat freely among themselves or if they did were promptly called to order by the President's gavel, and hence the speaker of the moment was always audible. At this meeting both Huygens and Sorbière were elected Fellows. Monconys was never elected although he seems to have understood the experiments shown and discussed them at more length and more intelligently than did Sorbière. As Oldenburg later told the Montmorians, Sorbière's election was made at least partly out of respect for his membership in the Montmor Academy, as well as for what he said about 'his zeal for the advancement of solid and useful science'.

The problems that could face a Secretary as the Royal Society's fame spread were many, especially as not all learned men who heard about it properly grasped its purpose. Thus in the summer of 1663 the Society received a letter from a German physician and theologian named Leichner who seemed to think that the Society was more closely linked to the monarchy and the state than was at all the case and also assumed that it was concerned with more general subjects than was the case either, writing about the state of theology in schools and universities and asking the Royal Society to join forces with him to promote the subject.[37] Wilkins, Wallis, Pell, and Hooke (a somewhat odd combination) were asked to read and comment upon the letter, as the procedure required, but it was, as it became ever more common in the future, a draft letter written by Oldenburg a month later which, after being approved at a meeting, was sent signed by Oldenburg alone. Here he simply declared that

> the Royal Society says it is not its concern to have any knowledge of scholastic and theological matters, for its sole business is to cultivate knowledge of nature and useful arts by means of observation and experiment, and to promote them for the safeguarding and convenience of human life.

Nothing could better show Oldenburg's secure grasp of the principles on which the Royal Society had been founded, principles which he himself now held most firmly and which he was to enunciate when the occasion demanded in future years. It should perhaps be noted that correspondence with Leichner, as with Hevelius and other learned Germans, was always in Latin; this was not because Oldenburg had forgotten his native tongue but rather that, as was true in diplomacy, German was still at that time quite unsuitable for the discussion of abstract and technical ideas (unless in alchemy).

All his work for the Royal Society kept Oldenburg very busy. He had weekly meetings (on Wednesdays) to attend, and these he rarely missed.He had to write up the minutes, letters had to be drafted and often copied out, he had to oversee the amanuensis who copied Oldenburg's minutes into the Journal Book, copied incoming letters into a Letter Book (unless Oldenburg did so), sometimes copied outgoing letters, and copied papers read at meetings into the Register Book, when they were made available by the author and the Society had so directed. When the amanuensis failed to do any of his tasks Oldenburg often took them over. Either the amanuensis (after 1663 properly called the clerk) filed away incoming letters and papers, or more probably Oldenburg did so, and they both had charge of the Society's 'Books' (the archives). The surviving record of the Society's work at this time should, in theory, be remarkably complete. That it is not so was generally not Oldenburg's fault, for not every Fellow handed his papers in after they were read at meetings, and similarly not everyone (and this included Hooke as Curator of Experiments) who performed experiments at meetings wrote them up when requested to do so—a failure familiar to organizers of meetings today.

Busy although all this made Oldenburg, it paid him nothing. Hence it must be the case that Boyle, always his patron, provided him with enough remunerative work to support him. He wrote weekly to Boyle in Oxford, conveying the substance of what occurred at meetings of the Society as well as news gathered from his correspondence, especially news from abroad. Boyle found it important that Oldenburg should keep in touch with his old friends of the Montmor Academy, that he told them what was happening in England and in return receive scientific and political news. (He could thus report in what high esteem the French academicians held Boyle.) Oldenburg saw to sending copies of Boyle's books to Paris when requested to do so; these were the English editions when the recipient was known to read that language (as few of the French did) and otherwise Latin editions as these became available. As already mentioned, Oldenburg was soon acting as Boyle's publisher, dealing with the printers, reading proofs and writing prefaces in return for which he must have been paid, probably with the profits of the sales of the books. He also began steady work translating many of Boyle's books into Latin for publication in England and distribution on the Continent, for Boyle disliked having his books published in Amsterdam in unauthorized translations, which he regarded with suspicion as full of errors and perhaps deviations from the original; very possibly too he wished to secure the potential profits to Oldenburg.

Oldenburg by this time (1663) was hoping to derive some income by means of his already extensive correspondence, as suggested by vague references in letters to the profit which might be made by purveying news on a regular basis to subscribers to his service—clearly he did not as yet think of a journal but rather something more nearly approaching some of Hartlib's schemes. He certainly had such a plan as regards Ireland which he had first discussed with Southwell in 1662/3, but which he seems to have given up within a few months.[38] He had always encouraged his friends in Paris to send him political as well as scientific news, both of which he passed on to Boyle. Although there is no trace of his passing it to anyone else at this time, it is not impossible that he did so. In the autumn of 1663 he began a virtually uninterrupted and very regular correspondence with a man who was to supply him with immense quantities of political news and general gossip, by no means always reliably accurate but always of interest.[39] This was Henri Justel (1620–93), a Protestant scholar who held the post of secretary to Louis XIV, an office inherited from his father who had purchased it in the then usual way; this was an official office, not a close personal position, which probably involved only nominal duties. Justel possessed an excellent library and in the 1660s and 1670s was host to a group of learned men who met regularly at his house. (In 1681, as religious intolerance towards Protestants increased markedly in France, Justel was to sell his library, emigrate to England, become F.R.S. and keeper of the royal library at St James's Palace, probably not an onerous position.) Justel had a magpie mind, highly disorganized: when he wrote his sentences bore little relation one to another, his handwriting was execrable, and his scientific news inaccurate (he had no direct access to it) but his political news was more reliable and always interesting. Much of the political news which Oldenburg planned to utilize for monetary ends probably came from Justel, the rest being derived from Holland and other parts of the Low Countries. Amsterdam was a fruitful source and here an important correspondent was Peter Serrarius, an enthusiastic chiliast and general theological writer who regularly supplied Oldenburg with news of the latest milleniarist ideas currently being promulgated, a subject in which Oldenburg, Boyle, and very many others displayed a lively interest, and which was also thought to be of political importance.[40]

Oldenburg's private circumstances were to change greatly in the autumn. On 20 October 1663 he obtained a licence from the Canterbury Faculty Office to marry 'Dorothy West of the parish of St Paul Covent Garden . . . aged about 40 yeares and a mayden of her own Disposing', as it declared, Oldenburg himself, so he deposed, being a bachelor 'aged about 43. yeares' (which fits the assumption that he was born in 1618 or 1619). The marriage itself took place two days later in St Mary-le-Savoy in the Strand.[41] It is not apparent why this mature couple sought to be married by the relatively expensive method of special licence rather than the cheaper if slower method of having the banns read in church. Very little indeed is known about Dorothy West besides what is stated on the marriage licence which reveals that her father was dead. Oldenburg later told Boyle[42] that she had inherited four hundred pounds, held on her behalf by her trustees. Of this sum the married couple received half on her marriage, half of that being spent in setting up a new household in Pall Mall very near to Oldenburg's old lodgings and somewhat closer to Lady Ranelagh's. (A year later Beale was to address a letter to Oldenburg 'At Mr Storey a Stone Cutter in the Pell Mell in St James his fields', the modern St James's Square).[43] The money went on furnishings, 'the fine for the house' (the down payment on the lease which was for £30 a year), and immediate living expenses. The rest of her dowry remained in the hands of Mrs Oldenburg's trustees about whom we know more than we do about her: they were two baronets, Sir Brocket Spencer, a Hertfordshire gentleman and Cambridge graduate, and Sir John Cotton, a man of affairs and also a Cambridge graduate from either Cambridgeshire or Huntingdonshire (there being, curiously, two baronets of this name with similar careers and similar ages). Presumably these trustees were friends of Dorothy West's father, who must have had a reasonable standing in the world, well to do but not obviously rich, possibly also a Cambridge graduate. He is otherwise not identifiable nor is it known whether he had other children; it is only reasonable to guess that he or some other members of the family may possibly have been in some way connected with John Dury or his wife but impossible to learn how Oldenburg met and wooed his future wife, marriage to whom gave him a comfortable household at last and a modicum of prosperity.

Soon it brought him an enlarged family as a result of the death, probably early in 1664, of Mrs Dury, when her ten-year-old daughter, Dora Katherina Dury came to join them.[45] She was, at her father's request, made a ward of the elders and ministers of the Dutch congregation of the Church of Austin Friars in London, but the Oldenburgs effectively took over her guardianship. Exactly how the Oldenburgs came to be chosen is not clear, but probably they were chosen by the child's father. It is all rather odd, for Dora Katherina Dury had well-connected maternal relatives, including remote connections among the large Boyle family and she had a surviving half-brother, any one of whom might have been expected to show some interest in her welfare. The child possessed property from her mother, both personal property and land; if Oldenburg was in any sense formally her guardian this might have improved his finances, for guardians were entitled to use the income belonging to a ward, but as he never mentioned this when reckoning his income at that time it seems likely that his guardianship was only such in name, not in law, or that the money all went to her father, for as was to appear after his death, Oldenburg did sometimes transmit money to Dury in years to come.

After marriage Oldenburg still felt desperately in need of more income and slowly began to take steps to acquire it. He began a new venture in the summer of 1664 directed towards something more formal than personal correspondence and began to think of setting up some mechanism for a news sheet which he might be able to sell to subscribers. He then asked Boyle to look out for 'any curious persons, that would be willing to receive weekly intelligence, both of state and literary news'.[46] He reckoned that 'ten lb. a yeare will be the most expected; 8. or 6. will also doe the business', evidently hoping to secure enough subscribers to make up a substantial sum, as even ten or a dozen would be expected to do. He mentioned the subject again to Boyle in November[47] when he wrote of being

> offred . . . a new correspondence at Paris for all the news and Curiosities of France and Italy [by] a person of quality and philosophically given, which maketh me unwilling to decline the offer, if I had but leasure and means to entertaine it, as I ought,

adding that the proposed correspondent sought in return nothing more than news of books published in England and accounts of the activities of the Royal Society. It is tempting to guess that this correspondent, never named, was Adrien Auzout, from whom many letters survive from the next year, 1665;[48] certainly it seems that Oldenburg did accept the offer after telling Boyle about it. Auzout, whom Oldenburg must have met in Paris since he was an active member of the Montmor Academy, a man of whom Boyle should at this point have heard, had many connections with Italy which would explain why he could be expected to have news of activities in both France and Italy. He was an able although by his own admission a lazy natural philosopher, active in both astronomy and optics (the effective inventor of the filar micrometer for use with telescopes) as well as the deviser of some ingenious experiments with the Torricellian tube. In 1666 he was to become one of the first members of the newly founded Académie royale des Sciences.

Parenthetically it must be remarked that it is ever more difficult to understand how Oldenburg had the means at this time to support his ever-increasing correspondence, for postage was by no means cheap. This explains why Moray in 1665, sending Oldenburg a letter for Huygens and perhaps others,[49] apologized for 'load[ing] your packet with other letters than your own' and thereby causing him 'trouble & expence' necessarily involved in receipt of such a 'packet'. Sometimes foreign correspondents could frank (prepay) the postage part of the way, for example when the letter was sent via Antwerp, but then Oldenburg must have felt obliged to reply in kind and in any case there would always have been some postage due on reception. That Oldenburg paid for letters addressed to him personally is obvious; less obvious is the question of who paid for the letters addressed to him as Secretary of the Royal Society. Presumably he did not have to pay for letters addressed to him care of Lady Ranelagh, as did those now very frequent letters from John Wallis, while one can hazard the guess that Boyle's letters to him came in the same way. But some not inconsiderable expense for a man with an uncertain in-come was inevitable, and it must have been a great relief to Oldenburg when, as happened in a few years, a remedy was found. Meanwhile he managed as best he could.

At the end of November 1664 Oldenburg told Boyle that his new correspondent had sent him word of 'a dessein in France to publish from time to time a Journall of all what passeth in Europe in matter of knowledge both Philosophicall and Politicall' by means of news of books, accounts of new experiments and discoveries, news of practitioners of both the arts and the sciences, lists of libraries and academies, and disputes in the world of learning. This was the first news to reach England of the proposed founding of the *Journal des Sçavans*, the first number of which appeared early in 1665.[50] It was never so ambitious a publication as the prospectus might suggest, being primarily devoted to notices and brief accounts of new books in all fields of learning. True it did from time to time include extracts from foreign journals and short papers with accounts of experiments and inventions, particularly those made by members of the Académie royale des Sciences. Yet it was not, and was never intended to be, a journal devoted more than incidentally to natural philosophy or technology. As Oldenburg told Boyle in the same letters in which he mentioned the journal's foundation he had been

> sollicited to contribute what I can concerning England, and what is found there, as to excellent persons, things, books . . . to be paid in the like coyne from France of what passeth there and in Italy etc. concerning these particulars.

And he added that he was

> very unwilling to decline this taske, but yet how to undertake it, being so much already charged upon me, I doe not yet know,

a somewhat pathetic declaration.

Although it is clearly not true, as his French contemporaries claimed and some later historians have asserted, that Oldenburg copied the idea of a scientific journal from the *Journal des Sçavans*, since that was plainly intended to be a different kind of journal, yet it is very probable that the news of the projected French journal stimulated him to put into practice the intentions which he had been excogitating about his own 'news sheet'. That it was not a direct copy is obvious from the very different plan which he developed only two months later. The first issue of the *Philosophical Transactions* began as it was to continue, as a *scientific* journal, and it rapidly took on the character which it has maintained ever since as a monthly journal of great importance. Curiously, there is nothing in Oldenburg's correspondence about his plans or intentions. Either he discussed it face to face with Boyle, or made up his mind too rapidly for discussion, or wished to keep his plan secret until it was accomplished. Whatever was the case, the first public statement of his new project was on 1 March 1664–5 when the Royal Society's Council minutes recorded that[51]

> It was ordered . . . that the *Philosophical Transactions*, to be composed by Mr Oldenburg, be printed the first Monday of every month,if he have sufficient matter for it, and that the tract be licensed by the Council of the Society, being first reviewed by some of the members of the same.

(The reason for the last proviso was because the Society was the licenser.)

Brouncker as President was requested to license the first issue 'to be printed by John Martyn and James Allestry', printers to the Society. The full title of Volume I (for 1665 and 1666) was 'Philosophical Transactions: Giving Some Accompt of the Present Undertakings, Studies, and Labours of the Ingenious in many Considerable Parts of the World', each issue or number being headed simply 'Philosophical Transactions'. This full title emphasizes the intention that it should contain not merely an account of what was said and done in the Royal Society or even in England but everywhere that natural philosophy was pursued. The term 'the ingenious' indicated that it was not a literary, legal or theological journal (subjects possibly covered in the *Journal des Sçavans*) but was confined to natural philosophy, inventions, natural history, travels, and similar subjects. The first number, dated 6 March 1664/5, consisted of sixteen pages and contained accounts of contemporary telescopes being made in Rome and then much admired, Hooke's observation of a spot observed on Jupiter, a long account of the recent comet by Auzout, a resume of Boyle's *History of Cold*, then in the press, an English account of a monstrous calf, an anonymous account of German and Hungarian minerals (in fact sent by Abraham Koch or Cronstörm, a Swede who wrote to Oldenburg from Liège) a description of whale-fishing as practised near the Bermudas and an Engish account of the use of pendulum watches at sea based on the trials made by 'Major Holmes'[52] as well as a brief obituary of the French mathematician Fermat (d.1665). Subsequent issues were filled with similarly diverse matter from home and abroad, mainly presented in the form of letters.[53] Such letters were sent either to the Society or, increasingly, directly to Oldenburg and generally intended to be read to meetings of the Society and/or published in the *Philosophical Transactions*. There were accounts of suitable foreign news relating to natural philosophy and inventions and occasionally translations from foreign journals. Beginning with the second number there were accounts of books, the first of these being a summary and laudatory review of Hooke's then recently published *Micrographia*. As time went on the formula remained unchanged. Oldenburg himself chose and edited the extracts of letters, papers, and brief accounts from foreign journals and wrote most of the book reviews, where desirable asking for help from others—so Wallis wrote most of the reviews of mathematical works. Oldenburg also wrote prefatory dedications to each volume, Volume I being, very properly, dedicated to the Royal Society itself. He also wrote the prefaces which began each volume, occasionally taking (or being given) advice from others, especially from Beale. Oldenburg naturally proofread all the sheets and he compiled the index with which each volume concluded. It was all a considerable undertaking, but at least he always had plenty of material available. The first volume, for reasons which will appear in what follows, was made up of twenty-two numbers running from March 1664/5 through February 1666/7. Later volumes generally spanned the calendar year beginning, as the English custom then still was, with March. Ideally they contained twelve issues, although for various reasons, mostly not the fault of the editor, this was not always possible. (Volume II was the worst example, for circumstances caused it to contain only nine numbers.)

Oldenburg initially hoped that he might make £150 a year by the sale of his journal, which, it must be emphasized, was his alone, although licensed by the Royal Society. In fact he was never able to make anything like so much and was soon fearing that £50 would be as much as he could rely upon. The financial details are uncertain but it appears that Allestry had initially agreed to give him £3 per printed sheet. Oldenburg was to pay half the cost of the illustrations and the print run to be 1000 copies, although in fact Allestry sometimes varied its size. There is no firm record of the way in which the profits from sales of the issues were managed nor is even the selling price known.[54] After a year (March 1665/6) Oldenburg was complaining to Boyle that the printer, convinced that there was no market for such a journal (seventeenth-century printers were notoriously reluctant to print mathematical or even scientific books) had printed too few copies so that the first issue was already sold out. Oldenburg even came to believe that the printer had gone so far as to discourage would-be buyers. And certainly he was unduly dilatory in giving Oldenburg the money due to him from sales.

As all this suggests, the journal was almost instantly much sought after. Unfortunately for Oldenburg he was forced to give away many copies, chiefly to foreigners like Huygens and Hevelius, while not unnaturally he had to give copies to the successive editors of the *Journal des Sçavans* in exchange for copies of their journal. From these he not infrequently printed extracts, as they did in turn from his *Philosophical Transactions*. The critical success of the *Philosophical Transactions* was great, it soon acquired an extraordinary reputation at home and abroad and this redounded to the already high credit of the Royal Society. It remained Oldenburg's private venture although it was universally taken to be the Royal Society's official organ, since not unnaturally the formal title was too long for ordinary use. The world insisted upon believing that the journal recounted what happened in the Royal Society, in spite of its far more international contents. Oldenburg firmly disapproved when the first translation of any of the volumes into Latin, eagerly anticipated abroad, was entitled 'Transactions of the Royal Society by Henry Oldenburg', at which he was openly dismayed.[55] At the end of issue number 12 of 7 May 1666 he did his best to disabuse the readership of this idea by writing rather petulantly

> Several persons perswade themselves, that these *Philosophical Transactions* are publish't by the *Royal Society*, notwithstanding many circumstances, to be met with in the already publish't ones, that import the contrary

adding that it was truly the case that he, a private person, was wholly responsible for them. He was to repeat this assertion in years to come without ever convincing readers of its truth, and it was one of the reasons why he disliked the various Latin translations which later appeared. His claim to be entirely responsible for the contents of the journal was effectively true, for although he received some assistance occasionally from individual Fellows when circumstances demanded and always had the encouragement and approval of the Royal Society, he was the editor and chose what to print and when, although of course he often responded to requests that something be printed or withheld.

Responsibility for the *Philosophical Transactions* did not affect Oldenburg's work for Boyle by which he supplemented his small income. He was the publisher of Boyle's *Experiments and Considerations touching Colours* of 1664 and the same year translated and published the Latin edition destined for sale abroad, signing the preface of the English edition with his initials. His letters to Boyle chart its progress through the press: thus we know that he had sent to Boyle corrected proof sheets of all but the concluding experiment by August 1664, and these must have been the proofs of the Latin edition for the custom of printers was then to publish in the autumn while dating the title pages with the next year's date.[56] He was the next year publisher of Boyle's *New Experiments and Observations touching Cold* (1665), the experiments described therein having been made during the previous, very cold winter. Although the printer had begun to run off the final sheets in September 1664, progress was very slow, to the disgust of both the author and the publisher, being partly held up by the cold weather during the winter of 1664–5 when the presses could not print, and partly by the dilatoriness of the printer, so that it was not finished until March 1664/5 according to Oldenburg's dating of the preface. By November 1664 Oldenburg had begun to translate an early section of the book into Latin on his own initiative, warmly welcomed by the author. The first part, 'New Thermometrical Thoughts and Experiments', was sent to the printers by Oldenburg by the end of the year simultaneously with the later parts of the English text. But although some of the text was in print by the end of the next summer and the whole of the translation was finished by the end of the year 1665, it is doubtful whether the text was ever officially published. No trace of any Latin version of the book has ever been found either in England or on the Continent, so that it is unlikely that the printer sent the sheets to be published in, say, Amsterdam, unless the parcel was lost in transit, the other possibility being that the sheets were lost in the Great Fire of London, but neither Boyle nor Oldenburg ever mentioned the Latin version again, nor have bibliographers so far found even one copy.[57] Boyle paid Oldenburg for all his translations at a fixed sum per sheet and, as already noted, presumably allowed him any profits from the books of which he was publisher, as the custom then was. He also probably paid him for proofreading and his general assistance. As will appear, Oldenburg regularly undertook this kind of work in the future and it must always have provided him with a reliable if not constant income, small but useful.

Work for Boyle and the publication of the *Philosophical Transactions* were both financially profitable even if not so much so as Oldenburg might have wished. Work for the Royal Society was satisfying but unremunerative yet, since it gave him status among learned men both at home and abroad, it was well worth his efforts. It was precisely during 1664 and 1665 that Oldenburg's reputation became established to the point where he represented the Royal Society in the eyes of the world of natural philosophy. It was a rapid rise from the obscurity of only three years earlier, and his professional life was now firmly established. It was to be in no way his fault that the next few years were filled with difficulties, for most of these were the result of public disasters of the times in which he lived.

4
The difficult years 1665–7

The hard winter of 1664–5 brought with it a series of catastrophes for England and immense difficulties for Oldenburg personally, most of them arising directly from the national disasters. Those years were so full of dire events that many invoked the milleniarist tradition of prophetic numerology that turned the year 1666 (1000 + 666) into the herald of, possibly, the end of the world foretold in the Book of Daniel, and at least presaged great turmoil and chaos in the social and political world. Certainly England was to be beset with enough catastrophes to satisfy the prophets: a lengthy and on the whole disastrous war, a devastating outbreak of plague, and the Great Fire of London. And all these in different ways affected Oldenburg himself if not disastrously at least for the worse.

As if national disasters were not enough,the period opened for Oldenburg with a private calamity. At the beginning of February 1664/5, after less than a year and a half of marriage, Mrs Oldenburg died of some unspecified illness. She was buried in their parish church of St Martins in the Fields on 4 February, to be mourned by her husband, his loss rendered the greater by its effect upon his financial status. At her death her widower received the remainder of her dowry of which a sizeable part was necessarily spent on her funeral, the rest vanishing, as Oldenburg later ruefully noted, on household expenses.[1] It is tempting to wonder whether Dorothy Oldenburg, by then in her early forties, might not have died in pregnancy or childbirth, although there is no existing reference to such a thing. However, if this had been the cause of her death it could explain when and why Oldenburg wrote (and carefully copied out) a long essay,[2] ostensibly a letter, addressed 'Dear Child' and signed 'Yr Affectionat Father H.O.' It is headed 'Admonitions and Directions of a good Parent to his Child, especially a Son', indicating that it was written before the expected child's birth.

To understand this little work it must be remembered that Oldenburg had grown up with strong Calvinist convictions even though over the years these broadened to accommodate the Anglicanism of Restoration England.

Certainly by the 1670s he could take the sacrament according to the Anglican rite. More important still was the fact that he was, all his life, deeply and truly pious and that his early training in personal piety never faltered. What he wrote here was a very personal statement equally applicable to an adult or a child and no doubt it described the way of life that he himself strove to follow. It is by no means dissimilar in tone to the reflective piety expressed by many literate Englishmen of his day when intense soul-searching was regarded as an important religious exercise. A Calvinist touch is perhaps noticeable in the opening sentiment concerning the obedience which the child owes to the parent as a reflection of the obedience which the parent owes to God, a sentiment which reappeared strongly in evangelistic Victorian Anglicanism. Most of the essay is concerned with daily life and the need for the constant exercise of piety. First, continually remember God the Creator, beginning each day by

> considering soberly, what you are to do in it, in order to your giving a good account there of when it shall end, and when all your days shall doe so too.

Similarly, end each day by going over what has been done and seeing whether it has been done 'soberly . . . righteously . . . and godlily'. Begin and end each day with a reading from the Bible and meditation upon the meaning of the passages read, accompanied by prayer. Remember that prayer is only meritorious if felt deeply and truly, for words alone cannot ever be efficacious. Further, whenever opportunity serves, listen to preaching and talk 'with good and experienced Christians'. Take the Sacrament 'when you are fit for it', fitness meaning the presence within of Spiritual Life which is not just governed by age. Strive in daily life to be just, merciful, truthful and temperate, seeking always good company, such company consisting not only of persons devout and moral but also of persons helpful

> in acquiring knowledge of the works and creatures of God, I mean of Natural good things, as Physick and Natural Philosophy; or in Artificial good things and Mechanical Ingenuities [since] all good is of God . . . and therefore we are to discern it and make use of it in the several degrees, in which he hath scattered it up and down through the Creation.

This expression of what came to be called Natural Religion, the argument that through the study of nature we discern the majesty of its creator, was one with which Boyle and many devout natural philosophers of the period were in sympathy and it was to be strongly held and developed in the next century. To include physic (medicine) was not usual, for physicians were commonly regarded as at best sceptics in religion, at worst atheists. Oldenburg was probably thinking of the wonders of the human body, its construction and workings, as recently discovered by anatomists and physicians rather than the practice of medicine as he knew it, but it was even so an original precept. Finally, like Polonius (and there are here to the modern reader many echoes of his advice to Hamlet) Oldenburg counsels the avoidance of extravagance either in dress or in the general expenditure of money. He also advises the bestowal of charity upon the poor, at least 'according to your power'. (This last strikes the modern reader as a slight descent to the practical after the generally lofty tone of the body of the essay.) It was all truly sincere in intention and we may believe that Oldenburg did, whenever possible, follow his own advice beginning and ending each day with prayer, meditation on his course of life, and reading the Bible, as very many of his contemporaries strove to do. One must hope that in doing so he found some consolation for the loss of his 'dear wife' as he called her.

Not only was Oldenburg a widower in somewhat straitened circumstances but he had a child ward living with him as well as, presumably, at least one servant. More pious and more conscious of the proper proprieties than Hooke (who at this time not only had a young niece living in his bachelor establishment at Gresham College but was soon to make her his mistress), Oldenburg obviously felt that it was not at all proper or right for Dora Katherina Dury to continue to live in his household now that he was a widower. Fortunately, provision had been made for just such an event: when Dury had asked the leaders of the Dutch congregation to look after the education of his daughter they had presciently replied that 'our Deacons if need shal require (as by the death of Mris. Oldenburg) will take care of the good education of jour daughter in jor absence', adding carefully 'not doubting that jou will provide competent means for it'. This Dury had in a manner done, for he had handed over the papers relating to his (really his wife's) estate.[3] Now Oldenburg wrote to Caesar Calendrin, the minister of the Dutch Congregation, asking him what he and other responsible members of that congregation would do to keep safe both Dury's papers and his wife's 'plate and jewels' (which did survive intact for many years). He also requested that they now again decide 'by whom and where this child might receive a vertuous and religious education, becoming her sexe, and her parents, and her duties to God'.[4] This was obviously a major difficulty for Dora Katherina was at most only eleven years old. Oldenburg at this time made no suggestion about her care, perhaps too bewildered after his wife's unexpected death, nor is there any record of what happened to the child for the next three and a half years. (Oldenburg kept Dury's 'papers' and probably the 'plate and jewels' in his own house.)[5] Otherwise we only know that Dora Katherina grew up safely, possibly in or near her own property in Kent, for there is no mention of her in Oldenburg's correspondence until 1668.

The first of the public disasters of this period began at the end of February 1664/5. Although at first it disturbed Oldenburg little personally it was in 1667 to affect him severely and dramatically. This was the outbreak of the Second Anglo-Dutch War, officially declared on 22 February and publicly proclaimed on 4 March 1664/5. The causes were again ostensibly economic, but involved also what would later be regarded as imperialistic considerations. Minor hostilities had already occurred from time to time, most of them provoked deliberately by the English, convinced that they must somehow reduce what they saw as the Dutch advantage in world trade, trade which, so popular opinion was convinced, properly belonged to England. As early as the summer of 1664 Oldenburg had passed to Boyle[6] news from Holland that Dutch East Indian merchant ships were being protected by the Dutch fleet and that English ships were preparing to sail for West Africa where the Dutch had secured a monopoly of trade. Everyone knew that the voyage of Robert Holmes equipped with 'pendulum watches' for longitude determination was a daring but ultimately unsuccessful attempt to break this monopoly. Oldenburg remarked presciently, 'tis thought, a warre cannot now be avoyded', adding that he had learned that in Paris it was there thought that the English were deliberately provoking the Dutch to ensure that war did break out,[7] which was almost certainly true. A week later correctly reporting to Boyle that Prince Rupert was to command a fleet intended for Guinea to try to drive away the Dutch, Oldenburg commented that this was 'a plain beginning of the Warre'. He perceptively added that since the presence of plague in the Netherlands had caused a cessation of Dutch international trade, they had plenty of ships available for their navy (which was in fact bigger than the English navy). More optimistically he noted that to set off against this threat was the fact that Dutch misfortune was increasing the profits of the English East India Company's cargoes.[8]

Oldenburg treated the threat of war calmly at first but by the end of

September 1664 while recounting to Boyle the triumphs of the English fleet under Holmes he commented 'This I look upon as the foundation of an implacable warre' and added the gloomy prediction that it would probably lead to an attempt by the Dutch to expel the English from the East Indies.[9] By early October Oldenburg, listening to gossip in a coffee house, had picked up the 'state News' that an English fleet under Prince Rupert was about to set sail at the same time as the Dutch African and East Indian ships were planning to sail through the Channel, a confrontation expected to provoke immediate hostilities (although in fact it did not do so).[10] Naturally many rumours flew about which Oldenburg regularly reported to Boyle, with English public opinion becoming more and more convinced that war was inevitable. The French, who at this point tended to sympathize with the Dutch, thought so too.[11] And so things continued for the remainder of 1664. Oldenburg was on the whole confident that the English had the better cause and would win the expected war, although he was aware of the tremendous preparations being made in the Netherlands to permit the Dutch to carry out what would obviously be an entirely naval war. He knew something of English preparations too, such as the appointment of the Duke of York, the future James II, to command the English fleet. Wiser than those in authority, he doubted whether the fact that he was accompanied by many courtiers would go down well with 'the Seamen, who love neither Ceremony nor imperiousness'.[12] (The presence of so many gentlemen unused to the sea and to naval warfare was to be a great hindrance to the English fleet. What Oldenburg almost certainly did not know was that the English, unlike the Dutch, had not undertaken a naval building programme in recent years and there was to be a decided shortage of spares during the coming years.) Oldenburg approved greatly of the King's firm speech to Parliament at the end of November when, Charles II proclaimed, 'he was resolved to make no peace, but upon a ferme bottome' and spoke 'of ye nation's honor and wellfare to that purpose' but he also noted the reluctance of Parliament to vote a necessary subsidy.[13] He showed himself, indeed, as he was always from now on to do, a perceptive but very loyal Englishman.

Once war was officially declared, on 22 February 1664/5, Oldenburg's political comments to Boyle virtually ceased. This was partly because there was little to report until the fleet under the command of the Duke of York sailed in the early spring and partly because, since Boyle was not living in Oxford during the early spring, no letters to him survive.[14] Surprisingly there is no mention in any of Oldenburg's correspondence of the first two great naval engagements at the end of the spring. These involved the capture of the Hamburg convoy by the Dutch on 20 May 1665, a great blow to the English since it was laden with urgently needed naval supplies, and the dramatic action of 3–5 June generally known as the Battle of Lowestoft.[15] This, fought off the Suffolk Coast, was an English victory, for the English beat off the Dutch with greater losses to the Dutch than to the English fleet. It would have been an even greater triumph for the English had a courtier on one of the ships not, quite improperly, called off the pursuit, perhaps because of the admittedly great English losses. The sound of the guns had been heard clearly in London, emphasizing the reality of the war, and Oldenburg must have heard them. But for some reason there is no mention of these exciting events in his surviving correspondence until a month later when he correctly reported that the Duke of York, fully and dangerously active during the Battle of Lowestoft, had now been 'prevailed with, to stay at home' and that Prince Rupert had been appointed in his place.[16] He also passed on the fact that Dury had written from Switzerland that there it was thought that the English fleet had had the worst of it so far, public opinion on the Continent being generally anti-English. Dury's opinion that the English had come off badly was not yet true, but it was soon to be true when the English attempt to capture a fleet of Dutch ships in Danish waters went disastrously wrong, the English ships being beaten off by Danish shore guns. This was just when Oldenburg was venturing on optimism, writing to Huygens[17] 'I hope for a good accommodation between our two nations'. Certainly the English had tried to put a brave face on the disaster by claiming a partial victory, reporting that three Dutch ships had been sunk with the loss of 1000 men.[18] In early September the English fleet really did do better, for cruising off Texel, the most southerly of the Frisian Islands, it captured both merchant and naval vessels, valuable prizes, and took many prisoners. In retaliation, towards the end of October the Dutch fleet cruised off the mouth of the Thames, but there was no engagement since, because of incompetence, the English fleet was not ready to sail. And so ended the 1665 campaign.

The late summer campaigns brought the reality of war closer to Oldenburg for an acquaintance, perhaps even a relation, had been captured and was now held a prisoner of the English.[19] In August Oldenburg wrote to Moray, as a public servant and a courtier, to ask him whether he could endeavour to secure the release of this prisoner whose family was suffering from his absence. Moray replied kindly and patiently but firmly that this would be quite impossible being

> entirely inconsistent with the prudent management of this warre, to release officers that may be usefull to the enemy who stand in need of such to strengthen their hands against ourselves.

He explained further that Dutch prisoners were needed for the English to exchange against their own captured officers and as for the family's distress, presumably the Dutch would look after them and if it did not, the Dutch population would become disaffected and the Dutch navy unable to recruit more seamen. This was a decidely cynical view, of course, and it must be said that the English were not known to look after the families of captured seamen, or even officers. However, plainly Oldenburg knew a convincing argument when he met it and recognized that any further attempt to get this man released would be futile; his fate cannot be known, for he is never mentioned again in the surviving correspondence. Moray's hard-headed and hard-hearted argument testifies to the fact that in the seventeenth century as in the twentieth war was waged where possible against civilian populations as well as against military and naval forces. It also shows a recognition of the fact that in a republic, such as the Dutch had, it was essential to ensure that public opinion in the civilian population remained in favour of the war. Some months later (October) Moray wrote,[20]

> I doubt not but all is done that can be done in sea matters. I am very sensible of your suffering with others, but all I can contribute towards relief is my best prayers & those are not wanting.

If Oldenburg's life seemed relatively little affected by the war during the summer and autumn of 1665, the case was quite otherwise with the second major disaster to affect both combatants during this year. Then plague swept through both England and the Netherlands. This was true plague, the Black Death which had been endemic in both countries since the fourteenth century, sporadically breaking out into epidemic proportions. The whole of south-eastern England was to be affected in 1665 and 1666 and the loss of life, particularly in London, was very considerable. Not only was the mortality high but the psychological effects were even more terrible than the mortality. For since no one knew the source of the contagion, nor were most people even sure whether plague was truly contagious, no one knew what steps to take to arrest its deadly blight. In fact, plague is carried by rat fleas, both rats and fleas being highly susceptible to the disease; moreover the common human flea can become a carrier if it bites a plague victim. The disease results from the injection of the causative virus into the body. What makes it so very hazardous is that normally fleas leave dying rats and turn to humans for food, carrying the virus with them from one human to another. There are two forms of plague. In the pneumonic the disease can not only be carried by fleas but is directly contagious like most infections of the lungs and respiratory tract and is transmitted directly by the coughing or sneezing sufferer. The bubonic form is only spread by bites from infected fleas and is characterized by swellings (buboes) in the glands, especially those in the groin. The Great Plague of London was the bubonic strain, manifesting itself not only in swellings but also, in the initial stages, by violent headaches and internal haemorrhages. Hence, although it was not so directly contagious as contemporaries supposed, any headache or bleeding in an individual caused excessive fear in those who came in contact with him. Nonetheless, the cruel practice of isolating plague victims or suspected victims and their families by shutting up their houses and allowing no one to leave, a practice begun in June 1665 but discontinued in mid-September, was not entirely without reason although in the then state of personal hygiene fleas must have been so common in bedding and clothing as to make it very largely a matter of chance whether an individual was put in danger or not. The fourteenth-century principle—to leave a plague-striken city quickly, to go as far away from it as possible, and to stay away until the epidemic had died down—was still the best way of avoiding the disease. Very many did leave London, but more could not. Possible remedies were much in demand among those who stayed and the literate studied medical authors for their recommendations, as Oldenburg was to do, advised by Boyle in some cases.[21] Some few caught the disease and recovered, more who stayed in London met and even talked with those subsequently infected and managed to escape the disease, as did both Oldenburg and Pepys. But avoidance remained all important.

By the end of April there was plague in London, beginning naturally enough near the docks. Pepys recorded in his diary for 30 April 1665 that there were rumours that a couple of houses in the City had been shut up. Six weeks later (10 June) he learned more reliably that the first cases had only just appeared in the City although there were many cases in the over-crowded areas just outside the City boundaries, especially to the east. By the end of June he noted 'home by Hackney coach; which is become a very dangerous passage nowadays, the sickness increasing mightily', while two days later he saw two houses shut up 'which is a sad sight'. Certainly the plague was raging in the City by the end of June although not yet, rather surprisingly, in the down-river towns or in Southwark across the river which was, Oldenburg noted in early July,[22] 'a great mercy', and surprising since seamen mostly lived in those areas.

Pepys, living in the Navy Office, was in more danger than Oldenburg in the prosperous west of London, not overcrowded, esteemed healthy in part because of the open spaces and parks, without densely inhabited tenements. Pepys noted that he like others in London, presumably including Oldenburg, began to avoid acquaintances living in severely affected areas. Pepys, again like others, sent his wife and 'family' (servants) out of town, in his case to Woolwich where he could easily visit them. But it must be added that Pepys not only continued to work in his office in his usual way, but also travelled a good deal about London in public conveyances. Recklessly, he frequented some, if not all, taverns and enjoyed the society of friends, colleagues and pretty women as much as, perhaps more than, ever, only occasionally sobered by the sight of houses shut up and empty streets and markets. And in this he was by no means unique.

Oldenburg stayed and worked as normally as possible. Lord Brouncker, not only President of the Royal Society but a colleague of Pepys in the Navy Office, also stayed and worked although, as Oldenburg noted in July,[23] houses close to his official residence were 'infected'. The Royal Society met early in June but then adjourned to 28 July, after which it did not meet again until the spring of 1666. In July Hooke went to country houses near Epsom where he was joined by Wilkins and Petty; they worked together there and none returned to London until the Royal Society was ready to resume meetings once more. Oldenburg seems to have had little thought of leaving; in early July it is true he went so far as to decide that

> If it [plague] should come into this row, where I am, I think, I should then change my thoughts and retire into the Contry, if I could find a sojourning corner

but fortunately for him Pall Mall was never affected.[24] He continued to work, keeping up his correspondence as best he could, writing weekly to Boyle in Oxford (which unlike Cambridge largely escaped the disease) and tried to keep his head above water financially, no easy thing to do in those troubled times. There was, of course, no perfect safety and even after the Royal Society adjourned he was forced to move about, take letters to the post, meet people to transact business and so on.

He worried much about what to do with all the papers in his custody, especially those belonging to the Royal Society and in July told Boyle that

> All I can think to doe in this case, is, to make a liste of them all, and to putt them up by themselves in a boxe, and seale them together with a superscription; that so, in case the Lord should visit me, as soon as I finde myself not well,it may be ready to be immediately sent away out of mine to a sound house.[25]

The entirely rational fear is obvious and the pathos of a deeply conscientious man living in dangerous times, with no relatives and with all his closest colleagues out of touch, is evident. The danger was still there a month later when he told Boyle that he had taken those steps he had thought necessary, informing him that

> I have now putt all my little affairs and my papers in order, severing what belongs to the R. Society, to yourself, and to Mr Dury's Child and estate etc. from mine own.[26]

Not unnaturally, as the situation in respect to the sickness remained unchanged, Oldenburg continued to worry. In mid-October he was sufficiently concerned to propose drawing up a will, with, he suggested, several of the senior members of the Royal Society as executors.[27] Among these was to be Moray who, when told of the plan, commended Oldenburg's 'Christian & prudent reolutions' but suggested that either Brouncker, still in London, should be the sole executor or that Oldenburg should specify that any one executor be empowered to act in the absence of the others 'if so be it please god to call you while all of us you joyne with him, are barred from London' by the continuance of plague.

Oldenburg steadily took what care he could to avoid anyone who might have been exposed to plague, being certain as he wrote that, in spite of what some astrologers like John Gadbury alleged, the sickness was indeed highly contagious. (Gadbury seems to have said, somewhat ambiguously, that plague was 'no more infectious' than 'smallpox, Lues Venerea [syphilis], Scurvy, Fistula's, Strong breathe etc.', according to Oldenburg). Against Gadbury's muddled understanding of what was and was not a contagious disease, Oldenburg could only declare, 'it is a mysterious Disease, and, I am afraid, will remaine so'.[28] He much disapproved of those who carelessly and needlessly exposed themselves to possible infection, criticizing Moray's French protege de Son (or d'Esson) who had been in England since the previous year:

Oldenburg frequently visited him at this time on behalf of Moray, and both men were much upset to discover that de Son had carelessly exposed himself, and so possibly Oldenburg, in September 1665, even though de Son had since then carefully fumigated his house.[29] (It would be interesting to know what Oldenburg would have thought of Pepys's way of life had he known of his activities.) Oldenburg was evidently quite prepared to throw himself on God's mercy, but he not unreasonably could not accept the apparent belief of de Son and others that one just had to take one's chance. Cautious man that he was, he believed in taking reasonable precautions in case, as he was inclined to believe, plague really was contagious. In August, when the epidemic was at its height, he wrote to Hooke,[30] telling him that he was wise to stay away, even though he added,

> I find by all my own and other men's observations that very few of those houses whose inhabitants live orderly and comfortably, and have by nature healthy constitutions (you must take all these together) are infected, and I can say (God be praised for it) that as yet not one of my acquaintance, except an under-postmaster, who lived closely and nastily, and had all sorts of people coming to his house with letters, is dead; so that, generally, they are bodies corrupted, and persons wanting necessaries and comfortable relief, that suffer most by this contagion

Perhaps he thought that Hooke's notorious ill-health made it particularly wise for him to avoid London; he can hardly have meant his remarks to be taken as ironic.

In spite of Oldenburg's apparent naivete about the causes of infection by plague, he was not far out in his diagnosis. Infestation by fleas must have been greater and more widespread among those 'that lived closely and nastily' than among the well to do and respectable classes, and it is hardly surprising that the immensely crowded slums of riverside Rotherhithe saw some of the highest mortality. Hygiene was notoriously neglected in the seventeenth century even by the upper classes. Baths were hardly known, while that great gentleman Boyle used to wash every morning by dipping a piece of linen in water and wiping it over his face and hands. Pepys believed that on the rather rare occasions when he washed his feet it brought on a fit of the stone. Oldenburg had perhaps learned better hygiene in Holland, but even so the Dutch were also suffering from plague at this time. Certainly he saw how much worse off the poor living in overcrowded conditions and 'nastily' were, but he could not have known that it was because they were, inevitably, as greatly afflicted with fleas as their houses were with rats. Without any knowledge of the cause of the disease, any proposed remedies, like those sent to Oldenburg by Boyle and by the Italian alchemist Borri can have helped only psychologically.

Oldenburg appreciated the kindness with which Borri sent him a quantity of his own prophylatic enclosed in a letter; it had, so Oldenburg reported, an extremely strong smell which he described as 'comfortable' (i.e. comforting) but although he was willing to taste it cautiously he did not try taking it.[31] Instead he sent a small sample of the supposed remedy to Boyle and Moray to try to analyse but neither seems to have got far towards doing so. A widely recognized prophylatic was tobacco: when Pepys first saw two houses in Drury Lane shut up because their inhabitants were ill, he noted that he was 'put . . . into an ill conception of myself and my smell, so that I was forced to buy some roll-tobacco to smell and chew' while Oldenburg adopted the practice of smoking a 'companiable' pipe of tobacco every evening.[32] (It is not recorded whether he normally smoked.)

As the epidemic continued life tended to go on with increasing normality for those who escaped it. Pepys enjoyed himself enormously this autumn and winter although the more serious-minded like Oldenburg worried more and more as plague continued to claim more victims and the threat of contagion hung over all. The severity of the epidemic seemed to be decreasing in the autumn[33] as it was expected to do in colder weather and in the early months of the winter everyone was optimistic that it might end.

Although for many Londoners fortunate enough to escape the disease and remain healthy life did continue fairly normally there was perpetual alarm for anyone suffering suspicious symptoms like headache and fever, or when coming close to those so afflicted and there were still severe disruptions to normal life throughout all classes. Oldenburg himself suffered as well from the fact that so many of those with whom he worked were absent from London. In late June 1665 the Court and the courtiers moved in a body from Whitehall up river to Sion House (across the river from Kew), in early July to Hampton Court, from there to Salisbury and thence, in late September, to Oxford where it remained until the end of January 1665/6. This took several important Fellows of the Royal Society away from London, notably the courtier Sir Paul Neile and, of particular consequence to Oldenburg, Moray, then Deputy Secretary for Scotland. Moreover, business and trade generally were severely disrupted, both because potential customers left London and because workmen living in poor and overcrowded houses on the outskirts of the City died in large numbers. The publishing trade in particular was badly affected on both counts and this jeopardised the publishing of Oldenburg's *Philosophical Transactions.* Issue number 5 dated 3 July 1665 appeared as normal. But it carried an ominous warning on the final page stating

> The Reader is hereby advertised, that by reason of the present Contagion in London, which may unhappily cause an interruption as well of Correspondencies, as of Publick Meetings, the printing of these *Philosophical Transactions* may possibly for a while be intermitted; though endeavours shall be used to continue them if it may be.

Oldenburg acted quickly to keep the journal going by a scheme to have the printing done elsewhere than in London. He consulted Moray[34] (then with the Court at Hampton Court)[35] rather than Boyle, probably because Moray, after all also a very important member of the Royal Society and keenly interested in the publication of the *Philosophical Transactions* both held an official post and was more of a man of business than Boyle. Moray quickly pointed out potential difficulties because Martin and Allestrey in London were the Society's official printers and had been named in the licence to print for the first issue as granted by Brouncker as President of the Royal Society. Moray advised Oldenburg first to consult Brouncker, whom Oldenburg might not have previously approached because he was then ill, and secondly, if Brouncker agreed to the change, as in fact he did, to inform Martin and Allestrey of his intentions before approaching any other printer. Presumably Oldenburg followed this advice, although there is no mention of such actions in his correspondence with either Boyle or Moray until the autumn. By then Moray and the Court had arrived in Oxford where Moray found lodgings in the last week of September with Wallis in Catte Street near All Souls.

The settling of the Court in Oxford for the late autumn and early winter of 1665–6 considerably changed both the social life of the town, as Boyle was frequently to complain,[36] and the intellectual life of members of the Royal Society. As Boyle told Oldenburg at the end of September[37]

> there being now at Oxford noe inconsiderable number of the Royal Society insomuch that the King seeing Sir Rob. Murry & mee with some others was pleasd to take notise of it, I did not know why we might not, though not as a society yet as a company of Virtuosi renew our meetings.

This proposal was agreed to and the Oxford Fellows of the Society began to meet on Wednesdays at Boyle's lodgings

> where over a dish of Fruite we had a great deale of pleasing Discourse, & some Expts I showed them . . . . Mr Oldenburge was mentiond and drunk to . . . & . . . wishd here

especially by Moray, Petty, Wallis, and Boyle himself. This must have given Oldenburg pleasure to read and no doubt he also wished that he might have been there. On 11 October ten or a dozen Fellows met in Wallis's lodgings, when Boyle again showed some experiments and the contents of two of Oldenburg's letters were discussed, very much in the manner of a normal meeting of the Society even though their Secretary was not there to read letters.[38] Further, Moray, noting that he had already hinted at the possibility in a previous letter now not surviving, reported that at dinner in Wallis's lodgings, presumably before the afternoon meeting, he had discussed with those present including Seth Ward (then Bishop of Exeter), Boyle, Wallis, and others the notion of having the *Philosophical Transactions* printed in Oxford from copy supplied by Oldenburg in London when 'they all were instantly for it'.

Those present immediately named as printer Richard Davies, who had handled many of Boyle's books, sent a message to him and interrupted the meeting to discuss the proposal with him, demonstrating the importance they gave to the *Philosophical Transactions*, only so few months after its inception. Davies readily agreed to undertake the printing of the journal, asking only that he might be permitted to pay Oldenburg less than the London printers had done because, so he claimed, he could not expect to sell as many copies from Oxford as London printers might expect to do. He professed himself willing to accept any other conditions that Boyle and Wallis might lay down, but in fact they accepted his terms on Oldenburg's behalf. Moray bade Oldenburg let Boyle know what he thought of the proposal, although it seems to have been Moray and Wallis who were most active in the affair. Oldenburg must have been pleased to have matters settled with so little trouble to himself although he can hardly have been pleased at the potential loss of revenue. However he obviously in the event thought that it was better to settle for less and keep the journal going than to have it cease, when the revenue would have ceased altogether. Moray directed Oldenburg, if he agreed to the terms proposed, to send material for a new number as soon as might be. He was to send it either to him when he would pay the postage or to Boyle or to Wallis, if neither Moray nor Boyle were in residence in Oxford. (This is a little confusing, since Moray was still lodging with Wallis when in Oxford so that all post addressed to Moray would have been delivered to Wallis's residence in any case.) Moray assured Oldenburg that Wallis would look over the copy for the press if need arose and proofread the text when set, as he was faithfully to do. Wallis was extremely attentive and acted throughout the period when the *Philosophical Transactions* was printed in Oxford as, effectively, assistant editor, always in consultation with Moray and Boyle.

The first number of the journal to be printed at Oxford, number 6, was dated 6 November 1665 and carried the imprimatur of the Vice-Chancellor of the University. To anticipate, nos.7 (4 December 1665) and 8 (8 January 1665/6) were also printed in Oxford. Sadly, the Oxford printing caused difficulty and trouble for all concerned. There was no problem about content, for Oldenburg still had to hand a constant amount of potential material—from Hevelius, from Auzout, from Huygens, from various English correspondents, from Boyle, and from books he had read. But with the best will in the world it was difficult to maintain the usual high standard of supervision when his journal was being printed in Oxford and he was in London. For although letters travelled quickly and easily (often rather faster than they do today) parcels were always a problem. By mid-October 1665 Oldenburg had all the copy for number 6 ready to dispatch 'to be reviewed by my Oxonian friends at their leisure, before their entring the Presse' except for a French description of a burning mirror which he had sent earlier to Boyle to read and of which he had kept no copy.[39] He therefore had to ask Boyle please to return it to him so that he could translate it into English. (The only non-English extracts ever printed in the journal were those in Latin, far more readable to an educated English audience than French or Italian.)

It was all very troublesome and Oldenburg cannot have relished having editorial control taken out of his hands, as was to happen occasionally. Boyle decided that a reference to his own 'stereometrical' or 'hydrostatical' balance (for measuring the volume of an irregular solid) should be omitted 'because there was some mistake in the Business', namely that Oldenburg had said that the account was to be published in full in Boyle's forthcoming *Hydrostaticall Paradoxes* (1666) which was in fact not the case.[40] Oldenburg took this interference, as Boyle noted with approval, 'in good part'; as a replacement Boyle inserted a brief account of a physiological curiosity, namely of the occurrence of 'white blood' or 'milk' in the veins of a man.[41] Moray also reported this to Oldenburg in telling him that both he and Wallis were proofreading the sheets for number 6, while once the issue was printed off it was Moray who saw to sending copies to Oldenburg with critical comments on the printing and the paper, which he thought not so good as those of Allestrey.[42] (Oldenburg's carefully trained eye found many errata most of which, as Moray later told him, had been corrected by Wallis and remained only because the printer had ignored the corrections.)[43] Moray and Wallis continued to oversee the printing of the next two numbers,[44] for September and January, and to make minor editorial adjustments, all to Oldenburg's apparent satisfaction.

And so the *Philosophical Transactions* continued to appear regularly, although at a lesser profit to Oldenburg. For financially things were far from satisfactory. Davies was dilatory in forwarding both copies of the journal and the money he owed, the latter especially troublesome as Oldenburg needed it in order to keep afloat. Boyle partially solved the problem by sending Oldenburg an authorization to draw on Boyle's private funds in London, while Oldenburg in turn was to authorize the printer to pay the sum due directly to Boyle.[45] This certainly speeded up Oldenburg's receipt of his money, but it was not as generous as it might seem, for Oldenburg only got what was rightfully his, and Boyle probably lost nothing. The printer's excuse for not sending Oldenburg the money due was that he was unable to sell many copies of the journal. This Oldenburg found so discouraging that he told Boyle that if he had not already sent off the manuscript for the January issue he would have given up all thought of having the journal printed in Oxford.[46] No wonder he was glad when Boyle advised him in December 1665 that he should try to see whether the journal might be printed once again in London. (Boyle assumed with others that the cold wintry weather would reduce the incidence of plague.)[47] As it happened the sickness did, briefly, diminish although it did not wholly cease, and life in London seemed to become slowly more normal.

By early January 1665/6 Oldenburg was negotiating with Allestrey, but it was, he found, unlikely that it could be on the former terms.[48] Prudently Oldenburg prepared material for a possible issue number 9, which, dated 12 February 1665/6, was indeed once again printed by Martin and Allestrey. But Oldenburg's troubles with Davies in Oxford were even yet not over: as late as the end of March 1666 Davies still owed Oldenburg money, even though Oldenburg was prepared to accept only £9, half the sum rightly due to him. What was worse, Oldenburg suspected that Davies had been telling the London stationers that he (Davies) was suffering greatly because there was no demand for the issues of the *Philosophical Transactions* which he had recently printed and thereby, as Oldenburg thought, discouraging the London stationers from selling copies in London.[49] This was so much the case, as Oldenburg ruefully told Boyle,

> that, if they goe on to be printed, I shall be the worse for it by 40. sh[illings] a month; which is a great losse to one, that has no other way of subsistence for serving the Society.

Twenty-four pounds in a year meant much to Oldenburg, and presumably he regarded the *Philosophical Transactions*, though it was his private journal, as being of service to the Royal Society, as it was.

In spite of plague and the bitterness and intensity of the Anglo-Dutch War, foreign correspondence between England and the Continent, even between England and Holland, was little disturbed. During 1665 Oldenburg continued to hear from Dutch correspondents, including Huygens, Spinoza, and Serrarius, the last named the more remarkable for much of his news was political or quasi-political and therefore potentially dangerous. Letters travelled very slowly but fairly normally between Hevelius in Danzig and Oldenburg in London. And there was plenty of exchange between Paris and London throughout this year, with Justel supplying news and rumours and recounting the French view that the English were too self-assured and so deserved to lose, while Auzout, Petit, and others sent purely scientific news, seeing no reason why war should interrupt their friendly intellectual intercourse.

As the cold weather increased in December 1665 the number of cases of plague did, as already mentioned, decrease sufficiently to induce optimism and the hope that the worst was over. Pepys brought his family back to London as early as January 1665/6 and evidently others returned as well. Oldenburg for one did not share the prevailing optimism and feared greatly for the future sales of his *Philosophical Transactions*, telling Boyle in mid-January that it was doubtful whether they could as yet be printed in London 'because of the plague keeping still on foot, and discouraging all sorts of people from setling to businesse', adding that 'it keeps on foot in the depth of winter, and hath increased the two last Bills [of Mortality]'.[50] This was true, and plague deaths continued throughout the spring and early summer of 1666,[51] but fortunately for Oldenburg his printers settled back to business successfully and printed his journal. The members of the Royal Society began to drift back to London, the Council met on 21 February 1665/6 and then decided that regular meetings should resume, as they did on 14 March 1665/6. In spite of the numbers of people not previously exposed who came back to London, it does not appear that many of them were affected by plague—certainly no Fellows of the Royal Society suffered, it seems.

So far so good. But other disasters were to threaten Oldenburg's activities. War with the French loomed after miscellaneous hostilities between the French and English fleets mostly apparently initiated by the English with some success—as Justel remarked early in the new year[52] 'the English have taken a few of our ships'. The French after some delay suddenly entered the war on the side of the Dutch on 16 January 1665/6. Then Oldenburg immediately realised how much a new war would damage his activities. As he wrote pessimistically to Boyle,[53]

> Now the French King hath declared warre against England (which he did, it seems, very briskly, when he sent word to the Q[ueen] Mother of England, that he must doe it within 2. dayes) we shall, I feare, meet with some interruption of our Philosophicall Commerce, as we cannot but doe with a totall one of Merchant-trade.

In preparation he hastily sent a message to an Englishman about to return home in the Ambassador's suite asking him to bring a number of books recently published in Paris. He devoutly hoped that, as usual in the seventeenth century, some communication would still be possible in spite of the hostilities between countries; as he wrote didactically to Boyle,

> Let Princes and States make warre and shed bloud; let us cultivate vertue and Philosophy, and study to doe good to Mankind.

A somewhat pompous declaration, but one which he endeavoured to keep, not without difficulty.

There had already been some problems in the autumn of 1665 over the transmission of parcels containing books and journals, both French and English.[54] At the end of the year there seems to have been some particular difficulty with the reception of copies of the *Journal des Sçavans* in exchange for copies of the *Philosophical Transactions*. Oldenburg then applied to Moray for assistance in resolving this problem and early in January 1665/6 Moray promised to see what he could do[55]

> if before that time Mr Boile or Dr Wallis have not done what I advise in it. For Mr Boile is in mighty credit with Secr. Morrice, & if that way take not he must try Mr Williamson.

It seems probable but not certain that Boyle did try what Moray advised and turned to Williamson, for by the end of February Oldenburg could write[56]

> I hope, Mr Williamson and I, have so ordered the matter in the point of Correspondency, that there will be no exception taken at it.

Joseph Williamson was then Keeper of the State Paper Office under the Secretary of State Lord Arlington. He was also Latin Secretary to the King, and organizer of a news sheet, later to be called the *London Gazette*. He had been a Fellow of the Royal Society since 1662, and although it is not known that he attended many meetings he was to serve on the Council in the coming year, to become President in 1677 for three years, and always to support the Society by paying his subscription with regularity. He obviously had facilities through the State Paper Office for sending and receiving letters and parcels from abroad, and he and Oldenburg soon came to a simple but ingenious solution for Oldenburg's problems.

The mechanism was this: Oldenburg's foreign correspondents were to be instructed to address their letters not to Oldenburg personally nor to the Secretary of the Royal Society but simply to 'Monsieur Grubendol, London.' Letters so addressed were delivered to the State Paper Office where Williamson received them and paid the postage due; he then gave them unopened to a messenger to deliver personally to Oldenburg. When Oldenburg had read them he copied out any political news they contained and sent this in a letter to Williamson, for use either in State affairs or for his *Gazette*. (Many, although far from all, these letters survive.) Letters from such persons as Justel and Serrarius, both full of political gossip, Oldenburg probably promptly destroyed as soon as he had dealt with them, a wise precaution in this dangerous time. This mechanism was designed not only to protect Oldenburg personally from possible charges of corresponding with the enemy and also freed from the financial burden of paying postage on what were often very bulky letters. Probably also Williamson paid Oldenburg something for his useful transmission of news from the enemy. Parcels from Hevelius were still to be sent direct to Oldenburg at his house for there were no hostilities between England and Danzig. Moreover, when Hevelius sent books, Oldenburg usually exchanged them for others or sold them and sent the proceeds to Hevelius, so this was private, not Royal Society, business. Undoubtedly much correspondence, like that with such French friends as Auzout and Huygens (who had become resident in Paris as a member of the Académie royale des Sciences in 1666), might be suspected of containing political news and in this case too the Grubendol address was a protection as well as a considerable financial saving for Oldenburg.

In fact the natural philosophers of Paris, like Oldenburg himself, carefully kept their distance from politics, never writing any thing about it and opting firmly for the neutrality of the intellect. As Auzout put it,[58]

> It will be annoying if the war hinders the commerce of sciences, there being so many excellent things we have cause to exspect from your Illustrious Society . . . But I think that at least we can maintain some intercourse through the post.

Oldenburg in turn, writing to Huygens, expressed the same sentiment:[59]

> In time, I hope, all nations, however little civilised, will join hands as loving friends, and combine their intellectual and economic resources to banish ignorance and inaugurate the rule of true and Useful Philosophy.

Brave words, but the governments involved did not concur. However the arrangement worked well for over a year and was to continue after peace was eventually declared, to the no small puzzlement of contemporaries as well as of subsequent historians. During the continuance of the naval war, when English dominance of the seas was by no means always certain, it provided Oldenburg with a feeling of security no doubt precious to him. Besides, it must always have been of financial benefit to Oldenburg as well as useful to Williamson in several ways.

What with the French entry into the war and the Dutch fleet having been much strengthened over the winter the outlook for the English was by no means good. Nor can Oldenburg have been cheered by Justel's continued insistence that the French counted upon certain victory over the English, although Justel's belief that divine intervention on the side of the French would ensure that this was so was not quite so assured as he made out.[60] In fact hostilities did not really break out until the early summer of 1666 when, at the end of May, the French fleet was erroneously reported to be at the west end of the Channel and the Dutch fleet, reliably, to be approaching from the east. The English fleet commanded jointly by Prince Rupert and Monck, now Duke of Albemarle, divided to face the two threatening fleets, a manoeuvre which was to prove the more disadvantageous because the threat from the French fleet did not materialize. The subsequent engagement with the Dutch begun on the first of June was to be known as the Four Days Fight; during this the sound of gunfire was heard in London and was, as Oldenburg put it 'rude and obstinate'. It was a destructive and ultimately inconclusive engagement: the English fleet was much smaller than that of the Dutch and the worse damaged but since the English fought bravely on, the Dutch in the end disengaged and withdrew. No one at the time could quite agree whether the battle was a success or failure; Oldenburg put the situation succinctly,[61] writing

> as the Engagement will not easily be parallel'd, so we shall find that the successe all things considered, [is] a Great deliverance and a Dear Victory to us.

Dearly bought it was, but it could be called a victory since it was not a conclusive defeat. Oldenburg consoled himself with the remark

> Mean while, God be thankd, both our Generalls are well, having done prodigious things and the king in a condition, to sett out very speedily another Fleet; which that it may be done so, as to prevent our Enemies, is the highest concernment in this businesse.

The English fleet was, with difficulty, reformed, strengthened and re-equipped and six weeks later was, surprisingly, ready to put out to sea again. The result, since the two fleets were quickly in sight of one another, was a serious engagement on 25 July, the St Jameses Day Fight. Unlike the previous battle, this was a clear victory for the English, only spoiled once again by their failure to pursue the retreating Dutch. The only surviving comment by Oldenburg on this battle was in a letter to Williamson quoting Justel's somewhat braggadocio claim that the English success would stiffen the resistance of their enemies. On this Oldenburg sardonically commented 'that did not appear on St Jameses day'. He must have shared the general euphoria, naturally increased by the news of what became known as 'Holmes's Bonnfire', when Sir Robert Holmes on 9 August 1666 landed on the islands of Vlieland and Terschelling off the north coast of Holland and burned both a large part of the Dutch East India Fleet with its cargo, and the town of Terschelling.

When Oldenburg heard the news he promptly wrote of it to Wallis (and perhaps to Boyle as well)[63]; Wallis's reaction was that he hoped that the action would lead to peace, for

> a good peace . . . well established would bee much more acceptable than the news of Desolations, though of our Enemies, though this, as the case stand, be good news too.

This action is generally regarded by historians as the last campaign of the year, but of course Oldenburg and his contemporaries could not know that. To them there were many threats: as Oldenburg told Boyle in mid-September, the French fleet was then off the Isle of Wight and the Dutch fleet not far away. Only a sudden very violent and prolonged storm beginning on 17 September was 'like to have disappointed all the 3. Fleets', which it did.[64] And a week later he was to tell Boyle that a battle was expected soon because the Dutch fleet was off the Kent coast 'in Margat-rode'.[65] Again the Dutch manoeuvre came to nothing for storms, increasing for some weeks, were so violent that no engagement between the fleets was possible. No doubt many English, perhaps Oldenburg among them, attributed the providential storms to divine intervention.

And divine help was badly needed, for London lay helpless under yet another disaster, the Great Fire of London which the Dutch, not surprisingly, saw as divine retribution for Holmes's bonfire, and which in Oldenburg's view, explained why the Dutch ventured so near to London later in September. The story is well known: on 2 September 1666 early in the morning fire broke out in a bakery in the City. This in itself was nothing out of the ordinary, for city fires were common everywhere in England, when houses were mainly timber-framed, roofs were commonly of thatch, city houses were built high, close together, and often so over-sailing on the upper stories as nearly to meet over the usually very narrow streets. Candles and domestic fires were a recognized hazard, while bakeries necessarily stored large quantities of kindling and faggots to heat their ovens, and there were many stables filled with hay and straw. Fire-fighting equipment was minimal. Pumps were inefficient even when water was available, as often it was not, and the only really effective method of stopping the spread of a fire once it had taken well hold was to pull down or blow up houses and buildings to create a gap wide enough so that the flames could not jump across it, a desperate measure which naturally the authorities were reluctant to take. What was truly catastrophic about this fire, what caused it to be remembered as the Great Fire of London as it had been the Great Plague of London, was that, fanned by the extraordinarily high winds of that September and after an exceptionally dry year, it was to rage for four days and nights, destroying all in its path. Such buildings as escaped (like the Navy Office) were nearly all saved as a result of the drastic expedient of creating gaps all around them.[66] The fire finally died out only when the wind temporarily dropped. By then thousands of houses, nearly one hundred churches including the greatest of them all, St Pauls, and many public buildings, in all probability four-fifths of the City of London, had been destroyed with most of their contents, although, mercifully, there was very little loss of life.

But as with many modern catastrophes, the very magnitude of the disaster generated a collective will to overcome its effects, no doubt made a little easier by the cessation of plague. Just over a week after the fire began Oldenburg could write,[67]

> I cannot omit to acquaint you, that never a Calamity, and such a one, was so well bourne as this is. 'Tis incredible, how little the sufferers, though great ones, doe complain of their Losses. I was yesterday in many meetings of the principal Cittizens, whose houses are laid in ashes, who in stead of complaining, discoursed almost of nothing, but a survey of London and a dessein for rebuilding.

And he added that he hoped that some members of the Royal Society would be active in the surveying of the old City and the design of the new, as indeed they were to be. He had suggested this to Brouncker as President, who was prepared to propose it at the next meeting of the Society. However, the Society had other pressing business to attend to, namely the finding of a new meeting place, for Gresham College was one of the few large buildings left standing in the City and was likely to be requisitioned, as it was, becoming the meeting place for the Mayor and Aldermen of the City. Other Fellows were keenly interested. Wren very quickly drew up a plan which he showed to the King, a plan which Oldenburg had seen by 18 September.[68] Hooke was active as well, on the 19th showing his plan to the Society. (As is well known, plans for a total re-design of the City of London came to nothing, for everyone was too anxious to get the City rebuilt as quickly as possible to wait for an agreed plan. The most that could be done was to insist that the rebuilding be done using far less wood and that at least some of the original streets be straightened).

The Royal Society managed to hold its Council and ordinary meetings in Gresham College on 12 and 19 September, on the latter day in Dr Pope's lodgings. But it was obvious that other accommodation would be required and a committee was set up to consider where meetings might be held in the future. This committee, consisting of the President, Moray, Oldenburg and three other Fellows, met at Arundel House, the home of the Howards in which Charles Howard, F.R.S. 1662, a younger son of the Duke of Norfolk, offered rooms. Arundel house was in the western part of the Strand, on a large site now occupied by King's College and Arundel Street. Arundel House was an old fashioned huddle of buildings around a courtyard, whose rooms were less suitable for the Society than those they had left in Gresham College and the location obviously less convenient for Hooke and the other Gresham professors, though more convenient for Oldenburg and Boyle in Pall Mall and for courtiers and government officials like Moray. As there was no better alternative available the Society accepted the offer and continued to meet there until 1673.

The worst direct effect of the Fire on Oldenburg himself was the probable fate of his *Philosophical Transactions*. As he told Boyle in early September[69]

> I doubt I shall find it very difficult to continue the printing of the Transactions; Martyn and Allestrey being undone with the rest of the Stationers at Pauls Churchyard, and all their books burnt, they had carried into St Faiths Church [destroyed by the Fire] ... besides, that the City lying desolate now, it will be very hard to vend them at present.

To anticipate, the situation was not to be quite so bad as Oldenburg had feared. Somehow the September issue, number 17 dated 9 September 1666, had escaped the Fire and duly appeared, although Oldenburg was forced to let Martin and Allestrey have all the profits, hoping only that it would encourage them to continue printing his journal. This in fact they did with help from other stationers and printers, including Moses Pitt and John Crook (both of whom sometimes printed Boyle's books). But Oldenburg understandably did not feel that he could afford to continue to let the printers take all his profits; as he ruefully told Boyle[70] 'I tell them that I should not want generosity to doe so, if I had the ability to bear that retrenchment'. A month later the situation was unchanged: as he told Boyle[71] the stationers and printers had lost so much that they

> doe very much scruple to print anything, except it concerne the present affaires of the warre and of the City: in regard whereoff, it will be very difficult to persuade them to continue the printing of the Transactions, unlesse I let them be printed without consideration for the charges and pains, I am at in the digesting of them; as I did the last: which my condition will not beare, however my soule be free enough to consent to it, if I could.

He had the small issue number 18 ready by 22 October, but had as yet no one to print it; however, Crook was persuaded to undertake the work (and that for the next issue as well) when Oldenburg promised him to try to get him 'some good vendible books, as occasion shall serve'. Number 22 for December was to be printed by Pitt—which suggests that the printers were sharing what little work was available—and in the new year Martin and Allestrey resumed their undertaking.

This strong probability that any profit he had been deriving from the *Philosophical Transactions* might cease severely threatened Oldenburg's financial status at a time when the result of the continuation of the war and the need to finance some of the rebuilding in the City was an immediate increase in the level of taxation. In September, Oldenburg recorded that he had to pay two pounds for 'chimney's [he had four], taxes and watching', and felt that if the level of taxation continued to rise he would be forced 'to run away'.[72] Plainly he needed more income. He had been hoping that the combined influence of Boyle, Brouncker and Moray would secure him the post of Latin Secretary to the King (presumably in succession to Williamson), and he diffidently asked Boyle to speak to Brouncker and Moray about this. Boyle sent him a testimonial to give to Brouncker but could not see why Oldenburg could not approach at least Brouncker directly, since Oldenburg had told Boyle that both Brouncker and Moray 'have an affection for me'.[73] Nothing seems to have come of this aspiration, although Brouncker was openly sympathetic and expressed his regret that he had not known earlier of Oldenburg's need for a paid position, promising that he would keep him in mind whenever a suitable position became vacant (but nothing ever came of this promise). Perhaps oddly, few of his associates realized that Oldenburg made so little as he did from his many occupations or that he required financial assistance. It was at his interview with Brouncker that Oldenburg spoke[74] of having 'declined several offers of conducting young Noblemen abroad, which would have been worth to me 100lb per annum, if not also some annuity ad vitam [for life], after the employment ended'. But as far as we know, no previous employer had ever in fact offered Oldenburg a life-long annuity, and certainly he would have been reluctant to give up his interesting life in London to return to the drudgery of bear-leading young men. But he continued to hope that, now that his circumstances were understood, something might be done by his influential friends to help him to supplement his income as he constantly told Boyle, trying to explain how much more efficient he could be in the service of the Royal Society if he were to be given assistance, especially as his correspondence increased as it did in November 1667.

In the autumn of 1666 plague decreased markedly in the City, (although it was still active down river) presumably partly as a result of the drastic cleansing produced by the Great Fire. This made it easier for the citizens of London to begin to rebuild their stricken city. (It must be remembered that it was the City, and only the City, that was affected by the Fire; the area west of the City, including the western end of the Strand and Pall Mall in one direction, Whitehall in the other having been quite untouched.) Meanwhile and simultaneously with the rebuilding of the City, the Anglo-Dutch War continued. In October 1666 England declared war on Denmark, to Oldenburg's expressed approval, the reason being that the King of Denmark was thought to have broken his word to the King of England the previous year by ordering the Danish shore guns to fire on the English ships while they were attacking the Dutch fleet off the Danish Coast.[75] Little came of this, except that, as Oldenburg foresaw, essential supplies for the Navy were now cut off. The whole affair was entirely a matter of national prestige, with no regard for economic policy.[76] After the battering received by both fleets by the autumn gales the two sides withdrew their ships for the winter.

Foolishly, with an eye to saving money, the English fleet was laid up in the Medway River and most of the ship's crews were laid off (but in large part not paid off) to the disgust of the more experienced professional officers like Holmes, helpless because the Navy was starved of funds. Some wiser heads pressed for the erection of defences in the Medway but little was done and that inefficiently. Oldenburg naturally knew little of all this, and in mid-Novermber could write cheerfully,[77] 'Our Winter-fleet is ready to goe to Sea, and to fetch home the Gottenberg-fleet, so necessary for our next setting out against our Ennemies', for these merchant ships would be bringing essential timber and cordage from Sweden for the Navy, now that trade with Denmark was an impossibility. It is doubtful whether the Navy Office had any money to pay for supplies as it had not been able for many months to pay the seamen nor those who had supplied goods so that merchants were highly reluctant to grant any more credit. In fact, the winter was so quiet on the war front that in Paris, according to Justel,'everyone' believed that there would soon be peace, a hope he reiterated three months later.[78] And in the spring of 1667 steps were indeed taken to begin negotiations for peace with the Dutch at Breda.

In spite of these approaches towards negotiation, by June 1667 the war had become fiercer than ever, and infinitely more disastrous for the English.[79] At the end of the first week of June the Dutch fleet again approached the Kent coast, rightly causing great alarm. The train-bands (militia) were armed and a chain-boom was belatedly built across the Medway to protect the fleet lying at anchor at the great naval base at Chatham. Undeterred and totally unsuspected by the English, the Dutch fleet first attacked the Isle of Sheppey at the mouth of the river and the next day, forced the chain-boom across the Medway, sailed for Chatham and, virtually unopposed, setting fire to many ships and taking the *Royal Charles*, the largest ship in the English fleet, as a prize to the Netherlands. The English could do nothing. Further attacks were to be beaten off, but the initial defeat caused the English, from the king downwards, to be convinced that the catastrophe was caused by betrayal. The king felt betrayed by Parliament which had not granted him the funds he had requested to improve the fleet, as Evelyn noted in his diary, while many ordinary people suspected betrayal by Catholics and spies. Yet others, who knew the state of the Navy as Pepys did, went further, blaming the defeat partly on incompetence including the failure to prepare a sufficient number of fire-ships and to strengthen defences, and partly to the fact that the sailors, unpaid, deserted to the Dutch in large numbers leaving the fleet not properly manned. There was no question of the extent of the disaster. On 28 June Evelyn recorded in his diary

> went to Chatham, and thence to view not only what mischief the Dutch had done; but how triumphantly their whole fleet lay within the very mouth of the Thames . . . a dreadful spectacle as ever Englishmen saw, and a dishonour never to be wiped off!

Evelyn promptly sent all his valuables away from his house at Deptford, reckoning it too near the river to be safe, as Pepys and many naval officials had already done.

Pepys's view, that the disaster was at least in large part the result of widespread incompetence on the part of the naval commanders (in fact, the whole of the civil administration of the Navy was in some sense more or less responsible, and knew it) seems to have been that of Oldenburg as well. Everyone involved looked for a scapegoat hoping to save his own skin. The official finger pointed at Peter Pett, F.R.S., Commissioner of the Navy at Chatham whose defences were so poor, so he was imprisoned in the Tower on 18 June.[80] In October a Committee for Miscarriages was apppointed and its members closely questioned, Pett among others (including Pepys, who was able to extricate himself by means of his command of detail and his knowledge of his job, conscious that on the whole he had always worked conscientiously).[81] Pett himself was nearly impeached but somehow managed to escape and to fade away into obscurity. In June Arlington, Secretary of State, conscious that his intelligence service had been at fault, had ordered a round-up of possible and likely suspects who were to be clapped in the Tower along with Pett. One of these unlucky suspects, imprisoned by a warrant signed by Arlington on 20 June, was Oldenburg, accused as he came to understand it, of 'dangerous desseins and practices' on the grounds of his extensive foreign correspondence. At one stroke he was deprived of his liberty and his livelihood, to be allowed very few visitors and denied the use of pen and ink. Pepys, learning of this four days later, noted in his diary

> I was told yesterday that Mr Oldenburgh, our Secretary at Gresham College [i.e. the Royal Society], is put into the Tower for writing news to a Virtuoso in France with whom he constantly corresponds in philosophical matters; which makes it very unsafe at this time to write, or almost do anything.

Oldenburg's state of mind can be imagined, especially as he was left in complete suspense. About a fortnight after his incarceration Williamson wrote him a brief note apologizing for his inability to help Oldenburg in any practical way except 'perswade Patience'. Evidently he did not feel he could argue with Arlington, his superior. Williamson did offer to send Oldenburg, if he wished, any of the letters addressed to him from abroad, noting that he had opened those from France and kept them.[82] Oldenburg was able to reply in pencil on the back of Williamson's note, which he said he 'had by the particular favor of the Lieutenant [of the Tower]'. There he begged Williamson to try to secure him the privilege of having visitors and to tell Arlington that he hoped that

> his Lordship will have experience in time, when this present mis understanding shall be rectified, of my integrity and of my zeale to serve his Majty, the English nation, and himselfe to the utmost of my power.

And he begged Williamson, when it was

> seasonable, to cast in a word of the narrowness of my fortune for to ly long in so chargeable a place, as the Tower is.

(Prisoners who wished to live decently normally had to pay to secure reasonably comfortable lodgings and always to pay for food, washing, and so on from outside, as well as to fee the turnkeys). He added that if Williamson were, as he had offered, to send him letters addressed to him and now in Williamson's hands, it would provide 'a welcome diversion'. There is no surviving reply from Williamson; all that is certain is that ten days later Oldenburg still had no access to writing materials. But on 15 July[83] he did have a visitor, identity unknown, who at his request wrote on his behalf a letter to Seth Ward, now Bishop of Salisbury. In this letter, he explained his circumstances, as he understood them, namely that he was 'committed for dangerous desseins and practices' and understood 'that that's inferred from some letters and discourses of mine, said to contain expressions of that nature'. He himself believed that in fact these letters only contained the expression of his

> reall trouble . . . to see things goe no better for England, than they did; and that I thought, there were oversights and omissions some where, which might prove very prejudiciall to the honour of the King and the prosperity of the nation . . . Thus, this expressed from me some words of complaint of neglect and security on our side; which having given offence, I am ready to beg his Majesties pardon for upon my knees.

And he added in mitigation, and truthfully, that almost since the begining of the war all his French and Dutch letters had been delivered in the first instance to Arlington's office 'where they might be opened at pleasure'. This was indeed the case although of course his arrangement was with Williamson, not with Arlington and besides, the accusation rested on words used by Oldenburg himself in an outgoing letter, not on anything said by his foreign correspondents. Evidently some of his letters had been intercepted in the post, although he had not been aware of this.(Nothing to show this now exists in the State Papers.) Oldenburg, a foreigner and with a known correspondence with the enemy, was an easy suspect and perhaps that alone accounted for the interception of an incriminating letter; there is also the possibility that he had spoken indiscretely in public. Whatever the cause of suspicion, there he was firmly imprisoned and Williamson could not help him. What Ward might have done for him had he received this letter is impossible to say, but it stayed in the State Paper Office, presumably taken from the visitor who had written it as he left the Tower. As far as is known, Oldenburg received no help from Ward nor, equally probably, from Lord Anglesey, President of the Council of State whom he had also asked to help and to visit him he being, as Oldenburg had told Ward, 'pleased to be also my friend' and, one would imagine, an even more powerful person than Ward. No trace of any letter to or from him now survives nor of any letter to or from Boyle, although Oldenburg seems to have written to him. But if these letters had reached their intended recipients it probably would have made no difference. The putative existence of such letters merely shows what a number of powerful friends poor Oldenburg did have, powerless though they were under these conditions.

Fortunately the Lieutenant of the Tower continued to stay as friendly as he could in the face of his orders and gave Oldenburg permission to have pen, ink, and paper exactly a month after his imprisonment. This privilege he used to write a 'Humble Petition to the King'and a very humble petition it is.[84] This suggests that Arlington had alleged that he was acting by the direct authority of the King who had, he must have claimed, been offended by Oldenburg's words, improbable as it now seems that Charles II had any personal knowledge of the affair. It was enclosed in an equally humble letter to Arlington himself, in which the poor prisoner insisted he was not 'guilty of any evill intentions and desseins' and had only used the 'inconsiderat and foolish words' complained of

> from affection, and from a reall and great discomposure of mind, I was then in, to see the affairs of England not succeed so well, as I wished, and to find the honor and safety of his Majty and his kingdoms endangered (such was my impertinent presumption to think) by the Invasion of an insolent Enemy.

For the moment, this appeal had no effect, and almost certainly Arlington never submitted the 'Humble Petition' to the King, for it stayed in the State Paper Office.

The raid on the Medway and the subsequent blockade of the Thames by one hundred ships of the Dutch Fleet which stifled all trade into London while the presence of the French fleet off the south-west coast of England interfered equally with trade and post in the West Country[85] changed the point of view of the government. It had as decisive an effect as the Dutch could have intended. As a result the English, after a couple of months of indecisive negotiation at Breda, at last agreed at the very end of July to a Peace Treaty.[86] After this, the severity of Oldenburg's incarceration was evidently a little relaxed. On 8 August Evelyn, as his diary records, was able to visit Oldenburg 'by an order from Lord Arlington, Secretary of State, which caused me to be admitted'. There is no record that anyone else secured permission to visit the unfortunate prisoner. But as Oldenburg later learned,[87] 'Not a few came to the Tower, meerly to inquire after my crime, and to see the Warrant', but 'when they found, that it was for dangerous desseins and practices, they spred it over London, and made others have no good opinion of me'. Oldenburg felt that, as he told Boyle, 'I have learned, during this commitment, to know my reall friends', adding touchingly, 'God Almighty blesse them, and enable me to convince them all of my gratitude'. He concluded 'Sr, I acknowledge and beg pardon for the importunities, I gave you at the beginning'. Evidently, Oldenburg did write, probably more than once, to Boyle and, if he received these letters, Boyle had destroyed them promptly as a precaution in those suspicious times. Boyle and Moray at least must have tried to exert influence on Oldenburg's behalf (it would have been out of character if they had not, nor would Oldenburg have written what he did to Boyle unless he knew that the latter had continued a 'reall friend' to him). But they must have found Arlington adamant for the time being, although their efforts probably now did hasten Oldenburg's release. There is otherwise no particular reason why his liberation should have occurred when it did on 26 August, nearly a month after the signing of the Treaty of Breda. It would have been wholly characteristic of the times if influence on behalf of a prisoner had not worked to encourage his release, for prisoners without influence might expect to be 'forgotten' for long periods.

When he was finally released, Oldenburg's first action, as he told Boyle, was to wait on Arlington, 'kissing the rod' and stoutly protesting his loyalty to England and the English. He then fled from London to enjoy what he called 'the good Air of Crayford in Kent'. Crayford, south of Woolwich and the Thames and not far from Dartford Heath, was very possibly where Dora Katherina Dury was living, for her inherited property lay nearby and it is certain that Oldenburg from this time onwards spent his holidays thereabouts. Hooke, reporting Oldenburg's release to Boyle, remarked 'He will not, I hope, leave off his Philosophical Intelligence and correspondence', a sentiment that should be remembered in view of Hooke's later contrary opinion.[88]

Oldenburg himself was determined to continue as before and once back in London wrote to all those whom he called 'my corresponding friends' to inform them that he was prepared 'to fall to my old Trade',[89] that is, to resume correspondence and to incorporate any replies he received either into the meetings of the Royal Society or to print suitable extracts in his *Transactions*. But as he feared, most of his usual correspondents were slower to reply than he had hoped they would be. Only Boyle and Beale wrote quickly, while no foreigners responded, at first perhaps, as Oldenburg himself thought, because they feared for his future safety. One may guess too that many people were on holiday, for September was then a great holiday month, especially in England. Oldenburg himself declared that he was unconcerned about any potential dangers arising from his foreign correspondence, only, wisely, planning to be very careful that his letters were confined to

> matters of a philosophical nature, or if there be a mixture of civill things, to such of them, as cannot be misinterpreted, or suspected of any ill dessein.[90]

This was the easier to do in the autumn of 1667 as England was, temporarily, at peace.

By the end of September, Oldenburg's correspondence did resume as before. Justel wrote from Paris, still with a mixture of political rumours and gossip from the intellectual world, the former, to which Oldenburg presumably did not reply directly, being transmitted as before to Williamson. The mathematician René François de Sluse wrote from Liège, having only belatedly received Oldenburg's letter of 6 June. Domestic correspondence also picked up. So, slowly, in the autumn of 1667 Oldenburg's life returned to normal. Once again he was fully occupied with correspondence, his work for the Royal Society (its meetings, which had been suspended during August and September as was to become henceforward customary resumed again in October) and the editing of his *Philosophical Transactions*, of which number 27 appeared under the date of 23 September, being 'For the Months of July, August and September'. This produced some confusion for, to Oldenburg's indignation, someone, probably Wilkins,[91] had had an issue printed under the date of 22 July as 'Numb. 27' and in the usual format although without the customary heading of 'Philosophical Transactions'. This contained only a letter by a Parisian surgeon Jean Denis describing his experiments on the transfusion of blood from an animal into a man as a French invention. (In fact the English had begun experimenting with the transfusion of blood from one animal to another several years earlier, as had long been known to the Fellows of the Royal Society (below, Chapter 5).

Oldenburg was much annoyed both at the production of a pseudo-issue of his journal and at the printing of Denis's false claim for priority. He therefore began his own, true *Philosophical Transactions* number 27 with a strong disclaimer, stating emphatically that he would never himself have published Denis's letter without plainly pointing out the English priority in 'prosecuting' such experiments. His annoyance was increased by the fact that many, including at least some who should have known better, ascribed the false number 27 to him, blaming his foreign origins for what they saw as an act of betrayal of English rights.[92] During the 1660s and 1670s the English were decidedly and manifestly prone to xenophobia and to the belief that all foreigners, but especially the French, were perpetually trying to claim for themselves ideas and inventions which certainly belonged to the English as the first discoverers. (Both Wallis and Hooke were badly and openly afflicted with this tendency.) Oldenburg might think of himself as virtually English, and certainly he had married an Englishwoman, but he was not an English citizen and however well he spoke English many of his contemporaries never forgot that he was really German and were all too inclined to believe that he was not sufficiently protective of English self-esteem. In fact, only sometimes, and only when writing (in Latin, always) to Germans did Oldenburg ever speak of himself as a German. But whatever he thought, others disagreed, and this was to lead to obloquy in years to come. But for the moment Oldenburg was, fortunately for him, able to satisfy everybody by disclaiming false number 27 and then publishing so much material on the English experiments on blood transfusion and related matters as to ensure that the whole learned world knew at last what had been done in England and when.

PART II
The correspondence: method and content

5
The promoter of philosophical intelligence 1665–70

In 1668 Joseph Glanvill, a Fellow of the Royal Society since 1664, published *Plus Ultra*, a defence of the new experimental philosophy against the claims of the older scholastic learning, arguing that by its means alone knowledge had advanced prodigiously since ancient times. There is nothing very remarkable about this argument, by this time a familiar, even a well-worn line of praise for the Royal Society and its outlook and for the strain of natural philosophy which it espoused. What is especially to the point here is Glanvill's encomium of Oldenburg who, so Glanvill declared,[1]

> renders himself a great Benefactor of Mankind by his affectionate care, and indefatigable diligence and endeavours, in maintaining Philosophical Intelligence, and promoting Philosophy.

This remarkably laudatory statement is important because it is a public one, making known to readers unfamiliar with the detailed workings of the Royal Society what Oldenburg's fellow members knew very well, even if they did not always appreciate what labour it required to achieve the result.

Although much of Oldenburg's foreign correspondence, like his publication of the *Philosophical Transactions*, was at least partly intended to supply him with much needed income, his chief stimulus in both endeavours was his profound belief in the aims of the Royal Society of which he was proud to be a loyal member and a devoted officer. He had, it must be remembered, acquired the art of writing and maintaining informational correspondence over the ten years previous to his appointment as Secretary and had long tried to disseminate the kind of natural philosophical principles which were to be the ideals of the Society. These he truly believed in. He could not adequately advance them by direct investigation of nature, being rather a virtuoso than a serious natural philosopher. Hence he chose to advance them by means of the skills which he knew himself to possess. These skills were partly administrative, for he was to be always an effective and hard-working Secretary. They were partly journalistic as he rapidly became a highly successful editor of his *Philosophical Transactions*, which equally rapidly acquired a genuinely international renown. They were partly literary as expressed in the complex network of correspondence which he had established even before his appointment as Secretary and which must have been at least partly responsible for his appointment. And they were partly linguistic, employing a talent he had used in both correspondence and translation of mostly scientific works from several languages.

It must be remembered that the officers of the Royal Society, were all, as befited their gentlemanly status, honorary and unpaid, the unique exception being Hooke, the first Curator. (Later Curators were not gentlemen.) This situation, for most of them, was fair enough, for their positions were not onerous and they had other sources of income. Brouncker, President throughout Oldenburg's time, worked in the Navy Office in the City, and was domiciled nearby. As President he was not bound to attend every meeting of the Society or even every Council Meeting, few though these generally were at this time, for the Vice-Presidents, chosen each year by the President, could serve as chairmen in his place. Brouncker in fact attended a majority of all meetings during his term of office, but by no means all of them. He was, moreover, a man of some means. The Treasurer's business did not require much attendance for he could, and usually did appoint a deputy to receive monies paid in by the Fellows, that is, admission money and quarterly dues, in payment whereof they were, it must be said, often dilatory or even delinquent. True, he needed to see that accounts were kept and audited, properly four times a year but in practice only once, just before Anniversary Day on 30 November. He was required to pay any bills ordered for payment by the President and/or the Council. But none of this business demanded frequent attendance at meetings.

The duties of the Secretaries were far more arduous and constant, and although the Charter specified the appointment of two Secretaries, duties were never assigned to both but to either or one of them. And in fact the 'other' Secretary during Oldenburg's lifetime, successively Wilkins, Thomas Henshaw, John Evelyn and Abraham Hill (the latter Treasurer from 1663 to 1665), seem not to have done much of the routine work, leaving it virtually all to Oldenburg. True in the summer of 1667 during Oldenburg's imprisonment Wilkins acted as Secretary in taking the minutes, but there is no evidence that any other Secretarial duties were fulfilled until the autumn, when Oldenburg was once again available to undertake them. With few exceptions all the Secretarial duties specified in the Charter and Statutes—attendance at meetings, taking the minutes, seeing to their entry into the Society's books, taking charge of 'all papers and writings belonging to the Society', carrying out any formal orders of the President, Council and Fellows issued at meetings, writing the bulk of the letters to be sent in the name of the Society—were carried out and conscientiously so by Oldenburg. The 'Books' were the Journal Books containing the minutes of the meetings, fairly written out, the Council Meeting Minute Book, the Register of Papers, that is, copies of papers read at meetings of which ideally both the original and the copy made in the Register Book were kept provided only that the authors would 'give in' their papers for this purpose. There were also the Letter Books containing incoming letters copied by the Clerk or Amanuensis (the name varied) and sometimes drafts of Oldenburg's replies, usually in his own hand. The originals of incoming letters and rough minutes of replies Oldenburg seems to have kept loose, as did successive Secretaries, for they were entered into Guard Books only in the eighteenth and nineteenth centuries.

All this took up hours of Oldenburg's time. Then there were the meetings, held either on Wednesday or Thursday at different times, from 3 to 6 p.m. The Council, when it met, normally did so an hour earlier on the same day, except occasionally on days when the Society was not meeting, when it might meet at any convenient hour. There were, by custom, no meetings on certain holy days like Ash Wednesday or exceptional fast days, and none ever until 1834 on 30 January, the anniversary of the execution of Charles I. After the first few years the Society suspended meetings from July to October, the period when Londoners then were accustomed to take their holidays. Oldenburg himself customarily took a short holiday in early September, but otherwise seldom absented himself except through illness. There had been, of course, no meetings between June 1665 and March 1665/6 because of the plague. This left Oldenburg with less to do in one sense as he had no meetings to attend, but more to do in another for the absence of meetings meant that his correspondence increased with those Fellows no longer in London, and made him responsible for dealing with all incoming letters addressed to the Society. He had as well to carry out any commissions for absent Fellows, and organize incoming letters for answer and subsequent filing. Possibly some of the Fellows who stayed may have continued to meet informally on occasion at coffee houses and taverns, but there obviously can no longer have been the weekly meetings at neighbouring taverns which had customarily taken place after meetings, all producing some saving of time. However in general correspondence had to take the place of informal discussions of business between Fellows.

The injunction in the Statutes to Secretaries to take charge of correspondence both domestic and foreign on behalf of the Society was one which Oldenburg hardly required, since even in 1662 he had had the basis of a flourishing international correspondence with many ramifications centred in his own hands. Begun initially as a purely private affair, first for himself, later for others as well, it had developed during his years of travel with Richard Jones in line with the interests of his friends Hartlib and his patron Boyle. From 1663 onwards the centres of interest of Oldenburg's correspondence shifted in line with those themes which it was natural and useful to convey to meetings of the Royal Society, to individual Fellows and, after 1665, to the learned public at large by means of publication in his *Philosophical Transactions*. (It goes without saying that Oldenburg edited out anything private in his incoming letters before making the contents public, ordinarily merely a simple matter of omission.) It was all work. No wonder that Oldenburg remarked plaintively to Boyle in 1667, a few months after his imprisonment when his correspondence had resumed its normal flow,[2] 'I am sure, no man imagins, what store of papers and writings passe to and from me in a week from time to time.'

He then reckoned that he had thirty correspondents, all writing more or less regularly and demanding regular answers, and this had been true for the immediately preceding months. In fact if all his correspondents, both regular and spasmodic, were counted, he had very many more persons writing to him, for in the preceding two years he had come to correspond with an ever-increasing number. In the years 1663–5 he had dealt with letters to or from more than thirty persons; in 1666–7 the number rose to forty-seven; while for 1667–8 there was a further increase to fifty-two. And these numbers respresent a minimum, for it is reasonable to suppose that by no means all his incoming letters have survived, and indeed there is evidence of some missing letters. After 1669 the total number of correspondents in any one year declined somewhat, but those who remained wrote on the whole with greater regularity. After 1667, no doubt at least partly because of the lesson learned from his imprisonment, Oldenburg was more careful than he had been in earlier years to keep more incoming letters and to have the more important and relevant carefully copied into the Society's bound Letter Books. He also from this time forth often noted on incoming letters when they were received and when answered, frequently even jotting a few lines to indicate the principal contents of his replies, while drafts of important Latin letters more frequently than before also appear in the relevant Letter Book.

In 1698, more than twenty years after Oldenburg's death, the naturalist Martin Lister paid a singular tribute to his industry, which he held up as an example. In commenting unfavourably on the failure of the Académie royale des Sciences to have founded an equivalent to the *Philosophical Transactions* as a 'register' of the numerous observations made by people everywhere, he attributed this failure to the lack of a 'great Correspondence'.[3] He remembered that

> I heard Mr *Oldenburgh* say, who began this Noble Register, that he held Correspondence with seventy odd Persons in all Parts of the World, and those be sure with others: I ask'd him what Method he used to answer so great variety of Subjects, and such a quantity of Letters as he must receive weekly; for I knew he never failed, because I had the Honour of his Correspondence for ten or twelve years. He told me he made one Letter answer another, and that to be always fresh, he never read a Letter before he had Pen, Ink and Paper ready to answer it forthwith; so that the multitude of his Letters cloy'd him not, nor ever lay upon his Hands.

As an appreciation of Oldenburg's industry and influence this can hardly be bettered, and although it may have slightly exaggerated Oldenburg's methodical approach to his business, for certainly there were times when he could not reply to a letter as soon as he had read it for a reply might require information from others. But it is a fair summary of what he tried always to achieve and which he accomplished more often than not. His was certainly a thoroughly business-like approach to the task to which he had dedicated himself since 1662 and had prepared for earlier.

Such a vast correspondence obviously involved much time, thought, and labour. Many letters demanded what Oldenburg called 'business', that is action on his part. Many letters contained enclosures or messages to be conveyed to others. Even more troublesome, many letters asked for news of books recently published, information which Oldenburg obviously could not have at his finger tips. More troublesome still was when he was asked to purchase and send books, which might or might not be easy to procure, and which involved him in expense which he could ill afford and payment for which was at best slow, at worst dilatory or even negligent. Moreover, books did not travel by ordinary post and some other means had to be sought. Thus parcels sent to Hevelius in Danzig required Oldenburg to find a willing ship's captain going more or less directly to Danzig (or at least to Hamburg). Parcels for Italy had to await a willing traveller, not always easy, and even those for France were best sent by hand unless they could go to a merchant somewhere, like the one at Dieppe with whom Justel had an arrangement. Very often correspondents, although they addressed their letters to Oldenburg, really wanted communication with other members of the Society, relatively easy to arrange for London-based Fellows, but troublesome in the case of provincial Fellows like Wallis, for it involved Oldenburg in further correspondence, often requiring paraphrase of information sent or inquiry sought. And, increasingly over the years, Oldenburg was to undertake elaborate networks of correspondence, whereby A wrote to B via Oldenburg, so that he had to write to both, using his judgement in conveying opinions where controversy was involved, and sometimes a third party might even be involved. (See below, Chapters 6 and 7.)

On this last account particularly the range of subjects with which he had to deal became ever greater and often ever more technical. Almost all his correspondence demanded that Oldenburg be prepared to respond to the varied interests of his correspondents. He needed to be able not merely to acknowledge letters but to discuss their contents in an informed manner. He needed also to exercise his judgement as to whether a correspondent should be encouraged or not. As already noted, those who wrote about theology or educational theory had to be politely told that the Royal Society did not concern itself with such subjects. In a number of cases as with dealing with Malpighi[4] in 1667–8, he was to encourage men to complete and send for publication work they were engaged upon, or to seek permission to publish their letters in his *Philosophical Transactions*. Moreover, correspondents frequently so reasonably requested the Society's opinion of their work that Oldenburg turned to individual Fellows working in the appropriate field to ask them to send him their private opinions for transmission, since the society always refused to commit itself to an official judgement. Frequently too over the years correspondents sent comments on papers which Oldenburg had published in his journal. Proper reception of such letters required him to communicate the comments to the author of the paper in question, the latter's reaction then dictating whether Oldenburg was to publish both comment and reply or merely communicate the result by letter to the commentator. It was all time-consuming. What demanded all the diplomatic skill which Oldenburg had acquired in the course of his life was the conduct of controversy with himself as intermediary. Of this many examples, ranging from the trivial to the truly important, constantly occurred.[5] Thus, with every continuing correspondence, Oldenburg had to exercise judgement. This was necessarily and deliberately a personal judgement but he had always to remember that he was, as Secretary of the Royal Society, an official mouthpiece of that Society, as he was generally regarded by his correspondents. When framing an opinion he was forced to consider the views of the Fellows at large, either by estimating of their probable reaction or by consulting individual Fellows so that he could be sure of having the Society's backing subsequently for what he wrote, either openly or tacitly. Much judgement was required when he decided to read a communication to a meeting of the Society and then report the reaction of the auditors. Very often he merely said what he had done and sent thanks on behalf of himself and the Society, the latter of which he was very frequently instructed to send, but quite often it required tact to express the sense of the meeting's reactions without giving offense to the writer. Clearly, foreign correspondence was particularly difficult.

Domestic correspondence, of equal importance and perhaps greater volume, was of several kinds. There were letters to established Fellows communicating news of what the Society was currently doing and thinking and what foreign natural philosophers of importance had written either to the Society's Secretary or to Oldenburg as a private person. There was the acknowledgement of letters received and the Society's reaction to them if they were read at meetings. There was the encouragement of isolated provincials, young and old. In the first category falls the bulk of Oldenburg's letters to Boyle in Oxford before 1668 - that is, the purveyance of scientific news. (After 1668 Boyle moved to London to live with his sister Lady Ranelagh as Oldenburg's neighbour.) There were letters to Beale in Somerset, whose correspondence combined his passionate interest in cider and horticulture with general points of curiosity and a keen interest in the affairs of the Society and to a hardly lesser extent a benign interest in Oldenburg's private affairs. In the same category belong the letters to and from such virtuosi as Nathaniel Fairfax[6] and Samuel Colepresse,[7] correspondence with both of whom began in the winter of 1666–7 and that with Joshua Childrey[8] which began a year later. Fairfax and Colepresse had different interests but were alike in being men isolated in rural surroundings who possessed inquiring minds whose interests

Oldenburg could and did to some extent direct to useful channels. Fairfax sent accounts of various curious medical cases, reports of natural history of a fairly credulous kind and odd pieces of country lore; Colepresse had a far sounder grasp of natural philosophy and sent reports of considerable interest; while Childrey had on the whole an antiquarian approach to natural history. Of Colepresse it can be said that he understood the Royal Society's aims and so had more to contribute to the serious purpose of these than the others. In 1666 he took the trouble to visit London, met Boyle certainly and Oldenburg very probably, and attended at least one meeting of the Society. His interests ranged from the behaviour of tides off the Devon coast[9] (and Wallis was then developing his theory of the tides and was able to make use of this information) to astronomy and medical curiosities of which the Society was always glad to hear accounts, especially when, as in this case, rationally described. Encouraged by his correspondence with Oldenburg and his contact through him with the Royal Society, Colepresse in 1667 went to Leiden, a common goal at that time for Englishmen seeking medical degrees. From Leiden he regularly sent Oldenburg serious and interesting accounts of Dutch work in various fields including anatomy, and a certain amount of natural history. Sadly, in the late summer of 1669 Colepresse died from, presumably, the 'malignant fever' which, so he told Oldenburg, had been hindering his activities.[10]

Fairfax also went to Leiden, in 1670, less to study than to acquire a quick M.D. degree. This was not from aspiration induced by Oldenburg's encouragement as it had been with Colepresse but shame at having betrayed ignorance of anatomy, especially in a passage which Oldenburg had printed in his *Transactions*. This embarrassed him so much that he felt he must apologize for it to Oldenburg since he thought his mistake might have brought the journal into disrepute.[11] One must hope that he stayed long enough in Leiden to acquire a better knowledge of the subject! He wrote many accounts of the kind of medical curiosities to which most virtuosi of the time were partial. The most reliable of these, and the most interesting to modern taste, was the account of a case of anorexia in a young woman, his patient. She, believing upon the death of her father that her income would no longer be sufficient to support her, determined to fast to death. According to Fairfax's detailed account she fasted for a month and a half except for a very little fruit and fluid. Then she suddenly gave up the attempt, enjoyed a dish of 'buttered peas' and returned to normal eating habits.[12] This tale is more credible now than it was then, unlike some other of Fairfax's medical anomalies. He reported some pretty tall tales of a man who ate toads and supposedly poisonous spiders without ill effects, tales which, although Fairfax was sometimes sceptical of their truth, trying several experiments to test the harmlessness of the spiders by observing their diet, on the whole he was uncritical, divided between simple but not unthoughtful experiments and complete credulity.[13]

In fairness it must be said that he was never so wildly credulous as Israel Tonge who in 1670 sent to Oldenburg a long account of a duel between a spider and a toad, a tale of whose truth he was firmly convinced. But even Tonge was sometimes a serious observer of nature. In previous years he had written sensibly about the rising of sap in trees, extracts from these letters having been published in the *Philosophical Transactions* along with those of the well-known naturalist Francis Willughby.[14] What Oldenburg made of all these tall tales is impossible to establish; certainly he continued to encourage Fairfax to write to him but equally certainly he never published any of his accounts of spiders nor any more anatomical accounts after his first inaccuracy. It is worth remarking that after he had taken his M.D. degree, Fairfax gave up correspondence with Oldenburg and settled quietly back into his country practice.

Childrey was also a virtuoso, but of a quite different kind. When he first wrote to Oldenburg he was no longer young and had already published a work on Baconian natural history, entitled appropriately *Britannia Baconica* (1660). His interests were mainly confined to observational natural history, such as details of agricultural practice, or accounts of mineral springs, all purely descriptive, not at all analytical nor theoretical. Oldenburg appears to have welcomed his letters and occasionally read extracts from them to Society meetings. But he never published anything from them in his *Transactions*, although he did publish Childrey's notes on tides, partly no doubt because of Wallis's interest.[16]

These are only three examples of the many provincial virtuosi with whom Oldenburg kept in touch at this time and later more or less intermittently, bringing London and Royal Society values to isolated clergymen, medical men and gentlemen. The case of John Beale[17] was different only in that he was an older man (he sometimes took advantage of their difference in age to patronise his 'worthy friend'), a man of established position and authority. In earlier years he had been in close touch with Evelyn, with whom he had collaborated in the production of *Pomona* in 1664, and with Boyle. In later years he did not venture to write to Boyle directly because of his appalling handwriting, but wrote through Oldenburg whom he trusted to convey the contents of what he wrote to Boyle. Like so many others, he felt isolated in the country remote from London, in his case in Somerset, and particularly welcomed contact with a wider world. He usually reported phenomena or discoveries related to the concerns of Boyle and others although frequently also on strange country matters. For all that he had a random and discursive train of thought his chief interests were closer to those of most Fellows of the Royal Society than those of, say, Fairfax or Childrey, and his letters did often, if not always, add to the observations of others. For besides what the original article in the *Dictionary of National Biography* called his 'hereditary interest in cider' which caused him to endeavour to do much to promote the spread of cider-apple trees in southern England, and his concern for agriculture, certainly a valid subject of interest to the early Royal Society, there were many topics more closely related to natural philosophy which caught his interest from time to time, all to be retailed at some length to Oldenburg. Thus on one occasion he described the light emanating from salted meat and rotting fish, a phenomenon which Boyle had previously investigated, and on another he reported that his defective eyesight was assisted by the use of cardboard tubes (also employed by Pepys). His accounts of weather in the West Country were added to the collection of accounts of meteorological phenomena elsewhere, to name but a few of his topics. Oldenburg frequently assured him of the Royal Society's appreciative interest in what he wrote, and printed extracts of many, although by no means all, his letters in the *Philosophical Transactions*.

There were many other Fellows scattered throughout England and Scotland who reported not merely random observations but systematic natural history, all adding to the number of Oldenburg's correspondents. Here Oldenburg could encourage such men by assuring them of the interest which the Royal Society took in their endeavours, telling them what others were doing, and attempting to steer their interests towards what the Society regarded as proper fields of investigation when this was required. Not that they always needed such direction. A productive network, for example, already centred independently around John Ray (1627–1705), rapidly becoming England's leading botanist, taxonomist and zoologist. Ray, although elected F.R.S. in 1667, did not at this time correspond directly with Oldenburg but did so through two of his former pupils, Francis Willughby (1635–72) whose estates were in Norfolk and Philip Skippon (1641–91) whose estates were in Norfolk and Suffolk.[18] Ray lived and worked with Willughby with whom and Skippon he had travelled on the Continent from 1663 to 1666. He only completely retired to a remote village in Essex some time after Willughby's death in 1672. Willughby himself was a constant and highly valued correspondent of Oldenburg's, providing considerable material for publication in the *Transactions*, sending papers to be read at Society meetings and seeking regular news of the Society's activities. He was also an able mathematician, sending useful comments on the Society's discussions of the laws of motion in 1669 (see below), but is best remembered for his studies on birds and fishes, studies left incomplete at his early death and which Ray was to finish for publication by the Society. Even in an example like this, where his correspondents were already engaged in productive research, Oldenburg could be useful in encouraging country dwellers to initiate or to continue to pursue research by informing them of relevant discussions within the Society and by telling them how much it was hoped that they would bring their studies to fruition. Country virtuosi, as distinct from such systematic workers as Ray and Willughby, might not make notable or lasting contributions to natural philosophy, but they had their uses and could be directed to consider matters of some importance which might lie within their competence, even if this was limited. Oldenburg persuaded other several country and provincial virtuosi as he had Childrey and Colepresse to send information on the behaviour of tides for the benefit of Wallis, for there were many places, as along the coast of Southern England and in the Bristol Channel which regularly experienced anomalous tides which local inhabitants knew of and their observations and recording of them could provide much needed information.

Wallis in Oxford was hardly isolated from intellectual contact but he could not regularly attend Royal Society meetings and to that extent depended upon Oldenburg for contact with the wider world of natural philosophy. He obviously also, as time went by, found it convenient to keep in touch with his own foreign correspondents through Oldenburg, at least from time to time. He habitually acted as adviser to Oldenburg on mathematical papers and books which Oldenburg was considering for inclusion in the *Philosophical Transactions*, and often wrote reviews. He brought Oldenburg into several mathematical controversies to which he himself was addicted. Thus in the early 1660s he involved Oldenburg a little in his vehement and amusing but not particularly important controversy with Thomas Hobbes which had dragged on since the mid-1650s, less a true controversy than an example of joint and mutually destructive criticism and quarrel.[19] (Hobbes was ultimately led to deny the rigour of mathematical proofs when they lacked empirical justification while Wallis naturally defended the independence of mathematics from sense experience, at the same time castigating Hobbes's mathematical incompetence.) Since Oldenburg published Wallis's attack in 1666 in the form of a book review in his *Philosophical Transactions* it is to be presumed that he accepted Wallis's arguments which Hobbes saw as a rejection of his point of view by the Royal Society. He replied to Wallis in a work on squaring the circle, and fortunately for himself Oldenburg was not further involved in Wallis's attack upon Oldenburg's former correspondent. Wallis was of course correct about Hobbes's mathematical errors, but it must be said that only a dedicated and fiery controversialist would have pursued the matter for so long.

In a mathematical controversy of 1668 Oldenburg did become involved. Then a French mathematician whom he had known in Paris, François Dulaurens, sent to the Royal Society (via Justel) a recently published book together with a challenging problem, as was not unusual at the time. The communication was, naturally enough, passed on to Wallis, who wrote long, repeated, pedantic and often fairly trivial criticisms of Dulaurens's work.[20] These, which Wallis sent to Oldenburg, the latter duly transmitted to Dulaurens, only taking care to soften some of Wallis's harshest expressions. Dulaurens replied patiently and with good will, saying that far from taking Wallis's criticisms ill, he was 'obliged' to him for having taken the trouble to respond so fully,[21] while he himself replied in a printed text rather than in a letter. (It was then common practice in France to print lengthy letters which could be circulated as widely as desired; there was, after all, no French equivalent of the *Philosophical Transactions.*) Wallis, unmoved by Dulaurens's soft words, continued his attack, but Oldenburg seems tactfully to have ceased to transmit Wallis's criticisms. It certainly must have removed some of the sting of such outspoken criticism as that of Wallis when it was transmitted by a third party who, as Oldenburg so often did, could manage skilfully to soften the language of attack. He was no doubt somewhat ashamed of the intensity of Wallis's attack on what Oldenburg was probably able to recognize as fairly minor points of difference between the two mathematicians. And in cases like this he must often have feared that he would be suspected of personal involvement when he was in fact merely transmitting another man's carping criticism, another reason for trying to mitigate its intensity. In later years he was neither so lucky nor so successful in doing so, and as a result was to be blamed for apparently stirring up controversy when he was merely an intermediary.

Far more frequently more pacific international exchanges passed through Oldenburg's hands and on the whole over the years he handled them with admirable tact. A good example occurred in 1668–9 when a discussion began at a Society meeting over whether the laws of impact motion (the theory of collisions) had been properly worked out. This discussion began on 23 October 1668 when Hooke proposed that the Society ought to investigate the problem. Brouncker then remarked that he thought that both Huygens and Wren had, independently, successfully developed a theory of impact motion. Consequently Oldenburg was ordered to write to Huygens to ask whether he did possess such a theory (of which he had not yet published anything) and if so, whether he would consent to communicate it to the Royal Society of which, he must be reminded, he was a Fellow, since the Society would like to examine any hypothesis he might have by experiment and 'to register it as the offspring of [his] mind'. This Oldenburg did in a most polite letter,[22] the outcome of which was that Huygens quickly accepted the Society's offer which he professed to regard as an honour. A few months later he sent a Latin paper, read at the meeting on 7 January 1668/9, as Oldenburg promptly informed the him.[23] Meanwhile, Wallis had sent a paper outlining his own theory and Wren produced his theory, both written in Latin and both read at the meeting on 10 December 1668, the two papers then being published in the *Philosophical Transactions* of 11 January 1668/9 (number 43). This publication resulted in detailed comments by Wallis's mathematical protege, William Neile, sent through Oldenburg, the result being a long-drawn out interchange of views in the first half of 1669.[24] Neile's paper was read to a Society meeting, as was that of William Croone on the same subject, and entered into the Register Book, although not printed. But the papers of Wallis and Wren were printed, and at this Huygens apparently was hurt. For while his paper had been duly read to a Society meeting and entered into the Register Book, which pleased him, it was not published in the *Philosophical Transactions* alongside those of Wallis and Wren.[25] (Whether its non-publication represented a private decision by Oldenburg or the preference of the Society's meeting at which it was read is not clear.) When Huygens learned that the papers of Wallis and Wren had been published, and without his, he promptly sent an account of his own hypothesis to the *Journal des Sçavans* for publication there,[26] for although the *Journal* was not originally intended to be a vehicle for papers, its editor was inclined to make exceptions in the case of short scientific or technical papers. In his account Huygens took care to stress his independence and his (quite genuine) priority in devising his theory, in spite of the fact that the very similar theories of others had been published first. Oldenburg must have been a little ashamed of his neglect in ignoring Huygens's claims: he tried to apologize by telling Huygens that he had not wished to publish the little work himself without having secured the author's express permission. Although other factors, like xenophobia, may have been at work this was not altogether a lame excuse for Huygens, like so many of his contemporaries, had never been inclined to publish, and perfectly well might have taken it ill if Oldenburg had published without permission.[27] Oldenburg tried to make amends by reprinting the paper published in the *Journal des Sçavans* in number 46 of his *Philosophical Transactions* (issue for 12 April 1669, pp. 925–8) with a brief English introduction and a longer, historical account in Latin, both written by Oldenburg himself. Huygens accepted the amends made and professed himself satisfied, remaining in the future on excellent terms with Oldenburg.[28]

No doubt Huygens found it easier to accept the English choice of procedure because at this time Wallis was entering a controversy between Huygens and the Scot James Gregory on the side of Huygens, a slightly unusual event since Wallis was normally always anxious to defend British priority against the claims of foreigners. This mathematical controversy, violent, bitter and complex, had begun after Gregory had sent Huygens a copy of his recently published *Vera circuli et hyperbolae quadratura* (Padua, 1667). Instead of replying privately, as no doubt Gregory expected him to do, Huygens published a highly critical review of Gregory's book in the *Journal des Sçavans* of 2 July 1668 (N.S.). Here he claimed that Gregory had made some errors (in this Huygens was partly correct), that he, Huygens, had anticipated some of Gregory's results (which again was at least partly true, although Gregory may not have known of it), and that Gregory, knowing of Huygens's work, was plainly guilty of plagiarism (which was false).[29] This controversy might seem to have nothing to do with the Royal Society itself or with Oldenburg, but unfortunately Gregory replied to Huygens in a work entitled *Exercitationes geometricae* (London, 1668) upon which Huygens in return commented in a letter to Wallis sent via Oldenburg. Huygens had assumed that the Royal Society would support Gregory, but this, Oldenburg was able to assure him, was not the case.[30] However, Oldenburg became personally involved when Gregory sent directly to him for publication in the *Philosophical Transactions*, as he hoped, an answer to a second comment by Huygens in the *Journal des Sçavans* of 2/12 November 1668 while Wallis, now himself involved in a different controversy with Gregory, began to comment on the whole affair in his letters to Oldenburg and Moray decided to intervene also.[31] Oldenburg correctly told Huygens that the Royal Society was keeping an open mind and it was the case that he himself had, as instructed by the Society's meeting of 1 February 1668/9, taken care to tone down some of Gregory's remarks and omit 'all, that might be offensive'.[32] The controversy began to simmer down during 1669, no doubt to Oldenburg's great relief, for it was nothing to do with him, although he had been to a certain extent caught in the middle. It had all been a sad contrast to the peaceful interchange over theories of motion, which had no doubt been rendered all the more peaceful because no one doubted the independence of all the participants, even though the theories of Wren, Huygens, and Wallis were very similar.

Other peaceable international exchanges were to be made from time to time and these Oldenburg seems on the whole to have enjoyed, although all too often those which began peaceably enough were to degenerate into acrimony. Astronomical observations were always a subject of international exchange and very frequently a source of controversy as they had been over the question of the exact path of the comet of 1664–5. Here the English were frequently asked to adjudicate between conflicting observations. This, it should be remembered, was much against the principles of the Fellows, so much so that except in special cases they were apt to be very wary even in cases of fact and virtually never conceded that they might frame judgements when the validity of theories was at stake, as Oldenburg had the duty of telling correspondents on many occasions. In the case of the comet of 1664–5, the problem was that Auzout assigned to it a circular orbit, while Hevelius not only rejected the circular orbit but reported that his own observations of the comet's positions differed from those reported by Auzout. The Fellows greatly respected Hevelius's skill as an observational astronomer, but were troubled by the fact that Auzout's observations and conclusions were supported not only by his colleague Petit (well known to Oldenburg and others) but by Giovanni Domenico Cassini of Bologna.[33] The controversy continued (slowly because of the difficulties of postal communication at the time) into 1667. On the Society's behalf Oldenburg wrote in cautiously polite terms to Hevelius and in rather warmer terms to Auzout. The astronomical Fellows were, on the whole, inclined to regard Auzout's observations as probably more accurate than those of Hevelius, since the latter used old-fashioned instruments without telescopic sights whereas, after all, Auzout had perfected the filar micrometer.[34] Through Oldenburg's skilful management, no bitterness ensued from this really profound difference of scientific approach. And this was even though Hooke was to be extremely critical of Hevelius's neglect of telescopic sights, which he himself firmly advocated on sound grounds.

However, Hooke did readily agree to assist Oldenburg in filling Hevelius's request for a good English telescope, since he did not have the leisure required to grind one for himself.[35] Oldenburg gladly offered to procure one for Hevelius, an offer which was gratefully accepted, Hevelius suggesting that he pay for the telescope with copies of his recently published *Cometographia*, which was done.[36] (In years to come Hevelius was often to send copies of his books to Oldenburg either to distribute to such Fellows as Wallis or to sell to pay for purchases on Hevelius's behalf.)

A more lively and only occasionally bitter controversy developed about the same time between Auzout and Hooke on the subject of lunar astronomy, the grinding of lenses, and the potentials of telescopes. These topics were all managed through Oldenburg who this time played an active part on Hooke's behalf. It all began in 1665 when Auzout published a little work in which he incidentally but very firmly criticized the proposal which Hooke had made in the preface to his *Micrographia* for a machine which would permit the easy grinding of spherical lenses, which he forecast would greatly improve their power.[37] The criticism was based upon two points. The first was that Hooke had announced his design without constructing an example of his proposed machine to prove that it worked (as was Hooke's habit). The second was directed at Hooke's optimistic suggestion that the construction of telescopes with very great focal length and spherical lenses might, so some people, as he said, claimed (but he did not positively do so) make it possible to see greater details on the planets and, in the case of the moon any existing animals and plants. Oldenburg carefully translated the relevant passages of Auzout's critical remarks for publication in the *Philosophical Transactions*. This translation was also intended for the use of Hooke, who read no French, so that he could reply, a reply which Oldenburg published alongside Auzout's original remarks.[38] Hooke wrote politely, thanking Oldenburg for sending him the translation and commenting with restraint upon Auzout's criticisms. When this reply was sent to Auzout he responded less temperately than Hooke had done, but sufficiently peaceably for Oldenburg to declare

> Surely, Sir, it is the right way to manage a correspondence between two worthy men and fine minds, when each expresses to the other his thoughts and discoveries in a frank and polite way, without offense given or taken, so that their minds may reciprocally stimulate each other and learn from each other, to the further progress of knowledge.

Hence he offered to act as 'mediator' between the two men.[39] As he put it here, all his life Oldenburg was to believe that the way to persuade natural philosophers, all too often reluctant to publish their ideas and discoveries to let the world share their thoughts, was in just this form of mild and controlled controversy. This was excellent in intention but not an easy thing to manage in practice, as Oldenburg was to discover more than once.

In this case things proceeded unevenly. Auzout had Hooke's letter printed in French for distribution without asking Hooke's permission, an action which displeased both Hooke who had written it and Oldenburg who had communicated it. Moreover Auzout, in replying to Hooke at some length, rather wilfully, as it seemed to the recipients, misunderstood some of what Hooke had written, as Oldenburg carefully noted in the margin of the translation.[40] Oldenburg's comments range from 'to this I say, He will needs make you say, what you say not' and 'What say you to this' to 'A handsom sting again will be necessary' and 'Me thinks, here you may tosse railleries with him' and 'To play again on him'. Once, on Auzout's brief remarks on how the earth would appear from the moon, he did favourably remark 'This pleases me'. However, when Auzout said that when he had had time to make systematic observations he could say more about the appearance of the surface of the moon, upon which he had commented in very reasonable style considering the then state of knowledge of lunar topography, Oldenburg felt it necessary to comment provocatively, 'Qy. Hath nothing been done herein by English men?' Clearly, Oldenburg was enjoying this particular controversy and felt, oddly considering Hooke's later penchant for fierce criticism of those with whom he disagreed, that Hooke might lack finesse and persistence, which he tried to supply. But he had to apologize to Hooke for what sounded like his own indiscretion when Auzout remarked that he wished he had a secret to exchange with Hooke to persuade him to reveal the undisclosed method of improving telescopes and their eyepieces which Hooke hinted he possessed 'seing that you [Oldenburg] hint unto me, that that is the way for it'. In exculpation Oldenburg could only tell Hooke, 'If I had imagined, he would have been so nimble to print, I should have used another expression'. He was clearly and rightly embarrassed by Auzout's remark, fearing that Hooke might take offence, and he deleted the sentence from his published version. Oldenburg continued to encourage Hooke to reply to whatever Auzout wrote, in this case, at least, accepting the view of so many of the members of the Royal Society that foreigners and especially the French were inclined to claim for themselves discoveries and inventions really initiated by the English. Quite rightly, he defended Hooke's genuine priority and his views about the best form of lenses. It had been a lively interlude, friendly on both sides, and Oldenburg had clearly enjoyed coaching his younger colleague, who now discontinued the controversy. But the whole episode may have contributed to increase Hooke's innate xenophobia.

It was not only disputes about matters of fact that frequently arose in the seventeenth century. These might often, as with Hooke and Auzout, involve questions of priority. This partly arose because of the absence of any publishing conventions, partly because publication more frequently than not was delayed or only occurred some time after the discovery had been made or even after it had been communicated to friends of the discoverer, partly because many men were reluctant to commit themselves to the finality of publication. It was also difficult to determine precisely when a book had been sent to press, for English books at least were often printed off in the autumn and dated as of the year to come. For example Hooke's *Micrographia* was officially published in 1665 but in fact copies were available in London in the late autumn of 1664. The Royal Society invented two methods of solving the dating of a discovery: it permitted any Fellow to 'register' an idea or invention by depositing a sealed paper in the archives to be opened only at the request of the author (a method used by Boyle, for example), while the custom of publishing letters in the *Philosophical Transactions* helped in dating, even when Oldenburg failed, as he mostly did not, to note the date at which the letter had been read to a meeting of the Society. A modified form of the latter method was of course open to the French if they could gain access to the *Journal des Sçavans* and this briefly became common in the 1670s. The very existence of these two journals encouraged the publication of new ideas in a brief or tentative form and by their authority served to validate an author's claims for priority, as of course letters also could do. For it must be remembered that letters to and from Oldenburg were not properly private but were at least quasi-public documents. (This had been true of letters to and from earlier networks of intellectual correspondence, like that of Marin Mersenne earlier in the century.)

As a consequence of this, both Oldenburg and his correspondents felt free to transmit the contents of the letters they received to others, unless the writer of the letters specifically requested secrecy or limitation of communication; hence information which Oldenburg received or transmitted counted as being almost published, very much like the printed sheets of Auzout and other Parisians, which were only privately printed. Rarely indeed did anyone think otherwise, recognizing, as most did, that it was an important part of Oldenburg's business as Secretary of the Royal Society to disseminate news of ideas and inventions. For he well knew, as did most of the Fellows, that the best way to discover what a foreign natural philosopher was doing and thinking was to offer him information about what English natural philophers were doing and thinking. Auzout was quite correct in believing that Oldenburg had implied so much in the case of Hooke who was so very inclined to hint at discoveries and refuse to communicate them openly (but might do so if he heard that someone else had the same or a similar discovery in mind). There was nothing extraordinary about Moray's instructions to Oldenburg during the Plague time to stimulate Hooke to develop his ideas in various fields by telling him that he would lose his priority if he did not get on with making his achievements public—which might mean developing them from a first, sometimes nebulous, idea.[41] (Whether Oldenburg did venture to rebuke Hooke in this vein cannot be known, but his treatment of him in the controversy with Auzout certainly suggests something of the same spirit.) To be fair, Hooke was far from being the only seventeenth-century natural philosopher to be slow about publication. In his case he had the excuse that he usually shared his ideas with his colleagues in the Royal Society and to his mind published them when he spoke of them at a 'public' meeting. But it was by no means then uncommon for natural philosophers to put their work on one side when more or less complete for many years for various reasons or apparently for none at all. Huygens was a notorious example, in both mechanics and optics. And so of course was Newton. A ready excuse was the then genuine difficulty in persuading booksellers to accept scientific and especially mathematical books. There was also a kind of diffidence, a reluctance to publish until everything had been developed to perfection, while of course there was then no professional compulsion to publish. Oldenburg sought to fight against these tendencies to delay by providing encouragement and a mechanism for publication of short accounts, as it were 'work in progress', which at very little effort on the part of an author vastly assisted the relatively rapid dissemination of information.

The problem of priority remained and here Oldenburg's correspondence could be of supreme importance. A famous and important example of the way in which Oldenburg could make use of letters and/or publication in his *Transactions* to establish rightful English priority concerns the rival claims for priority in the introduction of the use of injection of substances into the blood of animals and the experiments in the transfusion of blood from one animal to another which were a natural extension. Here the picture is quite clear for, although the world outside Britain was not aware of it, English workers, especially Wren and Richard Lower, had practised injection and were to practise transfusion before anyone else. Oldenburg's defence of their priority began at the end of 1664 and was to last for some five years, fully illustrating the dangers of non-publication. The necessity for the defence began after a German visitor to London learned by attending a meeting of the Royal Society of English injection experiments performed in Oxford some years earlier. When he returned to Germany he reported this to his medical friends with the result that Johann Daniel Major, a Hamburg physician, was moved to write to Oldenburg to state that he himself had some claim to having been the first to practise injection surgery which he used to hasten the absorption of drugs. (He was careful to state that he was very willing to believe that the English workers had known nothing of his work when they began theirs.)[42] His practise of this technique had been described in a book published earlier that same year and he sent a copy of it to Oldenburg for the Royal Society in order to substantiate his claim. There was no obvious way in which he could have heard of the English experiments, nor the English experimenters of his. In fact the English experiments, mainly made with poisons and on animals, which had been carried out in Oxford some dozen years earlier, had been described by Boyle in his somewhat miscellaneous work *The Usefulnesse of Experimental Naturall Philosophy* published in 1663, but since only in English probably quite unknown to Major. The latter clearly assumed that the experiments of which he now in 1664 heard news had been performed only recently, a mistake which Oldenburg could and did rectify. Oldenburg in return readily accepted Major's complete ignorance of the English work, politely assuring him that 'we do not at all concern ourselves with what you write about the invention of this injection method and the time of its discovery'.[43] This, perhaps surprisingly, was the reaction of the Fellows present at the meeting of 11 January 1664–5 when Oldenburg read Major's letter and presented his book. He was instructed then to thank Major and to tell him that 'one of their members' [Dr Timothy Clarke] was soon to publish an account of what the English had done in this field, as he did. It was easier to take the affair calmly at this stage because Major showed no great anxiety to insist upon his own priority and because his experiments were directed towards the possible use of injection in medical practice, whereas the English were concerned with experimental physiology. Had Clarke published his account at this time, Oldenburg's subsequent efforts to establish English priority would have been easier or even unnecessary. But Clarke failed to complete his review of the subject until 1668, when it was to be published (in Latin) in the *Philosophical Transactions*.[44]

Sadly, by no means all the members of the Royal Society were prepared to take such a calm approach when questions of their priority were involved. Even Boyle, usually anxious to emphasize his disinterested love of making discoveries in natural philosophy for their own sake, could tell Oldenburg that he wanted nothing known of the anatomical experiments on the heart and blood, involving transfusions, that he and Lower were making in 1665 'without being before hand registered by you together with the Time of their having been made or proposed' in order to avoid the possibility that 'they may beget such claimes & disputes as that which was formerly made here [only the year previous] of iniecting into the Veines of live Creatures'.[45]

Partly because of Boyle's concern, partly because of pressure from Clarke and others, in December 1665 Oldenburg published in the *Philosophical Transactions* 'An Account of the Rise and Attempts, of a Way to conveigh Liquors immediately into the Mass of the Blood' presumably written by himself with guidance.[46] This ascribed the idea of injection experiments to Wren, then in Oxford, and the practice of the method to 'at least six years since'; that is in 1659. The account insisted that it had initially been undertaken under Boyle's supervision, since when 'it hath been frequently practised both in Oxford & London, as well before the Royal Society as elsewhere' and especially by Clarke. (As this issue of the *Transactions* appeared when the journal was being printed in Oxford, the account could have been and probably was edited by Boyle, Wallis and Clarke.)

The controversial nature of these curious and important experiments, which in fact never led to any significant use in medical practice at the time, was increased in the late summer of 1667. Then Oldenburg learned of the injection experiments performed in Italy by Carlo Fracassati and published at Bologna in 1665 in the form of an 'anatomical letter' in a collection of other medically-related essays in the form of letters, two by Malpighi and one by Fracassati. At the time Oldenburg told Boyle that having heard of these (he did not say by what means) he had written for a copy. Only ten days later he was able to describe the work in *Philosophical Transactions* number 27 (the real number 27, dated 23 September 1667; for the false number 27 see above, p. 120). Even when he first mentioned them to Boyle he knew that the Italian experiments consisted of the injection of strong acids and alkalis into the veins of a dog, some of which experiments, not surprisingly, caused the death of the unfortunate animal by 'coagulation of the blood'.[47] Oldenburg asked Boyle to have the experiments repeated (showing on what familiar terms he was by now with his patron). Boyle then reminded him, in a reply printed in the

November issue of the *Transactions*,[48] that Boyle had informed the Royal Society some three years earlier of his own experiments (in fact, on 21 December 1664, as Oldenburg found when he examined the Journal Book). These experiments involved pouring strong acids and alkalis into samples of blood drawn from a dog, a far more humane form of experiment than those of Fracassati and equally informative. Boyle now generously commented that Fracassati might have had the idea independently, although Boyle's experiments had been made public at a meeting of the Royal Society so that he thought that Fracassati might have heard 'a Rumour of it'. (The fact that the members of the Royal Society all regarded communications to the Society's meetings as putting information into the public domain suggests that many more non-Fellows came to meetings as visitors than are known to have done so, as well as that those Fellows who were present at a meeting might be expected to tell a fairly wide circle of acquaintance what had occurred at meetings.) Now in 1667 Boyle suggested several new injection experiments, presumably hoping that they would be tried at or for meetings of the Society. (In fact he wrote 'If you try the injection . . .', but the 'you' was almost certainly collective, as there is no indication that Oldenburg himself possessed the anatomical and medical skills required to perform such an experiment.) Injection experiments continued to be performed in various places as news of them spread, accounts of many being sent to Oldenburg who generally published extracts of such accounts in his *Transactions*. There was great interest in the medical and physiological implications of such experiments, and fortunately for Oldenburg any possible controversy over priority in the invention of the method soon died down almost completely.

It is perhaps surprising that any controversy over priority in this case should have been of concern to the English, for already there had developed what was to be a violent and more significant controversy over the priority of the extension of injection experiments into the practice of blood transfusion. This occurred fairly quickly in England and only a little more slowly elsewhere, transfusion after all being merely the injection of the blood of one animal into the veins of another. When the Royal Society met for the first time after the suspension of its meetings because of the Plague in London, in March 1665/6, Dr Timothy Clarke reported to the meeting that during the previous months he had continued his experiments on both injection and the 'Transfusing of bloud out of one Animal into another', experiments which he had previously tried two years before. He was apparently still not completely successful, for the minutes of the meeting state that Moray then remarked that

Boyle had 'had hopes of mastering the difficulties that are met with in that Experiment'. (These difficulties presumably mainly caused by the rapid clotting of the blood being transfused.) Boyle must already have mastered these difficulties or at least devised a means of doing so for in Oxford late in 1665 Lower,[49] under his direction, had performed a successful blood transfusion between two dogs, as was reported to the meeting of the Royal Society held on 20 June 1666. At Boyle's request, Lower at the end of the year wrote a detailed description of the experiment which was promptly published in the *Philosophical Transactions*[50] as the lead article. This was followed two months later by 'Quaeries and Tryals' proposed by Boyle to Lower and in May 1667 by accounts of further blood transfusions between dogs[51] demonstrating how thoroughly the English had mastered the technique. The same issue of the *Transactions* also contained the first intimation of newly performed French experiments on animal transfusion in an account drawn up by Jean Denis who had performed the experiments, the account having been printed in the *Journal des Sçavans* and quickly translated by Oldenburg into English.[52] The priority of the English in the performance of such experiments seemed obvious. But English confidence was rudely shattered by the appearance of a printed letter dated 25 June 1667 in which Denis claimed that 'the Project of causing the Blood of a healthy animal to passe into the veins of one diseased [had] been conceived *ten years agoe*.' Since he only claimed 'conception' as early as 1657, while dating the 'performance' of the experiments to 1667, this in fact showed them to postdate those of Lower which had even been printed six months earlier.[53]

However, Denis did have something totally new to announce. He had transfused the blood from a lamb into the veins of a young man who had been so severely bled during a violent fever as to become almost comatose, the first case of human blood transfusion. As noted above (p.120) his account in English translation (probably not by Oldenburg) had been printed, without comment, in July 1667 in the false *Philosophical Transactions* number 27, probably by Wilkins, during Oldenburg's imprisonment in the Tower. Many who read it thought it had been planned by Oldenburg, and Clarke, Lower, and others blamed him bitterly for not denouncing the claims therein by Denis for priority.[54] All that Oldenburg could do was, as soon as possible, to publish a true *Philosophical Transactions* number 27, dated 23 September 1667 and carefully entitled as being 'For the Months of July, August and September'. It began with 'An Advertisement concerning the Invention of the Transfusion of Bloud' in which Oldenburg first explained that he had been prevented from publishing his *Transactions* during the previous months 'by an extraordinary Accident' and then went on to state firmly that 'if himself had published the Letter, which came abroad in July last, *Concerning a new way of curing sundry diseases by Transfusion of bloud*', he would then, as he now did, have corrected the claims by Denis to have been the first to perform such an experiment, stating the facts which could be confirmed by the printed accounts in the *Philosophical Transactions* published before Denis performed his experiment. Oldenburg wisely added that it was 'needless to contest' where and by whom the process had first been thought of, for what mattered was when it was first performed. He went on to begin the next issue of the *Philosophical Transactions*[55] with a careful and thoughtful summary of the conflicting claims for the invention of transfusion by various medical men on the Continent, noting potential dangers in *human* transfusion and once again emphasizing English priority in the first practice of animal, although admittedly, not human, transfusion. He ended, picturesquely,

> But whoever this Parent be [of the discovery] that is not so material, as that all that lay claim to the Child, should joyn together their endeavors and cares to breed it up for the service and relief of humane life, if it be capable of it; And this is the main thing aimed at and sollicited in this Discourse; not written to offend or injure any, but to give every one his due, as near as can be discerned by the Publisher.

This was to continue always to be his aim in future cases of priority disputes, as far as he could contrive it.

From this time on interest in England as on the Continent centred on the possibility of human blood transfusion, to cure madness or disease by replacing supposedly tainted blood with sound animal blood, usually from a sheep or lamb. After studying the work done abroad the Fellows of the Royal Society decided to attempt an experiment on a human being themselves. They chose a poor man who seems to have suffered from no specific illness, except 'a hypochondriacal disorder'. He received sheep's blood in November 1667, survived unhurt and even claimed to feel some benefit, being pleased to be experimented upon in return for a small gratuity.[56] He was clearly lucky to have experienced no lasting ill effects, for when Denis gave two separate transfusions to a madman in Paris hoping that the blood of a sheep, reckoned a docile animal, would relieve his furor the man died soon after the second, (Denis claimed poisoned by his wife). After this calamity transfusions were banned by law in France and promptly discontinued in England.[57]

In spite of this fatality controversy over priority did not die, as can be perceived from Clarke's somewhat belated, detailed account of English injection and transfusion experiments published in Latin in the *Philosophical Transactions* for May 1668.[58] Clarke was decidedly belligerent in tone. He quite irrelevantly and unfairly took the occasion to denigrate the work of the young Dutch anatomist Regnier De Graaf whose book on the organs of generations Oldenburg had reviewed favourably in his *Transactions*. The only reason for dragging his name into the discussion was that Clarke wished to insist that all foreigners were inclined to claim priority for what were properly English discoveries, in this case quite without foundation.[59] Not surprisingly, this attack prompted De Graaf to write in his own defence, so that for the next couple of years Oldenburg was forced to mediate between Clarke and De Graaf in an interesting correspondence which, unusually, included an exchange of specimens.[60] It must be said that this time Oldenburg was at least partly responsible for the continuation of the controversy, as Clarke plainly declared later in 1669.[61] Fortunately, Oldenburg's care in toning down any bitter words allowed him to escape all personal acrimony and De Graaf continued to be a useful if infrequent correspondent in years to come.

It must be said that the transfusion controversy was an over-long and not conspicuously glorious interlude in Oldenburg's attempt to serve the Society by informing the world of learning what its Fellows had accomplished in experimental philosophy and to defend their interests as they saw it. Really nothing except national pride was here at stake and it is possible that the whole incident could have been minimized had it not been for two things. The first was the unfortunate publication of Denis's first letter in Oldenburg's unavoidable absence. The second was the determination of such men as Clarke to insist on the importance of correcting foreigners about what he and others saw as infringements of the just claim of the English to be in the forefront of discovery and invention. Oldenburg could not afford to run counter to such determination. But it was not only what quarrelsome men like Clarke demanded in the way of recognition that created controversy, for it was a much more general attitude among natural philosophers. After all, Wren was one of those who complained strongly about the French claims in this matter and Wren never had been and never was to be one who boasted of his efforts. Normally he was like many others, including notably Huygens and Newton, who were inclined to be satisfied by making a discovery or working out a theory and little inclined to make their efforts known to the world at large, partly for fear of criticism and controversy. Yet such men were ever eager to defend their priority when some one else published what they had discovered earlier. This was only human: few people like to have others claim novelty for something they have proposed or even merely known years earlier and Wren was clearly not one who could accept having his ideas published by others, although he had not wished to publish them himself. It is worth noting here that, as so usually before and after this episode, Oldenburg took throughout the English point of view and was recognized as acting entirely in the English interest.

As the above examples show, maintaining a philosophical correspondence on the Society's behalf was no light task. An example not yet touched upon was Oldenburg's effort to gather information for the eventual compilation of a 'universal natural history'; that is, a description of the natural history of he whole world, in accordance with Francis Bacon's recommendation, to which the Society whole-heartedly subscribed.[62] This was a long-continuing project which Oldenburg himself seems to have found particularly congenial; it is tempting to guess that he hoped to be able one day to publish at least a part of this edited by himself. He had been gathering material for this project in a desultory way for many years during his travels, which is presumably why he had been appointed to the 'committee . . . for considering of proper questions to be inquired of in the remotest parts of the world' in 1660/1. When his travels ceased, he endeavoured to collect information useful for this purpose through correspondence, frequently opening a new correspondence by asking for meteorological, mineralogical, botanical, medical, and even ethnographical informations, whether at home or abroad. It was in large part on behalf of this idea of a universal natural history that Oldenburg tried over the years, with some success, to establish contact with remote areas of the globe. These included the Middle East, where English merchants and consuls in places like Smyrna, Aleppo, and Isfahan might be inveigled into writing accounts of the natural history of the areas, mostly in response to direct queries;[63] with Abyssinia, for which a learned Jesuit of Portugal known through Robert Southwell when he was Ambassador at Lisbon, was asked for information;[64] with Brazil, contact with learned Jesuits there being established through an English merchant, the brother of a Fellow of the Society.[65] So too Oldenburg's correspondence with North America—with the schoolteacher and mathematician Richard Norwood in Bermuda[66] and with John Winthrop of Connecticut, an early Fellow of the Royal Society,[67]—was largely directed towards learning something solid of the natural history of these remote colonies.

Yet another source of correspondence arose out of the ever-increasing fame of the Royal Society and with it the fame of its hard-working Secretary. Foreign travellers were ever anxious to visit the Society, to attend a meeting, and to meet its leading members, notably Boyle after he settled in London in 1668, and increasingly its renowned Secretary. Whole networks of correspondence might arise in this way. For example, consider the chain of connection with a number of German physicians, always eager to claim Oldenburg as one of their own. When the young Matthias Paisen (1643–70)[68] later a physician of Hamburg, travelled through Holland, France, and England between 1666 and 1668, he met many Dutch medical men. Among these was the young De Graaf, whom he then encouraged to initiate a correspondence with Oldenburg, telling him, so De Graaf said when he first wrote, of Oldenburg's 'very great courtesy to foreigners', of which he knew because he had met both Boyle and Oldenburg when in England. Arriving home, Paisen promptly initiated a correspondence on his own account which ceased only with his early death. Oldenburg used Paisen's good offices to convey a letter to Martin Vogel (1634–75), also a physician in Hamburg, a devout disciple of Joachim Jungius, mathematical philosopher, logician, and botanist, whose biography Vogel had already published. He had also edited some of Jungius's works and was to edit more. He was delighted to be put in touch with Oldenburg and remained a life-long correspondent, reporting on the progress of his Jungian studies, his history of the Lincean Academy (which greatly interested Oldenburg, but which was never completed) and his study of oriental languages, especially Turkish. For help in these studies Oldenburg enlisted the assistance of the Oxford Orientalists, whom he frequently consulted on Vogel's behalf. Here, as so often, correspondence with one man led to correspondence with others.

By this time, the activities of the Royal Society and the willingness of its Secretary to correspond with foreigners were well-known throughout Germany. Correspondence was particularly brisk with Hamburg, no doubt partly because it was a postal city. In 1666 Oldenburg was approached in terms of exaggerated flattery by Stanislaus Lubienietzki (1613–75) who wrote from Hamburg where he was living as a refugee from his native Poland for religious reasons. (Like many Poles at the time he was a Socinian and therefore liable to persecution now that the Jesuits had reconverted Poland to Catholicism.) He began his letter with flattery

> Your name, which all in the republic of letters know and honour, has reached my ears even though I concern myself little with such matters, and has filled me with admiration and respect. It was to you particularly, among the members of the Royal Society, that I wished to address myself,

modesty only, so he claimed, having previously restrained him from doing so.[70] At this time Lubienietzki was at work compiling what became his very large *Theatrum Cometicum* by the simple expedient of writing to all the European astronomers whose names he knew asking them for information about any and all comets ('since the Flood' as he put it) and he subsequently published all the replies, including those from Oldenburg. Although Oldenburg welcomed his letters, gave him news of the Royal Society's current interests, and encouraged him to continue, he supplied no cometary information. Undeterred, Lubienietzki continued to write until his book was ready for the press in 1667, after which the correspondence ceased. This is an excellent example both of the way in which foreign natural philosophers might learn of Oldenburg's role as the key to communication with the Royal Society and of the speed and care with which Oldenburg responded.

Oldenburg's correspondence was never static. There was always a solid core of 'regular' correspondents at home and abroad, but otherwise there were constant changes, the more so as he continually sought new correspondents, men whose names were made known to him or to his colleagues, in spite of the ever-increasing volume of correspondence which he already had in hand at any one time. The only criterion he demanded was that these new correspondents should be involved in important work in some field of relevance to the Royal Society and its aims. Sometimes such correspondence was intermittent or brief, but often it continued for many years with great profit to both sides, especially when the new correspondent proved worthy of election into the Royal Society as a Fellow.

A good example of the latter type was René François de Sluse, mathematician of Liège, known to Wallis and others of the English mathematicians by repute. Oldenburg initiated the correspondence in February 1666–7, writing in flattering terms of Sluse's published work.[71] He described the aims of the Royal Society and sought an exchange of letters which should inform the English of Sluse's research undertaken since the publication of his *Mesolabum*, which, he said, had been much appreciated in England, although copies were scarce.[72] Oldenburg also inquired about the natural history of the country around Liège, a well-known centre for mining, and about 'whatever novelties there may be of a mechanical kind', all queries quite usual with new correspondents. Sluse was later to reply fully to all these requests, satisfying the Society's mathematical interests in good measure as well as Oldenburg's expressed interest in natural history and the history of trades. In this first instance he responded only with polite acknowledgements, doubting, he said, whether a sustained correspondence was possible since it had taken three months for Oldenburg's letter to arrive.[73] (There was after all a lively war on, or rather a pair of campaigns: the Anglo-Dutch War and the French attack on the Spanish Netherlands which had devastated a part of Flanders.) In fact, Sluse's own letter reached Oldenburg in only nine days, and in future there were to be few postal problems between Liège and London. Oldenburg was quick to communicate the sense of Sluse's letter to the Society's meeting on 28 November 1667. The result was a request from John Collins, recently elected a Fellow, that Sluse be asked to comment upon a new algebraic method of dealing with fourth degree equations. There was also a query from Nathaniel Henshaw, an Original Fellow, about the manufacture of tinned iron in the area around Liège.[74]

This is a good example of a not uncommon method by which Oldenburg developed correspondence; that is, by communicating a letter to others either directly or at a meeting of the Royal Society. Since Oldenburg was not himself a mathematician, when his mathematical correspondence increased in coming years he was forced to rely heavily upon such men as Collins and Wallis for comments and assistance. Yet he could equally obviously understand the mathematics of his day to a reasonably high degree, for he could translate Collins's often badly organized and usually poorly expressed news and comments into correct Latin, which he could not have done had he not understood what he read. Sluse was to remain a faithful correspondent for many years and to become in 1669 the first to learn of the then little known mathematical skill and achievement of the young Lucasian Professor of Mathematics at Cambridge, Isaac Newton. (See below, Chapter 6.) Sluse's interests were in some cases very close to those of Newton and he received eagerly the accounts of his work written by Collins specifically for Oldenburg to transmit to him.[75]

As if all his purely official correspondence on behalf of the Royal Society was not enough, Oldenburg also found time to continue much only partly official correspondence with the Continent which he made as far as possible serve the interests of the Society, but which was mainly directed elsewhere. True, correspondence about theological matters, like much of his later correspondence with Spinoza or that about the state of Protestantism on the Continent or the new theological enthusiasms of the Jews of the Low Countries had no direct relevance to the Society as such, but it was of interest to many Fellows. These last included Boyle and Wallis to whom Oldenburg passed on what he learned of these matters. In 1666 there was a disturbance of international, although admittedly minor importance which impinged upon Oldenburg's non-scientific interests and activities, namely the prediction by certain milleniarists or chiliasts that the end of the world was about to occur, as some English religious sects active during the Civil Wars had also predicted. The prediction was of both political and religious importance, for as the millenial year arrived there were widespread disturbances in various religious communities of the Near East, so widespread as to alarm European countries. Oldenburg's most important source of information about this movement was Peter Serrarius who purveyed to him on a regular basis both political and religious news from Holland, he himself being a theological writer and a fervent chiliast who frequently wrote of various milleniarist predictions which Oldenburg, clearly interested himself, passed on to others including Williamson, potentially interested in any possible political disturbances which these predictions might create.[76]

In 1648 a Jewish mystic, Sabbati Zebi, Zevi, or Sebi (1626–76) had proclaimed himself the Messiah and the year 1666, as so often, to be the year of universal salvation. Zebi came from Smyrna where his father was agent for an English firm which perhaps explains how he came to be influenced, as he clearly was, by contemporary English Fifth Monarchy men, firm milleniarists. This explains why twenty years later the English government might be concerned at his preaching for milleniarism was by no means dead in England in the 1660s. Zebi acquired wide support among the Jews of Anatolia and his actions and predictions attracted great interest. At the beginning of 1666 he went to Istanbul (Constantinople the English still called it) to await the end of the world and its salvation. The Sultan promptly had him imprisoned after which, to the dismay of his followers, he was apparently converted to Islam. This produced much rumour and counter-rumour; it should have led to an abatement of the disturbances but did not entirely do so. Nor did the passing of the year of the predicted millenium mean the end of belief that it would come soon, and chiliasm continued to flourish.

Oldenburg, whether convinced or not, continued to be much interested in the movement as were very many others. In 1670 Oldenburg made detailed notes about various chiliastic views at a time when they were once again widespread in both France and Holland, and no doubt continued to keep himself informed about the spread of the doctrine, as were so many other devout and theologically orientated men and women of the time, not unnaturally in such a deeply religious age. It was perfectly consonant with Oldenburg's otherwise conventionally devout approach in his personal religion, although there is no indication that he ever firmly believed in the imminence of the day of salvation and the end of the world. In the atmosphere of the times and the known religious interests of Boyle, Wallis, Oldenburg himself, and many other Fellows of the Royal Society, there is nothing surprising in the undoubted fact that great interest in these movements was widespread.

Probably for Oldenburg, as for many others, there was nothing more out of the way in news of religious movements than in news of political movements of the normal kind. Some at least of the political news with which Oldenburg was supplied by Justel and which he then passed on to Williamson at the State Paper Office, was not so very different in character. This political news from Justel and others supplied Oldenburg with a source of much needed pecuniary profit, since Williamson presumably paid Oldenburg for his work in passing it on. Justel also supplied a certain amount of not always very reliable news about the natural philosophers of Paris. This, ironically, suffered from what was an important advance in French scientific organization already alluded to, namely the foundation in 1666 of the Académie royale des Sciences. This, especially after Huygens moved to Paris to be its most distinguished member, increased the amount of corporate news about natural philosophy. But Justel had no real connection with it and so could not report on its work. Fortunately for Oldenburg, in 1669 a reliable source of news of what was occurring in French natural philosophy appeared. This was supplied by Francis Vernon, secretary to the English embassy in Paris, for Vernon was soon intimate with many members of the Académie royale des Sciences and especially with Huygens.[77] (See below, Chapter 7.) Further, from time to time old personal acquaintances, many of them members of the new Académie, re-opened neglected correspondence to provide Oldenburg with matters of great interest to the Royal Society, and these Oldenburg did his best to maintain. All this compensated for the fact that Auzout, although a member of the Académie from its beginning, soon withdrew and emigrated to Italy, after which he wrote no more.

Although correspondence played such a major role in Oldenburg's duties as Secretary, it was obviously not his sole occupation in that position and his other duties, often very time-consuming, are worth reviewing. His weekly attendance at meetings caused him far more work than the mere taking of the minutes and writing them up fair for the amanuensis, when there was one, to copy into the appropriate Journal Book. Although, according to the minutes, he did not often speak except to report on the work of others, it must be noted that he was a modest man who did not make much of what he said and did, being decidedly self-effacing. Nevertheless, it is possible easily to deduce what an important role Oldenburg played at all the meetings he attended. He read letters and reported news and this he entered in the minutes. But more than that, he became the kingpin around which the affairs of the Society revolved, and he worked hard to keep the Society active. This emerges plainly in his complaint to Boyle in 1664[78] that

> our meetings are very thin; and . . . our committees fall to the ground, because tis not possible, to bring people together; tho I sollicite, to the making myselfe troublesome to others, not to say much of the trouble, which I create to myself, good store.

As time went on the trouble did not at all diminish, but the results were better and so the trouble more worthwhile. Committees it is true were always prone to languish, yet individual Fellows could be brought to work together on occasion to perform experiments or to consider new proposals for them. Many of the Fellows, even Hooke the official and paid Curator of Experiments, were inclined to neglect the performance of experiments which they had been assigned or had volunteered to do, or neglected to report on experiments they had done in writing for entering into the books, or to read the books assigned to them for review for the meetings or to seek out the information requested of them on specific topics. Yet they did not by any means always neglect these tasks, especially when Oldenburg tried to the best of his ability to make sure that they did do what they had engaged to do. He certainly ran about London talking to people about the Society's business, he wrote to those resident elsewhere to seek the information requested at meetings, he discussed Society business with Brouncker as President on a regular basis and he conferred with Boyle and with Hooke. Conscientiously and unostentatiously he 'ran' the Society, following the behest of Officers and Council and the will of the Fellows as expressed during meetings. No wonder that he became highly respected both at home and abroad, an indispensable servant of the Society which repaid his zeal by flourishing.

6

Scientific Diplomacy 1669–77 (1) Newton's Ambassador

During the 1670s Oldenburg's correspondence was to increase yet more in density as also, very markedly, in complexity. To the Continental world of learning he was, even more than before, the leading figure in the Royal Society's administration as we would call it, more dominant and more important than even its President, Lord Brouncker. Only strangers wrote to the President of the Royal Society in the first instance. Natural philosophers, physicians, learned men, and virtuosi nearly all had become accustomed to write directly to Oldenburg to tell him, and through him the Royal Society, what they and their compatriots were doing in their respective spheres. They all read or at least looked at his *Philosophical Transactions* as it appeared with gratifying regularity, even though most of it was still in English (the exceptions being mathematical and some medical papers). They also commented upon what they read and hoped to have these comments published. Most importantly, they sent accounts of their own original work which they usually hoped to see in print, as was often to be the case. More and more as time went on Oldenburg found himself acting as an intermediary in international exchanges between individuals who either could not (because of war) or did not correspond directly with one another.

It is obviously impossible to deal in detail here with all these matters. Nevertheless, the interconnections between such men were so vast in the 1670s and Oldenburg's role in promoting them so great that certain strands in his correspondence merit examination in some detail. They were of great importance to the development of the natural philosophy and mathematics of his time. They also bulked large in Oldenburg's life and show much about his methods of work. Of particular interest are those cases where controversy was involved, and the 1670s saw much dramatic international controversy which had important influences on the work of those involved. Oldenburg had, of course, long been involved in controversy, including international controversy, examples being the exchanges between Hooke and Auzout or the question of priority in the use of blood transfusion. Then, with his tactful help, the exchanges had mostly been conducted without great acrimony so that his role as mediator had been relatively easy and simple. Now in the 1670s controversy was to involve matters of far greater scientific and mathematical consequence, the participants were touchier, and it took all Oldenburg's skill to keep both parties calm enough to continue the dispute and to untangle the questions involved. It must be borne in mind that controversy was a common part of the scientific scene then, when new developments were coming thick and fast and there was not yet any well-developed set of conventions in respect to the publishing of ideas or discoveries or inventions. Besides, many men were content to make discoveries without going to the trouble of writing them up in finished form. There did exist means of asserting one's right to an idea without actually making it public. As already mentioned, one could convey to the Royal Society a dated, sealed letter, not to be opened until its author gave permission, which he was not likely to do unless work made public by others seemed likely to infringe his priority. This method was little used except by Boyle. Far more common was the use of anagrams as a form of cipher. Here the discovery to be registered was recorded in one or more short sentences, usually in Latin, enciphered by stating the number of times that each letter of the alphabet appeared in the statement. This is a form of encipherment virtually impossible to decode without the transmission of the sentence or sentences in clear later, and even these were often difficult to understand. Both Wren and Huygens in particular used this method.

Why, it may be asked, were such natural philosophers as Huygens, Newton, Sluse, Hooke, and many more so reluctant to reveal their discoveries when so many others were longing for publication? It was sometimes from a reluctance to commit themselves to an incomplete formulation of their ideas or discoveries, sometimes because they feared to arouse controversy. Controversy arose far too easily, witness that which had pursued Boyle in the 1660s when his conclusions, tentative though they were in his expression of them, were bitterly attacked by Linus and Hobbes who both firmly denied his belief that a theory must stand or fall by the results of experiments performed, however validly logical a rival theory might seem. (This in fact had been a controversy in which Oldenburg played no part, but it is worth mentioning because he did have an important part to play in similar controversies which were to plague Newton, for much the same reasons, in the 1670s.) Sometimes reluctance arose from the fact, as would appear to have been the case with Hooke, that natural philosophers when satisfied with an idea or theory they had conceived were reluctant to submit to the drudgery of trying to construct a model of an invention which seemed mentally practicable or of working out in detail a useful and seemingly saisfactory theory. Or sometimes, as with Huygens's work on optics, it was because their vision of their work was larger than what they could immediately accomplish seemed to be; thus Huygens never had a satisfactory theory of colours. And sometimes, as with Newton, they were content to have worked out an idea or discovery to their own satisfaction and had no real interest in sharing it with the world. In all such cases Oldenburg could and did encourage them, and beg for information. Often it seemed to him best to tell A what B was doing and vice versa, in the hope of stimulating both men to more work and more openness. Hence the promotion of controlled controversy was something Oldenburg regarded as worth striving for. Although he, of course dreaded the adverse effects which could arise from it he always optimistically hoped that he could turn away wrath with carefully managed soft words and calm reason. And indeed more often than not he was successful in this operation.

In the 1670s he played the role of instigator and mediator in a number of cases, both between English natural philosophers and foreigners and between English natural philosophers, occasionally even between two foreigners (as was to be the case with Huygens and Sluse, where he was fairly successful). As will be seen, he was unsuccessful, to his own cost since he became implicated, in respect to the dispute, rather than controversy, between Huygens and Hooke over the invention of the spring-balance watch. On the other hand very highly successful was his mediation between Newton and Sluse over a mathematical discovery which in the end they agreed presented no question of priority, for both had discovered it independently. And above all, and on the whole successfully, Oldenburg spent four arduous years handling critisms of and incredulity over Newton's optical discoveries, discoveries which he himself had managed to persuade Newton to publish. He played a leading role in transmitting general mathematical news of English activities in that field to the Continent. Newton for the first time, and largely with Oldenburg's encouragement urging him on, revealed at least some of the mathematical discoveries he had made in the preceding ten years. At the same time Oldenburg acted as an intermediary between Newton and Leibniz who, stimulated by learning new mathematical techniques under the aegis of Huygens in Paris, was developing the initial stages of his own contributions to the invention of the calculus, these being in parallel with, but differing from, Newton's own achievements in that field made earlier. All this Oldenburg did with great tact and diplomacy and it is fair to say that without him much of Newton's work in both optics and mathematics, all done before 1676, would never have been known outside of a very small English circle. Although he did not live to appreciate the full importance of his achievement, Oldenburg was later to be seen, for good or evil, as the catalyst who brought so much important English work into the open.

The most important of these exchanges and controversies for contemporaries as for moderns centred around the achievements of Newton. It must be kept in mind that until 1669 Newton's work was virtually unknown outside Cambridge even in England. There is even no mention of his name in Oldenburg's correspondence until that time, although English academics knew that he had succeeded Isaac Barrow as Lucasian Professor of Mathematics at Cambridge. Oldenburg with others first learned of Newton's astonishing mathematical prowess from John Collins, from whose accounts Oldenburg became aware in 1669 of Newton's importance, as he was then able to inform Sluse. Collins was a mathematical magpie; he was to publish several books on practical mathematics but these were of minor importance. His claim to fame must rest not on these but on his passionate interest in books and discoveries in mathematics and his coaching of Oldenburg on what Newton and others had achieved. It was undoubtedly for his enthusiasm that he had been elected into the Royal Society and it was for his indefatigable correspondence and collection of mathematical intelligence that he was valued by Oldenburg as well as by many mathematicians of his own time. He first met Barrow in 1662 and it was from Barrow that he learned of the rising genius of Barrow's much junior colleague, Newton. In 1669 Collins came into possession of Newton's most important early mathematical work, *De analysi per Aequationes*, copies of which Collins enthusiastically sent to various English mathematicians. He was so taken with this highly original work that he managed to meet Newton in November 1669 and to establish a correspondence with him. Further he persuaded him to undertake the writing of an introduction and commentary to Kinckhuysen's *Algebra* (first published in 1661), a commentary begun in 1670 and unfortunately never published, but noteworthy here because news of it was the first notice of Newton's interests to reach Oldenburg.[1]

Collins had become Oldenburg's mathematical adviser in 1667, soon after his election as a Fellow, sending him lists of books to enquire after from his Continental correspondents, helping him with reviews of new mathematical books for inclusion in his *Philosophical Transactions*, and supplying news for him to transmit to his mathematical correspondents.[2] Collins, himself poorly educated, always wrote in English, usually in a highly enthusiastic but unorganized style, sending a jumble of facts and news which it was Oldenburg's task to translate into coherence and, for Continental correspondents, into grammatical Latin. This latter he did with great skill, which confirms the belief that he understood what he wrote, at least to a very considerable extent; it is almost impossible to copy complex mathematics if it is entirely incomprehensible. And when Oldenburg did make mistakes they were clearly the result of inadvertence, perhaps from fatigue or interruption, never the result of obvious misunderstanding. As noted above, the earliest reference to Newton in Oldenburg's correspondence occurs in exchanges with Sluse with whom he had been in regular correspondence since 1667.[3] Then the mathematical content of his letters had derived from Wallis, Brouncker, and Wren, but by the autumn of 1669 Collins had replaced them as Oldenburg's chief source of mathematical news. Collins's rambling accounts were extremely useful as a source of the latest achievements in English mathematics.

At this time he was full of the accomplishments of Barrow, about to be revealed by the appearance of his optical and geometrical lectures (his Lucasian lectures) which Collins was seeing through the press. In the course of describing the contents of Barrow's books Collins summarised for Oldenburg's benefit what Barrow had told him about the discovery by a young Fellow of his own College, Trinity, (that is, Newton) of 'an universal Analytical method . . . for the Mensuration of the Areas' of very many geometrical curves, material which is to be found in Newton's *De analysi*.[4] Unfortunately Oldenburg's letter was lost in the post and although he sent a replacement Sluse seems to have read this a little carelessly, for he missed the name of Newton, only remarking in reply that he was anxious to learn more of the English method of finding tangents to complex curves. A year later Oldenburg, thanks to a couple of rough accounts by Collins, was able to give Sluse some examples of Newton's achievements, including his first work on infinite series and his solution to a problem first proposed by Collins about annuities.[5] Collins was to continue to supply Oldenburg with similar information to forward to Sluse. Once again, his letters demonstrate the fact that Oldenburg, as instructed by Collins, understood the information he was given to transmit. Further, he was able to translate Collins's disorganized accounts into clear and intelligible Latin with few mistakes—these being confined to such errors as anyone might be expected to make when copying out a long series of mathematical expressions, especially considering the nature of the text with which he was presented.

After this initial burst of information about the new English mathematical star in 1668, Collins continued increasingly to instruct Oldenburg from time to time. Thus in 1671/2 he told Oldenburg to inform Sluse that Newton had his 'Introduction to Algebra, his generall Method of Analyticall Quadratures, and 20 Dioptrick Lectures' almost ready to print. This information Oldenburg duly transmitted.[6] From now on Collins increasingly involved Oldenburg in the transmission of detailed mathematical information in 1672 and 1673, especially to Sluse. This was principally in connection with work on general methods for determining tangents to complex curves, problems which had been exciting interest in the mathematical world for some time. (It involved analytical work which was essentially an anticipation of parts of the integral calculus.) At the end of 1671 Sluse had mentioned to Oldenburg his intention of publishing an old work of his on this subject[7] and so, quite coincidentally, did the French Jesuit Pardies as part of a projected work (never completed since Pardies died in 1673) to be entitled *La Mechanique*. Oldenburg certainly told Wallis of this intention by Pardies,[8] and this apparently stimulated the former to send a summary of his own achievements to date for publication in the *Philosophical Transactions*.[9] All this forms an interesting example of the complex network of correspondence which Oldenburg successfully kept in a clear pattern, the more remarkable because he had many other equally complex topics in hand at the same time which he managed always to keep clear and distinct.

Not until January 1672/3 was Sluse to communicate his method of tangents fully and in detail, only to have Oldenburg in reply send him Newton's method of tangents which he had learned of through Collins.[10] In succeeding months Oldenburg, always acting under Collins's guidance, was able to inform Sluse of precisely what Newon had achieved and to inform Newton of what Sluse had done. It turned out that the methods of the two men were virtually identical, or, as Sluse put it,[11]

> I can say nothing else of the famous Newton's method than that it seems to me to be my own, which to be sure I have used for many years now. . . .

To this generous remark Newton responded equally generously, saying[12]

> By a former letter of yours I was a little dubious whether M. Slusius might not apprehend, by what you wrote to him concerning me, that I pretended to his Method of drawing tangents; untill I understood by M. Collins that you signified to him that you thought it here of a later date. For it seems to me that he was acquainted with it some yeares before he printed his *Mesolabum* & consequently before I understood it. But if it had been otherwise yet since he first imparted it to his friends & the world, it ought deservedly to be accounted his. As for the Methods they are the same, though I beleive derived from different principles. . . .

So, to the credit of both men, they agreed on the similarity of their methods and, with rare unanimity, agreed further that Sluse had made his discovery some years before Newton had made his, as well as that each had proceeded in perfect independence, neither being aware of what the other had done. So ended a potential dispute over priority in complete amity, no doubt to Oldenburg's considerable relief. To his considerable credit as well, for he could easily have managed the exchange otherwise than he did. Sadly, such a pacific exchange was to be rare in the future.

By this time Newton was almost wholly preoccupied with other matters, notably his dramatic discoveries in optics and their theoretical possibilities and implications. These Oldenburg was to relay to the world to the great benefit of natural philosophy but at the cost of much work and great difficulty to himself. It was a critical time for Newton and no less so for Oldenburg needed to use all his diplomatic skill and tact in coaxing a very reluctant Newton to write at length for the benefit of the Royal Society, which proved relatively easy. This material Oldenburg then, with Newton's permission always punctiliously sought, disseminated to the world by means of his *Philosophical Transactions*. What was more difficult was to persuade Newton to answer the hostile criticism and downright disbelief which his papers provoked at home and abroad. Without Oldenburg none of this could ever have happened and much of Newton's considerable achievement would have been lost to the world.

The history of Newton's optical work in the 1670s, made public after the revelation to the scientific world of his successful construction of a very small reflecting telescope, is both well known and much discussed by Newtonian scholars. Here it will be set out as it became known to Oldenburg and as he concerned himself with publicizng it. The question of when precisely Newton made his first model and the differences between that and subsequent models need not be dealt with here,[13] nor will the details of his later very important optical work be entered into, for neither concerned Oldenburg. What did concern him is the manner in which he first learned of Newton's work, his role in publicising it, in defending what he believed to be Newton's priority. After that he had to handle a long drawn out epistolary controversy, lasting from 1672 to 1677, which stemmed from Oldenburg's publication of much of Newton's views on light and colours. This was a controversy which had great importance in developing Newton's views as well as informing the world of them. For all that has been written on these topics, Oldenburg's role has not always been clearly understood nor have the relations over some five years between the Royal Society's Secretary and Newton been carefully examined, especially in connection with Oldenburg's total correspondence in this period.

As an instance, it is worth remarking on the intriguing possibility that Oldenburg's correspondence in 1671, before the world at large knew anything of Newton's optical work, might have had some role in stimulating Newton to return in that year to earlier optical investigations. In 1670 the still young Leibniz, then working in physics rather than in mathematics, wrote to Oldenburg at the instigation of a distinguished German diplomat, the Baron von Boineburg. He was about to use Oldenburg as a postal agent for letters addressed to Prince Rupert, sending letters from Leibniz with his own. Correspondence between Leibniz and Oldenburg continued since Oldenburg responded with his usual kindness to young unknown natural philosophers. In the autumn of 1671 Leibniz sent Oldenburg his recently published pamphlet on optics in which he briefly referred to his idea for a 'catadioptrical' telescope; that is, one using mirrors. Although Oldenburg does not seems to have read any part of this rather discursive letter to the Royal Society, he very probably either gave it to various Fellows to read or sent extracts from it in letters now lost. If this is so, it is not improbable that one of the recipients might have mentioned the Leibniz proposal to Newton, almost certainly without mentioning Leibniz's name, which would have been unknown to any of the English. Newton himself had, of course, already designed his first refracting telescope by this time but had put it on one side. Hearing of another's work, even though he probably already knew that Gregory had some years before also suggested using mirrors (in a design quite different from that of Newton) might have supplied the stimulus necessary to make Newton return to his own telescope. If this were so, it would have been the earliest, albeit indirect, contact between Leibniz and Newton, relations between whom were to involve Oldenburg so deeply in later years.

In considering the relations between Newton and Oldenburg, the reclusive Fellow of Trinity College in Cambridge and the Royal Society's Secretary in London whose business it was to disseminate information as widely as possible, it is worth recalling that Oldenburg always had to balance the preferences of all his correspondents with the demands of the Royal Society. This required the reading of papers at its meetings and, when these had been read, those present often gave its Secretary orders as to his subsequent procedure. For while it is true that the *Philosophical Transactions* was Oldenburg's own journal, not the Society's, and the choice of what to publish was usually his, yet he was often directed to publish what had been well received at a meeting of the Society in a manner which hardly permitted refusal. What remained entirely at his discretion was the manner in which he did this, how much the letters to be printed were edited, what comment or arrangement he provided, and generally the manner in which controversial letters and papers were published. His success in managing all this was notable, and it must be kept in mind that on the whole most correspondents, even Newton, touchy as he could be, were satisfied with Oldenburg's manner of publishing what related to their work.

To return to Newton's optical work. Fellows of the Royal Society, probably including Oldenburg, first saw an example of Newton's reflector in December 1671, conjecturally before the twenty-first when Seth Ward, Bishop of Salisbury, proposed that Newton be elected a Fellow, as he promptly was. Oldenburg certainly saw the telescope before 30 December, for on that day he wrote of it in some excitement to Wallis.[14] Oldenburg then on 1 January 1671/2 wrote of it to Huygens by which time he was able to describe the instrument briefly but clearly.[15] And on the next day, the second of January, he wrote to Newton himself, initiating what was to be for Oldenburg a life-long correspondence, of major importance both for Newton and for the history of optics and mathematics.[16] It was an encouraging letter, in which Oldenburg truthfully told Newton how much the Fellows of the Royal Society were impressed and interested by his invention. He briefly set out the reactions to it of those who had seen it, and emphasized their desire to have a diagram of it which could be sent to Huygens to supplement the verbal description provided by Oldenburg. (It should be remembered that Huygens was the leading representative of the French Académie royale des Sciences, as well as a noted practitioner of practical optics.)

Oldenburg stated plainly to Newton that the description and diagram were intended for the express purpose of preventing any foreigners who might see the instrument in London from claiming it as their own invention. That this was a real preoccupation and concern of many of the Fellows has already been noted. It is made explicit by the advice sent to Oldenburg a month later by Edward Bernard, classical and oriental scholar, later Savilian Professor of Astronomy, in which while thanking Oldenburg for an account of Newton's work he told him that, in his opinion, it would be wise if

> Mr Hooke & Mr Newton make all hast to print their Methods in Dioptriques least our Neighbours clayme great shares in the honor of their Inventions.

This is as clear cut an expression of the distrust of all foreigners, amounting to xenophobia, of even the most cosmopolitan Englishmen at this time.[17] When he received Bernard's letter Oldenburg had in fact already sent a draft description and a drawing of the telescope, both in his own hand, to Newton for correction and approval.[18] Newton then examined the drawing and returned it to Oldenburg after corrections; the corrected drawing with a Latin description was then sent to Huygens on 15 January.[19]

After this Newton responded readily to the interest shown by the Fellows of the Society in the details of his work and in particular the question of the best composition of the speculum metal about which Oldenburg had sent him queries. So successful was Oldenburg in his approach that, in a manner quite at variance with his earlier and later reluctance to communicate his work, Newton now volunteered to send an account of the optical work which had led him to the construction of his first reflecting telescope, if, he said, the Society would be interested in learning about it. Encouraged by Oldenburg, he sent what is now known as his first paper on light and colours.[20] Two days later, probably the day he received it, Oldenburg read the letter to the Society's meeting. Immediately afterwards he reported this to Newton, emphasizing the Society's great interest and approval and its desire for Newton to give permission for the paper to be published in the *Philosophical Transactions*.[21] Newton expressed pleasure at the Society's reception of his paper and not only agreed to publication but offered to send more experiments.[22] Oldenburg, clearly pleased, put Newton's paper at the very beginning of the next issue of the *Philosophical Transactions*, the number dated 25 March 1672. (It has been noted[23] that one sentence of Newton's original paper was omitted from the published version This is that in which he characterized his 'doctrine of colours'—that the coloured rays which appeared when white light was passed through a prism demonstrated conclusively that white light was composed of the coloured rays of the spectrum—as 'not an Hypothesis, but a most rigid consequence' of his experiments. It is possible that there was some reason for this omission, but it may have been the result of a simple error, which was certainly what Newton believed.)

During the first part of the year 1671/2 Oldenburg spread the news of Newton's new telescope far and wide, where it excited much interest. For although the idea of a reflecting telescope, that is one using mirrors rather than exclusively lenses, was not new, Newton's construction was novel. Moreover he had made not merely a model but an instrument which worked, and his method permitted a great reduction in the size of the tube without loss of power.[24] Writing to Huygens had sufficed in large part to inform French natural philosophers about it. This is shown by the news sent to Oldenburg from Paris at the end of February that the astronomer Cassini, who had left Bologna for Paris in 1669, and the physicist Mariotte, both members of the Académie royale des Sciences, were keenly interested in it, as they remained. Mariotte was to be highly critical of both Newton's 'doctrine' and of his experiments, not all easy for others to repeat.[25] Oldenburg wrote promptly to his correspondents in Italy, to Sluse in Liège, and to correspondents in Germany, taking always good care to stress Newton's priority of invention, all this before there was time for publication in the *Philosophical Transactions*.[26] All replied with praise.

About Newton's doctrine there was no less unanimity of interest once it had appeared in print, but far less agreement about its merits. Unfortunately, criticism began at home. Only a week after the Society had listened with applause to Newton's paper, the next meeting heard a paper from Hooke who, with Boyle and Ward, had been asked, in the normal way, to bring in reports on it.[27] Hooke was extremely critical of Newton's *Discourse*, insisting that Newton had not proved his principal claim about the nature of white light, that his own theory of colours was by no means disproved by Newton's experiments, and that he doubted the conclusions drawn from some of the experiments. The Society was embarrassed. Clearly, Newton must be sent a copy of Hooke's paper, but how could the Society's high praise of the previous week, the joint opinion of many Fellows, be reconciled with Hooke's forceful, even scornful rejection of its contents this week? True, Hooke had some right to speak dogmatically, for he had published his own theories of light and colours seven years earlier in his *Micrographia* which Newton had, in fact, read, although no one then could be certain of this.[28] But the manner of his speaking was the very reverse of kindly and as an established man, writing about the work of a younger man, it was a peculiarly savage rejection. The meeting collectively decided to publish Newton's paper forthwith[29] so that it could be made public, and to print Hooke's criticism only after Newton had been given an opportunity to reply,

> it not being thought fit to print them together, lest Mr Newton should look upon it as a disrespect in printing so sudden a refutation of a discourse of his, which had met with so much applause at the Society but a few days before,

as Oldenburg reported in the minutes. Presumably in writing to Newton in a letter now lost, Oldenburg expressed this very tactfully. What Hooke thought can be imagined, but he was powerless in the face of formal action by the Society.

This was but the beginning of a long drawn out and proliferating controversy between Newton, who was quite certain of his experiments and their, to him, necessary conclusions and those who, although admiring his work, found it impossible to repeat Newton's experiments, by no means easy ones even when he gave them clearer directions. Others refused to accept the conclusions which he drew from them and which to him were certain, preferring their own hypotheses to his 'doctrine'. Among those who involved Oldenburg in complex triangles of correspondence at this time was Huygens. He genuinely admired Newton's achievement, especially his telescope design, so much so that he had one made for his own use. Yet he was always to insist that Newton's theory of light and colours could be regarded as no more than an hypothesis, even if a probable one, and moreover an hypothesis which he could not accept.[30] In contrast, Pardies, uniquely among the Parisian scientists who corresponded with Oldenburg at this time, had by the early summer of 1672 accepted Newton's theory wholeheartedly after Newton had, through Oldenburg, patiently answered all Pardies's initial difficulties and criticisms.[31] These same criticisms were raised by many others, not always through Oldenburg. For example the rising young astronomer Flamsteed, soon to become a constant correspondent of Oldenburg, wrote to Collins of his difficulties; Collins firmly told him that Newton had answered all his objections when replying to Hooke, answers which Flamsteed could read in the letters published in the *Philosophical Transactions*.[32]

Hooke's criticisms, so bluntly expressed with lofty confidence by a man whose work Newton had admired, even if he had not accepted it, a man moreover senior to Newton by seven years and senior in standing in the Royal

Society by nine years, hurt Newton severely. Hence it is not surprising that he found it difficult to bring himself to read Hooke's paper carefully enough so that he could answer it in detail. It took all Oldenburg's powers of persuasion before he would undertake the distasteful task. Oldenburg suggested that Hooke's name be omitted from his paper when published, hoping that this would placate Newton, but as Newton himself shrewdly remarked, the authorship of Hooke's paper would be obvious to most readers from internal evidence.[33] Nevertheless he did finally on 11 June send a detailed reply, a careful, patient, and respectful refutation of Hooke's various criticisms.[34] Oldenburg read a part of this letter to the meeting of the Royal Society on the day after he received it; it was then copied and given to Hooke and others to read at leisure, before being finally printed in edited form as the lead paper in the *Philosophical Transactions* for November 1672. Oldenburg's editing consisted in removing Hooke's name from Newton's text, referring to him always as 'the Considerer', together with a marginal note saying that his

> Discourse was thought needless to be printed here at length, because in the body of this Answer are to be met with the chief particulars, wherein the Answerer was concern'd.

Newton, in giving Oldenburg permission to print his letter, had also given him permission 'to mitigate any expressions that seem harsh'.[35] It is reasonable to assume that the manner in which this was done was decided by Oldenburg. The decision not to publish Hooke's paper verbatim and the securing of his no doubt grudging consent must have been a diplomatic task of considerable difficulty, which would not have been undertaken by Oldenburg, even for his own journal, without the overt approval of at least some of the senior Fellows. All this exchange involved Oldenburg in considerable labour. He had to write to Newton, conveying the sense of his replies to Huygens, Pardies, and others. He had to see to the publication of Newton's text after securing his permission. And he had to try to keep on good terms with Hooke, with whom he was working closely on the day to day running of the Royal Society. It must also always be remembered that Oldenburg was, at the same time, writing on numerous other subjects to his many correspondents, some not without their own problems.

After six months Newton tired of controversy. As he told Oldenburg, it took up too much time and was in any case disagreeable, although he did add that he was still prepared to discuss optical problems privately. But he would refuse to sanction any further publication because, he felt, publication inevitably provoked controversy.[36] Oldenburg then politely replied by sending him only various items of foreign scientific news, although he also sent the latest comments from Huygens, no doubt hoping that this would count as private discussion.[37] Newton made no reply and Oldenburg then dropped the matter entirely. He did carefully see to it that Newton received a copy of the issue of the *Philosophical Transactions* which contained Newton's answer to the criticisms of the pseudonymous Hooke. Not until the end of the year did Oldenburg trouble Newton again, this time only to ask about the growing of cider apple trees in Cambridgeshire, a query perhaps originating with Beale and one to which Newton responded readily, as he was prepared to do to similar questions in years to come.[38]

Newton's resolution to permit no more publication was put to the test in the new year (January 1672/3) when Oldenburg sent him extracts from various comments upon the version of his reply to Hooke as it had appeared in the *Philosophical Transactions*. Among these were comments by Huygens in which he praised Newton's theory although continuing to call it an hypothesis.[39] Surprisingly, a couple of months later Newton patiently answered Huygens point by point in a letter of which Oldenburg sent a copy to Huygens but refrained from printing.[40] Huygens continued to disagree but as he declared that he was ready to drop the matter Newton was only too pleased to concur.[41] As a token of amity Huygens, in sending copies of his recently published and important treatise on mechanics, *Horologium Oscillatorium*, to Oldenburg for distribution to various English natural philosophers, included a copy for Newton carefully inscribed, a book which Newton read appreciatively and which still survives. In spite of this and in spite of the friendly fashion in which the potential controversy between Newton and Sluse had ended, Newton really for the time being wanted no more discussion of any of his work. As he put it to Oldenburg at the beginning of the summer of 1673,[42] he wished to be 'no further solicitous about matters of Philosophy', politely adding

> I hope you will not take it ill if you find me ever refusing doing any thing more in that kind, or rather that you will favour me in my determination by preventing so far as you can conveniently any objections or other philosophicall letters that may concern me.

And he ended his letter by thanking Oldenburg for having taken the trouble to have his annual fees as F.R.S. remitted, after he had threatened to resign his Fellowship on grounds of expense. (Although he did not know this, in fact it was usually done for provincial Fellows who could not attend meetings regularly.) The whole letter shows a patience very unlike the irascibility that Newton was to display in old age, when of course Oldenburg was no longer alive to act as an intermediary between Newton and those who disagreed with him. But more than that, it demonstrates the truly cordial relationship between the two men and the effectiveness of Oldenburg's endeavours to manage controversy without arousing the anger of the participants. To anticipate, Newton never in the future blamed Oldenburg for his part in the optical controversies, whatever some modern historians have assumed to be the case. It is not going too far to say that only Oldenburg could have managed the controversies, optical and mathematical, in which Newton was involved from 1672 to 1677 without incurring Newton's enmity, still less while staying on friendly terms with him.

For the rest of 1673 and the first half of the next year he scrupulously respected Newton's wishes, writing to him only to enclose a presentation copy of Boyle's latest work, his *Essay of Effluviums*, and to give him the latest astronomical news, namely the discovery by Cassini of additional satellites surrounding Saturn.[43] It was not Oldenburg's fault that controversy flared up again in September 1674, set off by a letter from Francis Line S.J., Professor at the English Jesuit College at Liège, a controversy which was to be pursued after his death the next year by his pupils.[44] A dozen years earlier Line had attacked Boyle for his explanation of the role of atmospheric pressure in creating the empty space at the top of a Torricellian tube filled with mercury and inverted, (this having been used as an experimental device since Torricelli had first publicized it in 1644).[45] This attack had been useful since it led Boyle to undertake a careful series of experiments after which he enunciated his eponymous law. Line's attack on Newton generated even more heat than that on Boyle but, in contrast, no significant amount of light. For Line insisted that Newton's papers contained not only fundamental errors of theory but also of experiment, especially claiming that Newton had confused the effects of direct sunlight with the effects of sunlight shining through a cloud (which of course Newton had not done). How Oldenburg excused himself for breaking his promise to respect Newton's withdrawal from controversy is not known, for his letter to Newton on the subject no longer exists, but he must have done so with even greater tact than usual, for Newton in reply almost apologized to Oldenburg for the trouble caused to him. Oldenburg seems to have tried to take a strong line at first in resisting Line's attack, telling the man who delivered Line's letter that any possible errors in Newton's work had long since been found out and corrected. But this failed to satisfy Line who persisted so determinedly that Oldenburg found that the only thing he could do was to send the original letter to Newton.[46] Newton was, not surprisingly, slow in replying but he did reply. Writing at the end of December, he said that he regretted that Line's letter had been sent to him, especially as in his view it required no answer. He then enunciated his famour disclaimer, 'I have long since determined to concern myself no further about the promotion of Philosophy'. For this reason, he said, it was that he had asked to be excused from what he called the 'honour' of submitting a discourse to the Royal Society for which he had been asked previously. Nevertheless, either because he could not leave the subject alone in spite of his protests, or because he genuinely wished to assist Oldenburg, he told him what he should reply to Line provided it was not, he stressed, as 'from me'.[47] Oldenburg promptly wrote to Line paraphrasing Newton's remarks,[48] and then printed both Line's letter and his own reply in the *Philosophical Transactions* for January 1674/5. This ought to have silenced Line, but sadly it did not; he not only wrote again reiterating his criticism but, in the autumn of 1675 when he found that his letter had not been immediately printed, complained bitterly, not realizing that Oldenburg would never have published a letter of criticism without being able to allow Newton to respond.[49] So, one imagines somewhat wearily and no doubt apologetically, Oldenburg sent Line's latest letter to Newton, who once again took Oldenburg's action in good part, only saying again that there was really no need for him to reply since Line was only questioning his 'veracity in relating an experiment', an accusation to which there could be no answer. Nevertheless, once again, Newton could not bring himself to leave matters there and gave Oldenburg details of a new experiment which Line was to be asked to repeat.[50]Line died, probably of influenza, before Oldenburg's letter could arrive, but this did not end the controversy for his pupils soon took it up as will be seen.

Meanwhile, in spite of what he had said so firmly in 1674, in November 1675 Newton, in the midst of writing about Line, suddenly offered to send for reading to the Royal Society a paper which, so he said, he had by him, 'if the custome of reading weekly discourses continues'. (Of course it did, not that Oldenburg would ever have failed to read any 'discourse' sent by Newton to a meeting.) Newton added that he had thought of writing a new optical paper for this purpose, 'but find it yet against the grain to put pen to paper any more on that subject'.[51] When Oldenburg read this letter to the Society he was, naturally, 'ordered' to thank Newton for the offer and ask him to send his discourse 'as soon as he pleased'. This Oldenburg obviously did for, after some delay for which he apologized, Newton sent off on 7 December 1675 his 'Hypothesis explaining the Properties of Light discoursed of in my several Papers' together with a 'Discourse of Observations'. In his accompanying letter Newton modestly apologized for any imperfections in the two papers, insisting 'I had formerly proposed never to write any Hypothesis of light & colours, fearing it might be a means to ingage me in vain disputes', but did so now because he thought it would make his previous work 'more intelligible'.[52] Sadly it was not to do so, and Newton through the medium of Oldenburg's correspondents was to find himself once again embroiled in controversy as he had feared might be the case. Oldenburg began to read Newton's papers to the Society on 9 December, only the day after he received them (indicating the importance he gave to them and the eagerness of the Society to hear more of the work of its Cambridge Fellow). It took several weeks to get through the reading of the whole collection but the Fellows always listened attentively, ordering Oldenburg to ask Newton's permission for copies to be made so that they could be read more carefully and one to be kept permanently in the Record Book.[53]

All this Oldenburg told Newton on 11 December. After Newton gave the required permission the original was returned to him, presumably at his request, so that only the copies remain in the Royal Society's archives. Oldenburg was also able to report that the experiment involving rubbed glass, which Newton claimed offered good evidence for the existence of an aether as transmitter of light rays (others probably agreed with him about this) had not succeeded when it was tried at a meeting, presumably by Hooke. Newton patiently replied with painstaking advice on how best to manage the experiment, advice which Oldenburg must have passed on, presumably to Hooke, with whom however he was not then on good terms.[54] Newton's patience was the more commendable because Hooke had strongly 'insinuated' (Newton's word) that most of Newton's 'Hypothesis' (Hooke's word) was to be found in his own *Micrographia* now ten years old. This Newton vigorously and reasonably denied. He freely acknowledged that he had made use of some of Hooke's published experiments, not unnaturally since they *were* published, but he insisted that he had given Hooke explicit credit for all of these. And, more to the point, he insisted that his interpretation of a sound aetherial hypothesis and Hooke's were entirely different.[55] There was indeed nothing novel at this time about proposing and utilizing an aetherial hypothesis to explain various properties of light; what was novel in Newton's work was the particular use he made of it).

Newton continued to be patient in replying to criticisms sent to him by Oldenburg, remarkably so indeed since he now had to deal not only with Hooke's objections but also with those of Line's pupils, John Gascoines and Anthony Lucas.[56] He even explicitly thanked Oldenburg for what he called his 'candor in acquainting me with Mr Hook's insinuations' and asked him to 'continue that equitable candor'.[57] This Oldenburg did continue to do, regularly reading Newton's replies to the Society and reporting any comments back to Newton. At this time Newton was, no doubt partly at least due to the tact with which Oldenburg wrote, very much inclined to be emollient and friendly. He even added a postcript saying 'If you have opportunity, pray present my service to Mr Hook, for I suppose there is nothing but misapprehension in what has lately happened.' Oldenburg must have passed the message on, for soon afterwards (on 20 January 1675/6) Hooke wrote to Newton in a most friendly fashion, telling him that he 'suspected' that Newton had been 'misinformed' as to what he had said, as he had, so he claimed, had 'experience . . . formerly . . . of the like sinister practice'.[58] Then and later Hooke, as here, quite unfairly blamed Oldenburg for any bad relations between himself and Newton and others, insisting that it was all Oldenburg's fault, and implying that he habitually misreported what had been said and deliberately stirred up trouble. It must be remembered that Hooke was in the midst of a tremendous quarrel with Oldenburg over the latter's siding with Huygens over priority (see below, Chapters 7 and 9). Besides, it seems likely that, since what Hooke actually wrote and said is fully recorded, he must have been one of those people who have no idea of the effect of their own spoken words on others, although they are very touchy about what others say and inclined to forget what they said in times of anger. In this letter Hooke insisted rather patronizingly that he found Newton's papers 'excellent', that he was always anxious to see good work, and that he found that Newton had gone farther in optics than he had done. But he then spoiled these conciliatory remarks (quite different in tone, be it noted, from anything he had said publicly) by adding that he had not read Newton's latest papers as yet, only heard them read, but when he had had time to read them properly he would be glad to correspond privately with Newton about them. Here again he was suggesting that what Oldenburg had written or might write was very different from what Hooke intended.

It is doubtful whether Oldenburg ever knew of this exchange which cast such a slight on his integrity, and indeed it is to be hoped for his peace of mind that he did not. Newton replied calmly to Hooke, without in any way accepting his criticisms of Oldenburg's part in their exchange and the correspondence between the two men then ceased for the time being. Oldenburg did believe in stirring up controversy by telling one correspondent what another said about his work, as he thought it a good way to make men work out their ideas on any topic carefully and fully. No doubt he appreciated that it was a dangerous practice, but he always tried to the best of his ability to report only the truth and to quote written words where this was possible. Unfortunately it was not always possible to do so in the case of Hooke, who at this time wrote nothing formally and did not even always 'give in', as the customary formula had it, the papers he read at meetings so that they could be entered into the Register Book. On the whole Hooke preferred to rely on what he had written ten years earlier, which he was sure was still correct whatever Newton might say. Fortunately for Oldenburg, most of those with whom he had to deal had easier temperaments than Hooke, who over the years was to remain one of his few enemies.

Newton visited London in the winter of 1675–6 and seems to have established even more cordial relations than before with Oldenburg. In the new year he sent some corrections and additions for inclusion in any subsequent publication of his work, at the same time thanking Oldenburg for all that he had done in regard to the controversy between himself and Line.[60] He took the blame to himself for Oldenburg's failure to print Line's second letter (Line had blamed Oldenburg for this). As he put it, Oldenburg had shown him the 'kindness' not to do so until Newton's reply was ready for publication with it. (It was in fact Line's pupil Gascoines, sending the news of Line's death, who had last complained about the non-publication of Line's last letter, when Oldenburg had generously replied that 'any fault' in the printing of letters in the *Philosophical Transactions* lay entirely with himself).[61] Line's second letter was printed in the *Transactions* for January 1675/6, no doubt with Newton's approval, for on his visit to London he had been given a copy of it together with the letter from Gascoines. On his return to Cambridge he wrote for Oldenburg's use a well thought out reply which included a defence of the faithfulness of Oldenburg's transcription of Newton's own words, Line having insisted that what Oldenburg had sent him did not agree either with the printed letter nor with his own experience. Newton had believed that if only Line would repeat the experiments described carefully he would become convinced that Newton's interpretation of them was correct, and now he believed that the same would apply to Gascoines.

Perhaps he might have been correct in the case of Gascoines but unfortunately for both Newton and Oldenburg the controversy was next taken up by another English Jesuit of Liège, Anthony Lucas.[62] Lucas was more emollient than Line had been but equally persistent. To anticipate, the exchange of views between Newton and Lucas via Oldenburg was to continue throughout the remainder of 1676 and well into 1677.[63] Presumably to Oldenburg's relief, Newton was then remarkably patient, only complaining because Lucas would insist upon returning again and again to Line's original criticisms, which Newton quite fairly believed that he had fully answered earlier. Oldenburg carefully abstained from printing any of Lucas's letters, to the author's annoyance, promising Newton that he would never do so until he could also print any replies from Newton. Newton never gave him the requisite replies, but to conclude the story, three months after Oldenburg's death Newton paid him a remarkable tribute, writing[64] 'Mr Oldenburg being dead I intend God willing to take care that [Lucas's letters and Newton's replies] be printed according to his mind.' And he then tried through Hooke and John Aubrey to write proper answers to Lucas 'to rid my self the easiest way of this frivolous dispute & stop their clamouring against Mr Oldenburg'.[65] After that, Newton very reasonably refused to have anything to do with the matter, and as it happened these last letters were never printed since Oldenburg's *Philosophical Transactions* effectively died with him.[66] These tributes by Newton to Oldenburg show touchingly that Newton had complete faith in Oldenburg's integrity and is clear evidence, if any further is needed after considering the letters the two men exchanged, that Newton did all along cooperate willingly with the Secretary of the Royal Society on optical matters, as long as that Secretary was Oldenburg.

Optics was far from being the only subject which occupied the correspondence between Newton and Oldenburg in 1676 and 1677, if the most public. Oldenburg was also to act as an intermediary between Newton and Boyle on more than one occasion. The first time was in February 1675/6 when he printed as the lead article in that month's *Philosophical Transactions* a paper by Boyle under the transparent disguise of B.R. (a not uncommon form of pseudonymity at the time) entitled 'Of the Incalescence of Quicksilver with Gold'. Here Boyle said that chemical writers disagreed about the existence of such a substance as 'a Mercury' which would grow hot when applied to gold as 'common mercury' (the metal) manifestly did not do when amalgamated with gold. (It must be remembered that for Boyle and other seventeenth-century chemists mercury and all other 'common' metals, including gold, were not simple, elemental substances but compounds. Hence there was for them no logical contradiction in speaking of 'a' mercury (as we speak of 'a' sulphide), nor in their believing that gold itself could be changed into a 'simpler' or 'purer' form. So since ordinary mercury was conceived of as a compound, there might well be a purified form of it which would react and not merely amalgamate with gold, as would be shown by its growing hot. If so this might well be the first step on the road to creating a more perfect form of gold, as indeed many alchemical writers claimed to be the case). Boyle was characteristically careful to insist that both the mercury and the gold required great care in preparation if the experiment was to succeed. But in a manner very unlike his usual insistence upon recording the exact description of his procedure in order to make his experiments readily reproducible by others, he here explicitly refused to give any precise directions for the supposed purification of the two metals, perhaps one reason why he insisted upon anonymity.

Now Newton had been studying alchemical writers for many years and, at the same time as his later optical work in the mid to late 1670s, had devoted much energy to chemical experiments, mostly with metals.[70] Neither Oldenburg nor most other members of the Royal Society were aware of this. The only exception was Collins who was probably not sufficiently interested to pass on the fact. It must have come as a surprise to Oldenburg, therefore, when Newton wrote to him as had by then become natural to express his keen interest in Boyle's paper.[71] He praised Boyle, whom he had by then met in London, for the reticence shown in not revealing all the details of the preparation of his mercury, which he thought wise because if the experiment were to produce the results described it was best to restrict the knowledge. He himself inclined to a naturalistic interpretation, comparing the action of the mercury to that of a strong acid. He thought that it would be best for Boyle to explore the matter further himself and thus to be certain of its possible consequences. Such a discovery, if valid, might be more complex and fundamental to the understanding of chemical operations than transmutation. Very interesting for the confidence which Newton had in Oldenburg's discretion is his conclusion, in which he stated that he had written only because he understood that Boyle wished for the opinion of others. (One must conclude, therefore, that he meant Oldenburg to show the letter to Boyle.) Oldenburg made a copy of the letter[72] and must have showed either the original or the copy to Boyle, but it seems certain that otherwise he must have obeyed to the letter Newton's parting injunction, 'pray keep this letter private to yourself'. Apparently Boyle sent thanks and a reply to Newton via Oldenburg although no trace of it now survives, for in the course of a brief note in May to Oldenburg Newton confessed, obscurely,[73] 'I perceive I went upon a wrong supposition in what I wrote concerning Mr Boyles Expt.' And with this now cryptic remark the discussion ended.

Oldenburg did not forget Newton's interest in Boyle's work. A year later he sent Newton a copy of the *Philosophical Transactions*[74] in which he had published the first part of a short work by Boyle of both chemical and optical interest. It was entitled 'About the Superficial Figures of Fluids, especially of Liquors contiguous to other Liquors'. Here, after a brief consideration of fluids in relation to his corpuscularian theory of matter, Boyle discussed capillary action as displayed by various fluids in 'narrow pipes' and the reflection of light from the interface between two differing fluids. In sending the journal to Newton Oldenburg must have asked for comments, for Newton soon replied.[75] He began by saying that he agreed with Boyle about the importance of his experiments for the development of 'mechanical systems of Philosophy'. But what had struck him particularly was Boyle's experiment in which a very light oil, lighter than alcohol, was poured on 'a deliquated Alcali, made of Niter and Tartar' (potassium nitrate and carbonate) dyed red with cochineal. At the interface between the two fluids Boyle had noticed 'a strangely vivid Reflection of the incident beam of the Light'. This, Newton remarked, was 'more in the way of things I have been sometimes considering' and he briefly related the phenomenon to the differing 'refractive densities' of the two liquids, which must affect their 'reflexive power[s]'. He then offered some suggestions for further experiments. But although Oldenburg must certainly have shown the letter to Boyle, nothing came of Newton's suggestions for further exploration of the phenomenon.

Busy although Newton and Oldenburg jointly were with the optical exchanges between Newton and Lucas, and busy as Oldenburg always was with his normal correspondence, they were both about to become involved in a totally different epistolary exchange. This was the important, complex, and ultimately explosive correspondence between Newton and Leibniz (always via Oldenburg).[76] Its origins lay in Collins's excitement at reading Newton's unpublished *De analysi*. (As noted above he had communicated a little of this to Sluse in 1670 through Oldenburg.) His interest and excitement did not decline and when in the spring of 1675 a young German mathematician named Tschirnhaus came to England on his way to Paris, Collins found this an opportunity to spread the fame of Newton's work. Tschirnhaus met Oldenburg in London and was given a copy of the *Philosopical Transactions* for

January 1672/3 from which he learned of Sluse's method for finding tangents, which 'highly delighted' him.[77] Oldenburg then introduced him by letter to Wallis, whom he met, and to Collins, whom he only met just before he left England. While he was in London he wrote out his current mathematical ideas in the form of letters to Oldenburg, who then conveyed their substance to Collins.[78] The correspondence between Tschirnhaus and Oldenburg continued during 1676 while the former was in Paris where he was in contact with Leibniz. In most of this Oldenburg was acting as an intermediary between Collins and Tschirnhaus, Collins being very anxious to disabuse the young man of his enormous admiration for the geometry of Descartes and using him also as a means of conveying news and information to Leibniz. This was how Leibniz learned more of Newton's work than he had previously known or been in a position to appreciate, now that his own mathematical development was rapidly progressing. What Tschirnhaus and Leibniz mainly learned was of Newton's method for determining tangents to complex curves (of course, very similar to that of Sluse although reached by a different method and differently expressed) and also something of Newton's work on infinite series, which was totally new to them. It was all a splendid example of Oldenburg's handling of a network of intellectual activity.

In the spring of 1676 Leibniz learned of Newton's use of infinite series to give the relation between the arc of a curve and the sine.[79] Leibniz thought this discovery was both ingenious and elegant and he asked Oldenburg for information about Newton's demonstration 'on the same theme'. Oldenburg read this letter carefully (once again displaying his own far from despicable mathematical competence), marking the points to which he wished Collins to respond. He also wrote to Newton, giving him what he called some 'particulars' of Leibniz's reactions to what he had learned of Newton's work and mentioning the desire of Leibniz for a demonstration.[80] Early in June 1676 Newton responded generously, composing a long letter, which he was later to call his *Epistola Prior* (*First Letter*). This was intended as a way of enlightening Leibniz of whose work he can at most have known very little. The treatise included Newton's achievements in the development of infinite series and of the binomial theorem, necessarily in somewhat summary fashion, but at sufficient length. What he wrote confirmed what Collins had been telling Tschirnhaus and through him Leibniz about the extent to which English mathematicians had outstripped Tschirnhaus's hero Descartes.[81] Oldenburg carefully copied this important letter and sent it off to Paris six weeks later. But not by post, for it was both too bulky and too important for that; instead it went by the hand of an unnamed German traveller, a normal method of conveyance for important parcels.[82] There was a couple of months delay in its reception caused by the inefficiency of this traveller who failed to get it delivered for some time, but as soon as Leibniz received it he read it eagerly and promptly acknowledged that Newton's work showed that he had indeed surpassed Descartes. At the same time he asked for more information about Newton's achievements.[83] He also sent an outline of his own parallel discoveries, not yet quite on a par with those of Newton which indeed he had not yet entirely mastered.

Meanwhile Newton was occupied with composing patient answers to Lucas's optical questions for Oldenburg to transmit, so it is not surprising that he did not complete his second, even longer, response to Leibniz's requests for more information for some months. When Leibniz visited London in early October 1676, sorrowfully going home to Germany to look for a job (see below, Chapter 8), he at last met Collins. Collins showed him copies of some of Newton's papers, about whose contents he was able to make some hasty notes, and also a letter from Collins to Oldenburg written at the end of September and intended for Tschirnhaus which Leibniz also found so useful that he made further notes on it.[84]

Leibniz was not able to see Newton's *Epistola Posterior* (*Second Letter*) because it was not then complete.[85] It was to be a further exegesis of Newton's methods and a response to Leibniz's comments on the *Epistola Prior.* Here Newton gave a brief autobiographical account of his discovery of his methods for the construction and use of infinite series and outlined some details of their use. However he said nothing here specifically about his method of fluxions. Writing it was a generous use of his time and, to his credit, Leibniz when he finally received it from Oldenburg was duly appreciative. The delay in transmission was not the fault of either Newton nor Oldenburg, but arose because once again the sheer bulk of the letter necessitated the finding of a complaisant traveller to convey it. Since no one offered himself before May 1677, the *Second Letter* only reached Leibniz in June. He was so excited by Newton's account of his work that he replied, via Oldenburg of course, in not one but two letters (on 21 June and 12 July 1677) amply conveying his intense interest.[86] Oldenburg ackowledged the receipt of these letters early in August, promising Leibniz that he would duly forward them to Newton for comment, but warning him that both Newton and Collins were then (9 August) 'out of town and also distracted by various other matters'.[87] In fact it was to be Collins, now back in London, who at the end of August sent Newton copies of Leibniz's two letters,[88] saying that 'The Originall of this I borrowed from Mr Oldenburgh to transcribe and send you in his retirement into the Country': that is, on holiday.

There is no record of Newton's immediate reaction, nor did he ever send a reply to Oldenburg, who died in early September. Hence the hope that Newton had expressed in the covering letter to the *Epistola Posterior* 'this will so far satisfy M. Leibnitz that it will not be necessary for me to write any more about this subject' was fulfilled, although hardly in the way he had intended or would have wished. Oldenburg being dead, the exchange of letters between these two leading mathematicians of the age ended there for the time being. When it was revived years later it ended, as is well known, in bitter controversy and enmity, peaceful exchange proving impossible without Oldenburg. He had managed the correspondence between Newton and Leibniz without either man's showing rancour. On the contrary both had expressed themselves as grateful to Oldenburg for his part in the exchange of information and both had gladly assisted him by writing out far more important and profitable mathematics than they ever would have done without him. He had here, as elsewhere, proved himself to be a notably able scientific ambassador.

7
Scientific Diplomacy 1669–77 (2) Huygens: Mathematics, Mechanics and Horology

Correspondence with Newton had employed Oldenburg on two fronts, the mathematical and the optical, but it was by no means his only fruitful correspondence nor the only one which generated controversy. A very important part of Oldenburg's correspondence in the 1670s centred around Christiaan Huygens. His appraisal of Newton's optical work has already been discussed in the previous chapter, but he also involved Oldenburg in several other controversial subjects which equally had real and important implications for the expansion of natural philosophy. Huygens was one of those often reluctant to publish and yet, again like so many others, was inclined to be outraged when others claimed priority for what he himself had discovered even though he had not made it public. Once again Oldenburg was required to use all his diplomatic skill in soothing wounded pride and in eliciting information in the face of reluctance to communicate, this time with mixed success.

For a number of years now Oldenburg had replaced Moray as Huygens's chief English correspondent and he took the position very seriously, particularly in the years from 1669 onwards. He then managed to handle successfully questions of priority in developing theories of motion in the complex discussion which had involved Huygens, Wallis, Wren, and others. (See above, p.137f.) He was at the same time concerned with Huygens's mathematical ideas and formulations, as will be seen, while five years later he was to be forced to try to mediate between Huygens and English writers on other questions of mechanics after the publication of Huygens's *Horologium Oscillatorium* in 1673. Then in 1675 a new horological invention by Huygens was to plunge Oldenburg so deeply into a devastatingly violent controversy between Huygens and Hooke as to make Oldenburg a victim of Hooke's fierce anger. It was all very different from his fascinating, important, and relatively smooth involvement with Newton's optical and mathematical disputes which he was handling at the same time. No longer was Oldenburg called upon only to play the role of ambassador. Now he was no innocent bystander or mediator, but was to be an embattled participant in a dispute which ultimately cost him the friendship of both parties.

Even in personal and professional relationships, Oldenburg found himself on a very different footing with Huygens from what he was with Newton. With the latter he was on a level of equality in social terms, and yet his connections had been and remained almost purely epistolary and business-like, the relations of the Secretary of the Royal Society with a distinguished provincial Fellow, cordial but never close. They met, briefly, on Newton's infrequent visits to London but apparently established no strong personal relations. True, over the eight years of their acquaintance some warmth of feeling appeared here and there in their correspondence, but it was necessarily distant. With Huygens Oldenburg's relations were far longer, going back at least to 1661, friendly but again never intimate. On Oldenburg's side there was necessarily a touch of deference, the Huygens family being such a distinguished one in Dutch affairs, intimate with the government and serving it. Huygens prided himself on his title of 'Lord of Zulichem'. His father made him a generous allowance so that, unlike his father and brothers, he never needed to be active in public service but like his contemporary Boyle he was 'a gentleman free and unconfined', able to devote himself entirely to natural philosophy. When the Académie royale des sciences was founded in 1666, Huygens became its leading member; he was then a pensionary of Louis XIV and played an increasingly important role in Parisian scientific life. As the Académie's most distinguished member he enjoyed immense prestige while being free to come and go as he pleased, so that he was not always resident in Paris. Yet at the same time he became so thoroughly a member as to be regarded by the English as virtually French, not to his advantage in the eyes of many Englishmen.[1] He came to regard himself as a defender of French natural philosophy, as a visiting Englishman put it after becoming very friendly with many of the French academicians, especially Huygens,[2] he

> hath a great Zeale for the honour of the French academie even to Jealousie & I believe . . . that hee is a little suspitious of the workings of the English nation, & thinketh that you would take the patterne by them.

Whether Oldenburg saw it this way is not clear, but it certainly suggests how carefully he would have to tread in order not to put himself in the wrong with either Huygens or the English, which on the whole he succeeded in doing for some years. Relations between the two men, although formal, were always to be friendly, rendered more so in the later 1670s by the genuine terms of friendship which arose between Oldenburg and Christiaan Huygens's father Constantijn, poet, diplomat, and statesman, who spent much time in England and was to correspond in warm terms with Oldenburg from 1674 to 1677, terms at least outwardly far warmer than those with the son. (See below, Chapter 10.)

Oldenburg had, it should be recalled, not only met Huygens during the latter's visit to London in the spring of 1661, but had called upon him in the summer of 1661 on his return from his last visit to Bremen. After this, there was a brief correspondence between the two men, ending with a letter of introduction from Oldenburg on behalf of a German traveller, something often to be repeated in years to come.[3] This was purely private correspondence. Huygens's principal source of English news for some years to come, even after he was elected a Fellow of the Royal Society in 1663, was Sir Robert Moray with whom he had become friendly in the 1650s during the years when Moray, as a royalist, was living in exile in Holland. There was then no occasion for Oldenburg to write to him in his own official capacity. He began to write frequently in 1665 and 1666 primarily as a postal agent for Moray, although occasionally on his own account. It was not until 1668, after Moray had left London for Scotland, that Oldenburg began a steady correspondence with Huygens to keep him informed of what the Society was doing and how he might, as a Fellow, become more involved in its affairs. Oldenburg first wrote at the express command of the Society, when there was beginning that discussion of 'the nature and the laws of motion', as the minutes describe it, featuring among others Wallis, Wren, and William Neile, to which Huygens was to be invited to contribute, as he did. (See above, Chapter 5.) Oldenburg's letter on this occasion was very formal and respectful and indeed his letters never entirely lost all of this tone, although he felt able to relax a little over the years as Huygens showed his appreciation of Oldenburg's work, but neither man ever wrote informally in their exchanges. The content of the letters which Oldenburg addressed to Huygens was not essentially different from that of those he addressed to any other Fellow of the Society. His letters were full of scientific news, attempts to encourage the recipient to communicate his own discoveries (it being taken for granted by Oldenburg that he was constantly making discoveries) and to criticize and comment upon the discoveries and opinions of others. The exchanges between Huygens and Newton managed by Oldenburg have already been described. There were to be several other such exchanges more or less controversial, which put a considerable burden of work upon Oldenburg as well as the necessity to strive to maintain neutrality between controversialists. In this he was more often than not to be successful. Certainly Huygens always thought he was fair.

Among the exchanges involving Huygens which were the most burdensome and complex for Oldenburg must have been the several mathematical controversies in which Huygens took a leading role. As noted above the Royal Society had in 1668 become concerned in a controversy between Huygens and James Gregory.[4] Wallis tried to act as an intermediary[5] and soon, with other members of the Society, Wallis came to feel that Huygens was in the right. Through Oldenburg he told both controversialists that this was the case, a conclusion naturally gratifying to Huygens and one which made him feel closer to the Society than ever.[6] This was a somewhat unusual case of the Royal Society's showing partiality to a foreigner. No doubt to Oldenburg's relief, the controversy ended early in 1669 and, since Oldenburg was here openly acting as a mere mouthpiece for Wallis and others, he escaped being implicated on either side, fortunately for him.

Oldenburg's role in drawing Huygens into the 1669 discussion and printing ideas about the laws of motion (all in fact very similar) by Wallis, Wren, Huygens and others has already been discussed. (See Chapter 5.) It should be recalled that Oldenburg was here for once not entirely successful in his relations with Huygens whose paper he only printed as a translation of what Huygens published in the *Journal des Sçavans*. Oldenburg soon managed to placate Huygens, to overcome his resentment, and to maintain an equable correspondence on more mechanics, optics, astronomy (Huygens's recent observations of Saturn which confirmed his discovery of its correct shape a decade earlier), discussion of the possibility of the determination of longitude at sea by means of Huygens's pendulum clocks (work which had been going on intermittently for nearly ten years), and many more subjects.[7] All these topics Oldenburg discussed with his usual aplomb and dexterity. Before 1670 it became normal for Oldenburg not only to communicate news to Huygens but to try to stimulate him to communicate his thoughts and discoveries for dissemination to the Royal Society and the world.

A second mathematical controversy involving Huygens was both calmer and more profitable to the world of learning. This was the long drawn out critical discussion, rather than a true controversy, between Huygens and Sluse with Oldenburg acting always as the tactful intermediary, touching the respective solutions by the two men of the famous problem in mathematical optics of Alhazen. This had been proposed and partially solved by the eleventh-century Moslem mathematician Alhazen: it demanded a mathematical solution to the problem of finding two points on any reflecting surface such that light falling on one point would be reflected to the other. Here Oldenburg, assisted as usual in such matters by Collins, was able to stimulate the two controversialists to continue to work at the problem. He was carefully to refrain from publishing anything until both were fully satisfied that they had reached their best possible conclusions. The interest of both Huygens and Sluse in this problem was aroused in 1669 by reading Barrow's *Lectiones Opticae*, published in that year, where it had been dealt with to some extent but, as Huygens was to say, with only partial success. (Collins had naturally had his interest aroused by the same source.) The exchange between Huygens and Sluse was initiated in a roundabout manner, when Huygens sent to the Royal Society some examples of his newly discovered invention for printing geometrical figures, one of which contained 'the construction for a problem which I solved lately and which our mathematicians thought pretty elegant'.[8] Oldenburg promptly sent the figure to Wallis who seems not to have been particularly interested even though, as Huygens had noted, Barrow had experienced some difficulty with its solution. About a year later Sluse, when reading Barrow, was led to remark in a letter to Oldenburg that he had some time previously found a way to attack the problem more easily.[9]

Oldenburg, undaunted by the lapse of time which might well have made him forget what had occurred in this matter so long before, and pursuing his usual method of stimulating research by telling one man what another had done, sent Sluse the solution which Huygens had sent to the Royal Society. As a result, as Oldenburg hoped, Sluse then sent his own version to Oldenburg, noting that he and Huygens had analysed the problem in the same way while using different methods for its solution, his own being, he believed, more general. After carefully securing Sluse's permission, Oldenburg sent a copy of Sluse's letter to Huygens,[10] copying out the mathematics himself to ensure accuracy. This was now some six months after it had been written and by this time, stimulated by hearing of what Huygens had done, exactly as Oldenburg intended, Sluse had pursued his researches further and found a yet better solution. Huygens replied (although he was really then more interested in the observations which he and Cassini had recently made on Saturn which confirmed his own hypothesis of the ring structure of that planet, an hypothesis proposed originally in 1659) agreeing with Sluse as to their approaches being similar but saying that he thought his own construction was 'more natural'.[11] Although Huygens then expressed his intention of writing directly to Sluse 'since he is the most knowledgeable and candid of all the geometers I know' he did not do so. In fact, he did not need to do so for Oldenburg duly reported to Sluse what Huygens had said. This in turn prompted Sluse a month later to send what he called his 'second thoughts' on the problem, ideas which Oldenburg also sent to Huygens although he was really then more concerned to tell Huygens the news of Newton's recently invented 'new kind of telescope'.[12] (See Chapter 6, above.) In February 1671/2 Huygens, while declaring himself 'apprehensive of getting myself too deeply involved in the study of geometry' because his health had been poor and he found the study of mathematics a strain, could not resist asking for a copy of Sluse's latest work.[13] Oldenburg immediately sent him a copy of what Sluse had written, together with a short note[14] in which he apologized for his brevity on the grounds that he was 'almost worn out from writing, after having myself transcribed Mr Sluse's letter about Alhazen's Problem, so that you might have it copied correctly'. (When sending Huygens a copy of Sluse's further thoughts on 10 September 1672 he also declared himself 'almost worn out').

All this is a vivid indication of Oldenburg's devotion to the art of learned correspondence and his appreciation of the care and attention required to transcribe mathematics accurately and correctly, as he generally succeeded in doing. After this Huygens expressed himself satisfied.[16] Oldenburg hoped to be able to print the exchange in the *Philosophical Transactions* fairly swiftly, provided the two authors gave their permission, but since both Sluse and Huygens continued to elaborate their constructions during the course of the next year (which of course involved Oldenburg in yet more copying) he was only able to do so in October 1673.

The whole exchange, conducted most amicably, was a brilliant example of what Oldenburg's methods could achieve, and if his correspondents sometimes complained a little at his needling of them, they very seldom really resented it. This particular example is the more remarkable because during its four years' duration so many other exchanges, mathematical and otherwise, were occupying Oldenburg continuously, so that his keeping the thread of this exchange unbroken is a tribute both to his methodical habits and to his memory. One can only assume that he had some kind of efficient filing system, sorting his incoming letters either by correspondent or by subject or by date. He certainly indexed each Royal Society Letter Book where the most important letters were entered in order of receipt. It is especially remarkable that he kept this particular exchange clear in his mind since at the same time that he was acting as intermediary between Huygens and Sluse over Alhazen's problem he was also the intermediary between Huygens and Newton in optical affairs, as well, of course, as dealing with many other topics for other correspondents.

By means of the diplomacy and care outlined above and always overtly acting as an agent of the Royal Society, Oldenburg led Huygens to feel himself a closely integrated member of the Society. (Under the terms of his election he was indeed as much a full member of the Society as anyone, for the concept of 'foreign Fellow' had not yet been invented.) How successful Oldenburg was in this endeavour emerged in 1670, when Huygens became seriously ill. For information about the effects of this illness one must read the letters sent to Oldenburg from Paris by Francis Vernon, a colourful youngish Englishman of, clearly, great charm, who in 1669, after an adventurous period of travel during which, among other vicissitudes, he was captured by pirates and sold into slavery, had been appointed to the post of secretary to the embassy in Paris under the Duke of Montagu. During the next three years Vernon became extremely friendly with virtually all the natural philosophers in Paris, especially with the various members of the Académie royale des Sciences, including Huygens, Cassini, and Picard. He was thus able to supply Oldenburg with detailed and accurate accounts of the important work being then done by the Academicians.[17] In the very cold weather of January 1669/70 Huygens told Oldenburg that he was not then well on account 'of the discomfort which I experienced during this severe cold', and this was confirmed by Vernon.[18] But it was more than a simple cold or mere discomfort from which Huygens suffered during the ensuing cold weather, for during the next month he was ill in bed, he received no visitors, and his state was regarded as critical, so much so that on 12 February Vernon received a message asking him to visit Huygens.[19] He found him very ill indeed and extremely weak, much depressed, and suffering from lack of sleep. He was gloomy about his prospects of recovery, fearful lest his illness might prove mortal. He expressed himself as very friendly towards both Vernon and the Royal Society, and, dismissing his attendant servant, told Vernon privately that he had sent for him

> to the end that while my memorie & strength serve mee for any such Purposes I may communicate something of my mind unto you [Vernon] & leave something in your hands

trusting that Vernon would faithfully obey his directions as to what to do with what he was to be given.

Huygens then handed Vernon 'a little Pacquet sealed' in which, he said, was to be found the solution to the anagrams concerning his 'Doctrine of Motion' which he had formerly sent to the Royal Society. This 'Pacquet' Vernon should, if Huygens died, open and give to the Secretary of the Académie. (If he lived, the packet was to be returned to him unopened.) Huygens also gave Vernon a printed sheet 'On Parhelia' to be sent to Oldenburg, who was to publish it in translation in his *Philosophical Transactions*, as he did a few months later.[20] Most important of all to Huygens was an unsealed 'bundle of Papers' which after his death was to be sent to the Royal Society for, particularly, Lord Brouncker who, so Huygens said, so well understood the 'Doctrine of Motion' that he hoped that Brouncker would rewrite what he sent in a better form, for publication. (Probably these papers contained an early draft of work only in fact to be published in 1703 in his *Opera Posthuma*.) According to Vernon, the chief reason why Huygens despaired of his recovery lay at the door of the Parisian physicians, for as he said, 'Oh if Doctor Willis were here I believe I should recover but these People have not a right conception of Physick'. Vernon also reported that Huygens denigrated the Académie, both on account of its internal disputes and because it depended on the whim of the King, at the same time praising the Royal Society as 'an assembly of the Choicest Witts in Christendome & of the finest Parts'. This explains why he was so anxious to share his work with its members, of whom he particularly praised Brouncker, Ward, Wren, Wilkins and, ironically in view of what was to occur a few years later, Hooke, whom he denominated 'a man of vast invention'. He evidently also praised Oldenburg greatly, but the line in which Vernon revealed that he did so is heavily crossed out.[21] This may have been done out of modesty by Oldenburg but equally probably years later by Hooke when Oldenburg's fierce enemy, who certainly went through all of Huygens's letters in 1677. This truly touching scene between Huygens and Vernon is necessarily only known from the latter's account and does not quite square with what he had written earlier about Huygens's suspicion of the English, but reflection induced by great weakness and fear of imminent death might well have made Huygens feel more charitable towards the Royal Society and closer to his English than to his French colleagues. There is certainly no reason to doubt the essential truth of Vernon's narrative for he seems to have been an honest and reliable witness of the French scientific scene and had no personal reason to falsify or exaggerate what occurred in his presence.

Huygens did not die and indeed fairly soon became convalescent as the cold weather departed. He then presumably asked Vernon for the return of his 'bundle of papers' and sealed note. But it was many months before he was well enough to travel home to Holland for recuperation and yet more until he was fit to work again or correspond. When he finally did write to Oldenburg it was with a simple letter of introduction for a German traveller coming to England, as so many did, together with a brief explanation of his silence. He needed, he said, complete rest after his 'illness, so little short of fatal', so much so that he had not dared to read the mathematical books by Barrow and Wallis which he had seen in Paris before his departure. He could only note that Barrow mentioned his own mathematical work, a characteristic touch of the interest which all scholars and scientists take to this day in citations of their work. He would, so he said, have liked to accompany his father Constantijn Huygens on his latest diplomatic mission to London, when he was accompanied by Prince William of Orange, later the English King William III, but he did not dare to venture on the voyage. A fortnight later, however, he was well enough to write a brief letter[23] which shows a returning interest in news of the world of learning and soon afterwards returned to Paris, but even then it was some time before he took up mathematics again. Nor did he reply to Oldenburg's letters,[24] not even to that of 22 July 1671 in which Oldenburg included Sluse's latest thoughts about Alhazen's Problem. This, he said, he only did, because he had learned from Constantijn Huygens, still in London, that his son 'had returned to Paris in good health'.

Not until 3 February 1671/2, two years after the onset of his illness, did Huygens resume correspondence with Oldenburg in his normal manner, stimulated by the news of Newton's telescope[25] and in March of that year, as already noted, he at last returned to mathematics and his interchange with Sluse about Alhazen's Problem, all nourished by Oldenburg. At this time, it must be remembered, Oldenburg was encouraging him to maintain an interest in Newton's optical work, an interest which he was to pursue for the next two or three years, sending to Oldenburg his opinions of it and his reactions to it, all of which demanded complicated replies. He then seemed in good health, but on 17 September 1672 he wrote apologizing for having failed to respond promptly to Oldenburg's faithfully maintained correspondence, showing how much he valued it and how much he appreciated Oldenburg's share in making it possible for him to exchange views with Newton.[26] This failure he ascribed, at least in part, to 'my ill-health, which ordinarily at this season brings on an attack', citing also 'other affairs'. These last included involvement with the laws of motion, optics, pneumatics, astronomy, and mathematics in an astonishing range of activity, all of which no doubt also did account for the slightness of his correspondence with Oldenburg in 1672. In fact he did not write to him again until the new year, writing twice in January 1672–3, being then clearly keenly interested in these 'other affairs', which included a new design for a barometer[27] about which Oldenburg had politely expressed his usual interest.

Sadly for Oldenburg, by the end of 1673 there was a real if temporary breakdown in communication between Huygens and the Royal Society which was ultimately to end all direct friendly relations between Huygens and Oldenburg, although not before Huygens had dramatically manifested his real feelings of friendship for Oldenburg. The break began insidiously after the publication of Huygens's great work on mechanics, his *Horologium Oscillatorium*. (This is an extremely important work for its discussion of a wide variety of mechanical problems, all dealt with in rigorously mathematical terms; but although much of this work went back to studies made as long as 1659, particularly those on pendulum motion, it still did not reveal everything that Huygens had done in this field over the years.) In May 1673 Huygens thoughtfully sent twelve copies o the book to Oldenburg for presentation to twelve English colleagues, whom he named, among them Wallis, Moray, Boyle, Wren, Newton, Hooke, and Oldenburg himself, presumably all inscribed as presentation gifts.[28] All recognized the importance of the work and all were grateful for being given it. Newton, whose copy still exists, wrote in an acknowledgement for Oldenburg to transmit to the donor[29]

> I received . . . Monsr Hugens his kind present, for which I pray you return him my humble thanks, I have view'd it with great Satisfaction, finding it full of very subtile and useful speculations very worthy of the Author. I am glad, we are to exspect another discours of the *Vis centrifuga*, which speculation may prove of good use in Naturall Philosophy and Astronomy, as well as Mechanicks.

This praise Huygens must have been glad to receive, even though Newton rather spoiled it by speaking of what he called 'an Illegitimat supposition' and noting that he himself had done something about 'The rectifying curve lines' which he thought was 'more ready and free from trouble of calculation than that of M. Hugens' which he would be glad to send to Oldenburg for transmission to Huygens if the latter wished him to do so. In copying this out Oldenburg wrote in the margin 'You have only to let me know your will about this' but Huygens showed no desire to receive the news.

Regrettably, worse was to come and Huygens to find himself in a veritable hornet's nest of English complaint and resentment. In the first place, to his great embarrassment, he learned from Oldenburg that he had, unaccountably, committed the solecism of failing to send a copy of his book to Brouncker. This is the more surprising as he greatly respected Brouncker, who was after all the President of the Royal Society, and who had, as Huygens had acknowledged earlier, done good work in some of the areas of mechanics discussed by Huygens in his book. In telling him this, Oldenburg displayed his customary diplomacy: when he acknowledged the gift to himself, he said 'When my copy has been bound I shall lend it to Lord Brouncker'.[30] Huygens was properly horrified to discover his lapse of good manners and claimed to be surprised, although his own list of those to whom copies should be sent also omits Brouncker's name, surely by inadvertence. He immediately promised to send a copy for Brouncker as soon as he could find someone to convey it,[31] as he did, although it is not certain that it was ever delivered. Brouncker was extremely good-tempered about the lapse and did not even take umbrage at Huygens's claim that Brouncker's early work on the mathematics of the isochronism of the cycloid was a mere demonstration of what Huygens had invented earlier although he did go so far as to have Oldenburg publish an anonymous note about it in the *Philosophical Transactions* of 19 May 1673.[32]

Others were neither so good tempered nor so complacent. Wallis very properly thanked Huygens for his gift, but could not resist defending the priority of his pupil William Neile, now dead, in the discovery of the method of rectifying the cycloid which Huygens had ascribed to a Dutch mathematician, Henricus van Heuraet.[33] What was worse, Wallis wrote a further defence of Neile for publication in the *Philosophical Transactions* for November 1673 with, in support of his claim, statements extracted from Brouncker and Wren (who had his own claim for priority).[34] It was in connection with this that Wallis accused Huygens of having become Frenchified,[35] telling Oldenburg (privately, it must be said)

> why Huygens should have put up pretences about all this can only be because he has been rather less than impartial towards us, being perhaps influenced in this, by the French, since formerly he used to behave more fairly.

In as late as the next spring Wallis continued to harp on this theme. When he then read more of the *Horologium Oscillatorium* he was moved to complain bitterly[36] that Huygens had ignored his (Wallis's) contributions, insisting that

> since he hath drawn in the Air of France, it is (it seems) below him to acknowledge any thing done by any body else: which while while he was but a Dutch-man he was wont to do.

As if these claims of Wallis were not enough, Oldenburg was also the recipient of Hooke's eager and earnest claims for priority, claims to which both Oldenburg and Wallis were at first sympathetic. Unfortunately it is now impossible to tell that Hooke knew precisely what Oldenburg had written about this and certainly later, and perhaps even then, he would not have believed that Oldenburg had indeed defended him. For in his first acknowledgement of the receipt of the book[37] Oldenburg had reminded Huygens (mildly, using the disarming phrase 'you doubtless remember') that Hooke had 'proposed and constructed a circular pendulum, applying it as well to a clock' in June 1666. To be exact, in the minutes of the Society's meeting on 13 June 1666 Hooke is stated to have 'exhibited a new contrivance of a circular pendulum applicable to a watch'. This was, it must be said, not necessarily quite the same thing as having applied it to a watch and although this may have been a genuine construction it may also have referred only to the drawing of such a construction. The point of the reminder was that in *Horologium Oscillatorium* Huygens had described the geometry of a conical pendulum and its application to a clock as being his own invention, as it almost certainly was, an invention made many years earlier although never before made public. Nor is there any certainty that he was aware of Hooke's invention of the same thing before he wrote the passage in the printed book, although, as he admitted, he had been told something about Hooke's ideas on his visit to England in 1663, now more than ten years earlier and after he himself had had the same idea. He now declared[38] that when he had asked to be told how Hooke had 'made the times of rotation of the pendulum equal' as was essential for accuracy, those to whom he spoke

> only told me vary brusquely that your gentleman did not want to have anyone lift their inventions, which is as much as to reproach me with being the one who tries to steal from them. There is nothing of which I feel myself less guilty nor less capable than that which you impute to me.

One can see Huygens's point and can also believe that Hooke then as later was inclined to be suspicious of all foreigners and equally was reluctant to reveal any of his best ideas, in spite of the apparent ease with which they came to him.

The full text of the letter which Huygens sent to Oldenburg is, very unusually, lost. Very probably it was given by Oldenburg to one of those mentioned in it (Moray, Wren, Hooke, Brouncker) who never returned it to him. But there is no reason to suppose that the surviving memorandum preserved in the

Huygens archive does not represent Huygens's feelings accurately, although he may have expressed himself somewhat more tactfully in the formal letter. That he was very angry and very hurt is clear enough. He had indeed reason to be so for, as he wrote in the summer of 1674 in an apology or defence (addressed to his father to be forwarded by him some months later)[39] he had made his own invention as early as 1658 and had told Wren of it in 1663, tactfully adding that perhaps Hooke had not known of it in 1666 when he made *his* invention (though probably Huygens did not believe this). Huygens also defended himself in a letter to Wallis in June 1673[40] and a year and a half later was still anxiously defending his priority.[41] Oldenburg was obviously in great difficulty, caught between two angry and equally worthy opponents. He certainly did not wish to offend Huygens. Nor did he wish to offend either the Royal Society or Hooke, the former his employer and to which he belonged as a Fellow, the latter his daily colleague.

He was asked by the Society to publish relevant letters from Wallis, Brouncker, and Wren in his *Philosophical Transactions*, and so he did in the autumn of 1673.[42] The best he could do for Huygens was to tell him 'I shall never fail to contribute all that I can to maintain goodwill between distinguished people everywhere', and to try to demonstrate this principle by publishing the exchanges between Huygens and Sluse and Huygens and Newton in the October and November issues of the *Philosophical Transactions* in 1673.[43] He told Huygens that he had also there inserted the priority defence written by Wallis 'in order to give his due to everyone, as far as may be'. But although he added optimistically,[44] 'I am convinced that your candour will take this act of justice in good part' he cannot have been surprised that Huygens did not reply.

Recognizing his mood, Oldenburg did not trouble Huygens with correspondence for some months, evidently hoping that time would heal this sad breach between the Society's Fellows at home and abroad. In March 1673–4 he sent issue number 100 of the *Philosophical Transactions* to Huygens 'to complete the century'. (Although there is no firm evidence that Huygens had received the immediately preceding issues, it is probable that he had done so through his father, for Oldenburg had been in extremely friendly correspondence during the intervening period with Constantijn Huygens, who supplied news of his son and who received news from London to transmit to him.) When he sent Huygens number 100, Oldenburg also sent a short covering letter[45] in which he gently pointed out that Huygens owed him many letters and expressed the hope that the receipt of the copy of the journal would 'arouse' him to reply. Perhaps a little naïvely, he added

> You seem entirely to have forgotten the interest which you have in the Royal Society or else you take it in bad part (which is what nevertheless I should not easily wish to think) that one or two of this body take the liberty of speaking frankly of some particulars which you gave to the public.

He then, truthfully as well as more tactfully, went on

> you may be assured that our Fellows do not fail to have the same esteem which they have always had of your merit, and which they demonstrate from time to time in a manner which cannot fail to please you.

With hindsight we might think that Huygens necessarily found this assurance difficult to believe, given the past, and in fact Oldenburg was more than a little disingenuous here, since Wallis was even then continuing to read *Horologium Oscillatorium* and to find objections to many statements in it, objections which he sent to Oldenburg who tactfully did not transmit them to Huygens.[46]

Events in 1674 brought a change after Hooke, in spite of his resentment of Huygens's failure to accept his priority the previous year, asked Oldenburg to send Huygens a copy of his Cutlerian Lecture entitled 'An Attempt to prove the Motion of the Earth from Observations'. He only showed some trace of his previous feelings by his request, duly transmitted by Oldenburg, that Huygens should acknowledge his receipt of the gift.[47] This did finally make Huygens break his silence, with a cordial, slightly shame-faced letter to Oldenburg.[48] This began, 'You show real goodness in continuing still to write to me and send me all that is new with you notwithstanding such a long silence', the reason for which, he said, was 'nothing but that I saw that my letters only put me wrong with those gentlemen over there . . . which I did not at all expect'. He acknowledged several books sent to him (one by Boyle, Hooke's Cutlerian Lecture, and several issues of the *Philosophical Transactions*). 'The Observations of Mr Hook' he found 'very good'. He had, he said, sent the promised copy of his *Horologium Oscillatorium* to Brouncker 'a long time ago by hand' and one to Sluse, but had heard nothing about either. (As already noted, nothing is known of the fate of the copy intended for Brouncker; that to Sluse was obviously delayed by wartime conditions, for the French were still active in the southern Low Countries.) He reported that he had had new thoughts about the solution of Alhazen's Problem which he had been, via Oldenburg, discussing with Sluse, in which connection he remarked that the extracts from his letters dealing with this problem 'were very badly translated' in the printed version in the *Philosophical Transactions*, adding handsomely 'I do not think that this was your Latin, because I know it of old'. (The reference is to a few sentences originally written in French in the middle of a letter to Oldenburg, although both Huygens and Sluse normally wrote all mathematical passages in Latin, which explains why this passage had to be translated.) It is not clear to what Huygens objected, and evidently Oldenburg himself did not understand the objection, for he merely answered politely that he was 'very grieved' at any errors and offered to print an errata sheet if Huygens would send suitable corrections. The errors, if they truly existed, cannot have been very serious for Huygens never sent any corrections.[49] To the body of the letter Oldenburg replied emolliently, evidently still hoping to persuade Huygens to resume correspondence on the old terms, even sending Brouncker's thanks in advance of receiving his copy of Huygens's book with his regrets for any misunderstandings. In spite of Oldenburg's efforts, Huygens did not write again during the remainder of 1674 although Oldenburg continued to send him issues of the *Philosophical Transactions* as they appeared, three in all, always with covering letters.[50]

When Huygens did write again at the end of January 1674–5 after eight months of silence it was because he had news of his own very different concerns.[51] He apologized handsomely for his silence, saying

> Although I only discharge at great intervals, the duty of thanking you for the continuation of your kindness, I beg you not to judge from this that I am not aware of it,

a disarming apology made more effective by his demonstration of his appreciation of the journals sent to him by discussing their contents. He also displayed his faith in the judgement of the Royal Society, in spite of his previously expressed doubts and difficulties, by asking for the opinion of the Fellows about the reliability of Leeuwenhoek's observations which the Dutch microscopist had been sending to the Society and which Oldenburg had been routinely printing in the *Philosophical Transactions* since the spring of 1673. (See below, Chapter 8.)

Huygens ended his letter with his real and important news, namely that he had 'recently' (in fact during the previous ten days) 'discovered a new clock invention' whose nature he was prepared to describe at this point only by an anagram as a postscript to the letter itself.[52] This proved to be the invention of a coiled spring balance and a practicable method of applying it to watches. Sadly, this brilliant and important invention was to create a new international dispute about priority and to involve Oldenburg in great difficulties, as could not, of course, have been foretold. As Oldenburg told Huygens in his reply,[53] the members of the Royal Society to whom he showed the letter displayed a 'great desire' to learn the secret of the anagram. (Like all such anagrams, it was quite insoluble since it merely gave the number of times the letters appearing in a usually cryptic sentence was used.) This was encouraging for Huygens, but at the same time Oldenburg had to tell him that Hooke had once again attacked him in yet another Cutlerian Lecture entitled 'Animadversions on the First Part of the *Machina Coelestis* of Hevelius', published in 1674, and to try to apologize for Hooke's rudeness in so doing. Although primarily addressing astronomical problems, Hooke took occasion here[54] to complain that in *Horologium Oscillatorium* Huygens had given

> a short description of a circular pendulum . . . without naming me at all as concern'd therein, though I invented it, and brought it into use in the year 1665, and in the year 1666

it had been described at a meeting of the Society and the fact of this transmitted to Huygens by Moray, (though Hooke did not say how he knew that Moray had done so). Now all this was true, except that Hooke does not really seem to have 'brought it into use' as he claimed, 'while Huygens, it must be remembered, could and did claim that he had made his own discovery eight years earlier than Hooke, that is in 1658, as was also true. Further, Huygens might easily, and no doubt had, genuinely forgotten in 1673 something of which he had been told in a letter written so long ago, and which, moreover, could have been of no great interest to him. But he was, inadvertently, laying up trouble for both himself and Oldenburg. What this trouble was will be discussed here only as it affected the relations between Huygens and Oldenburg; for its affect on those between Hooke and Oldenburg see below, Chapter 9.

Meanwhile, and before he could have received Oldenburg's letter, Huygens found out to his shocked dismay that his watchmaker whom he had instructed in the use of the novel balance springs had quickly made two watches of the new design and shown one to all sorts of people, claiming to be the inventor. Consequently there was no longer any point in secrecy and so Huygens now sent Oldenburg a clear description of his invention.[55] He had, as he informed Oldenburg, secured a Royal Privilege in France (equivalent of an English patent) and with extraordinary generosity and a decided indication of the warmth of his feelings for both Oldenburg and the Royal Society of which he was a Fellow added, in the words of the translation Oldenburg later provided for publication in the *Philosophical Transactions*,

> if you believe, that a Privilege in England would be worth something, and that either the Royal Society or You might make some advantage thereof, I willingly offer you all I there might pretend to.

He also gave Oldenburg full permission to print the 'explanation and description' of the invention. Oldenburg took full advantage of this permission by printing in the *Philosophical Transactions* for 25 March 1675 a translation of the account drawn up by Huygens for publication in the *Journal des Sçavans* and which had appeared in the issue for 25 February 1675 (N.S.).

All this was very interesting and very gratifying. Unfortunately when Oldenburg read the relevant portion of this letter to the meeting of the Society held on 18 February 1674/5, Hooke reacted extremely badly, claiming that the spring-balance watch was *his* invention. As he noted in his diary for that day, 'I shewd where it was printed in Dr Spratts book.[56] The Society inclind to favour Zulichems [that is, Huygens's]'. And indeed the meeting did incline to favour Huygens as the independent inventor who had, moreover, put his invention into practice. According to the minutes,

> It was ordered, that Mons. Huygens, notwithstanding, be thanked for this communication, and informed what had been done here; and what were the causes of its want of success.

(The final clause reveals what Hooke could never be brought to acknowledge, that there was a difference, as many of the other Fellows did recognize, between having an idea for an invention and bringing it to fruition with a working example, and that they did not all agree with Hooke's claim.) That the majority of the Fellows present at the meeting did tend to think that Huygens had a just claim to be the inventor of the spring-balance watch is not surprising, for what Sprat says, without ever naming Hooke, is that there had been among the Fellows suggestions for 'Several new kinds of Pendulum Watches for the Pocket, wherein the motion is regulated by Springs, or Weights, or Loadstones, or Flies moving very exactly regular.' Clearly, Hooke had in 1666 or possibly even earlier suggested that springs *might* be used to regulate watches, although in Sprat's account at least springs were not singled out any more than 'Weights, or Loadstones, or Flies'. Equally clearly, he never described any possible mechanism for doing so and certainly did not then make any going watches regulated by a spring balance. And Huygens had now done so. At the meeting at which Huygens's description of his invention was read, Hooke was prepared to say publicly that such watches had been made, although he did not say when this had been done. (It is tempting to wonder whether he did not then and there rush off to make one or at least a model of one, his mechanical ingenuity being so great that once he had learned that the idea worked he could easily have devised a method of his own to make a working watch.) Oldenburg cautiously reported to Huygens that the Fellows were very greatly interested in his news but would 'suspend their judgement' until they had seen a detailed description and drawing,

> principally in the face of Mr Hooke's having invented, some years ago, a similar thing, as he believes, which however did not then succeed entirely according to his hopes

but which he now thought he could improve.[57] Naturally Oldenburg thanked Huygens very warmly for the offer of possible patent rights, adding that 'The Royal Society will willingly leave me to enjoy any advantage which may result.' (In fact he could not resist immediately drawing up a petition for a patent for himself and a draft of a possible warrant granting it, documents which he seems to have submitted for possible Royal approval at the beginning of March.)[58]

It must be noted that throughout this affair it was almost certainly Brouncker who suggested leaving any profit from the potential patent in Oldenburg's hands and Brouncker who openly supported the claim of Huygens against that of Hooke, as the latter was well aware. Among the other prominent Fellows, some, including more notably Wren, preferred Hooke's claim, influenced no doubt by a combination of friendship and xenophobia. Brouncker added weight to his interest by asking through Oldenburg for Huygens to have a watch made and sent to him; he was explicitly prepared to pay for it, as he later did, and much of his reason for wanting it was so that he could show it to the King. (In spite of this, Huygens did not hasten to have the watch made for Brouncker, so that Oldenburg was twice forced to repeat the request, the last time in May.)[59]

Huygens's generosity in offering Oldenburg any possible patent rights backfired, for it was to embroil Oldenburg in a mighty, bitter, and prolonged quarrel with Hooke from which their personal relations never recovered, with Hooke literally pursuing the quarrel beyond the grave. (For more details, see

Chapter 9.) Here it will suffice to record that Oldenburg never received a patent, because Hooke succeeded in designing and having made a practicable spring-balance watch, similar to but differing in detail from that of Huygens. He showed several designs to the King himself and was able to have at least one made by September, when Oldenburg told Huygens that one of Hooke's watches was running well.[60] Hooke never secured a patent either, although he sought one. He was still hoping for one some months later and refused to let anyone but the King see his watch until a patent was secure even though it was said to work fairly well and boasted a minute hand, as Huygens's first watches did not. Meanwhile Huygens did have a watch constructed for Brouncker which was conveyed to him by the hand of a traveller, with detailed instructions as to how to regulate and adjust it (directions which Brouncker seems to have found difficult to follow); the price, which Brouncker duly paid, was 80 livres (francs).[61] Brouncker was so pleased with it that he asked for more watches, to be furnished with minute hands, a request with which Huygens rather slowly complied in December 1675.[62]

Huygens was very anxious to learn whether Hooke had really made the same horological invention independently, which he seems to have doubted, remarking that[63]

> The procedure of Mr Hooke seems to me neither good nor honest, wanting to make himself the author of everything new which comes along, and particularly as regards this invention, it is very ill-natured to say he had it long ago, having produced nothing until I sent you the anagram . . . which I told you contained a new invention relating to clocks. Those who are capable of finding first-rate inventions by themselves do not go on so.

In reply,[64] having first remarked of Hooke that 'he is a man of an extraordinary temperament', Oldenburg asked Huygens to

> write three lines to Lord Brouncker to point out to him that you knew nothing of Mr Hooke's invention before you sent the anagram you said contained a new invention for clocks; to which you might, if you please, add that a person who was the first to apply a pendulum to clocks, and discovered the cycloidal figure to regulate the vibrations, could easily devise some means of replacing pendulums by a suitable spring.

(This was of course by no means the first time that Oldenburg had advised a correspondent how best to reply to an adversary.) He then added, no doubt with complete honesty,

> Which I advise for no other reason than that I wish with all my heart, that everyone should receave his due, and that jealousy of curious minds should not interrupt their friendship, or that exchange which is so useful to the growth of the arts and sciences.

The last phrase was a sincere and often repeated expression of Oldenburg's faith in the value of correspondence. In reply[65] Huygens wrote

> I beg you to tell me if before receiving the anagram it was known that Mr Hooke claimed to have a new clockwork invention, for you never let me know of it except after my invention had been printed,

wisely seeking reassurance on this importance point. The deadlock is obvious and Oldenburg's position so delicate as to be almost indefensible. Huygens was justifiably proud of having made a discovery which proved eminently practicable and he was not a man to give way easily to another's claims of priority, especially to someone who had consistently questioned his independence of discovery. To him, Hooke's claims were ill-founded because he had not publicly produced his watch until he knew o the existence of that of Huygens; to Huygens Hooke had done nothing but, as he put it,[66]

> what the watchmakers here do, that is to vary the construction and application of the spring which regulates the balance, which is not difficult after one has seen it succeed in one way.

To Hooke it was certain that he had long before had the *idea* for a spring-balance watch and could have made one. The proof of this contention was that he had immediately made one as soon as he heard what Huygens had done and almost certainly before he heard Oldenburg read Huygens's detailed description. Indeed, the watch which he gave to the King seems to have been more reliable than the first watch that Huygens sent to Brouncker and moreover it had a minute hand which that sent by Huygens did not have, although his later watches did have them. Oldenburg was clearly caught between opposing forces, especially as both Hooke and Huygens were Fellows of the Society he served and of equal standing and seniority and with both he had had over many years mostly good and close relations. Huygens was prepared to have the case laid before Brouncker as President 'of whose justice and impartiality I am as much persuaded as yourself'.[67]

Hooke was quite fairly right to think that Oldenburg was partial as was Brouncker, at least to some extent. Where Hooke was totally wrong was in believing that Oldenburg had ever before 1675 written to Huygens about Hooke's invention. Even Moray, Huygens's principal correspondent in the early 1660s had only mentioned Hooke's horological work in vague and general terms. (Almost certainly Huygens had genuinely forgotten what he had been told, as why should he not have done?) But Hooke could not be shaken from his belief that Oldenburg had 'betrayed' him and the bitter quarrel which arose from this waxed ever more fierce during the latter part of 1675 and was to continue for the next two years. It is not necessary here to go into details except insofar as Huygens was directly involved. This he became after Oldenburg had appealed to him for support in the autumn of 1675 when Oldenburg reluctantly told him[68]

> Something has happened here which obliges me to write to you this. This is that Mr Hooke, having known that you had given me permission to make use of the benefit which you could claim of a patent for your watch in this country, has been so rash and shameless as to say publicly that you gave me this permission as a reward for having divulged to you his invention, adding that I am your spy here for communicating everything of note found out here, and that I want to defraud him of the profit of his invention.

Clearly Hooke, as Huygens thought the English were prone to do, had become convinced that Huygens had stolen his, Hooke's, invention, put up to it by Oldenburg. The result, as Oldenburg told Huygens, was that 'my reputation, which is dearer to me than life itself, suffers very much'. Evidently Hooke's friends joined him in believing that Oldenburg indeed 'betrayed' secrets to foreigners in his correspondence.

To counter these charges Oldenburg requested Huygens to write to Brouncker as President of the Royal Society to declare that Oldenburg had told him nothing at all about Hooke's horological inventions until after Huygens had published his own. Agitated though he was by Hooke's accusations, Oldenburg yet continued his normal correspondence with Huygens. The next day he wrote a letter of introduction for Evelyn's son. A couple of days after that he sent a copy of Hooke's Cutlerian Lecture entitled 'A Description of Helioscopes' pointing out the postscript in which Hooke, although in much milder words than he used elsewhere, attacked both Huygens and Oldenburg.[69] Huygens replied pretty promptly to Oldenburg's plea with a letter addressed to Brouncker as requested,[70] in which he completely exonerated Oldenburg from what he did not hesitate to call 'Hooke's slanders.' He declared that he had only given Oldenburg any possible patent rights because he

> believed that he very well deserved to profit in some way from the new inventions which he had taken so much trouble to publish in his *Transactions* to the great advantage of learning.

And he made it clear that he believed that Oldenburg needed any potential profit more than most. This declaration could hardly have been expected to pacify Hooke, as it did not, but it satisfied both Oldenburg and Brouncker. The latter thanked Huygens for his letter and as it were apologized for Hooke's behaviour, assuring Huygens that Hooke's accusations could not affect Huygens's reputation in England where it was always high.[71] Meanwhile, as if to prove the last remark, Brouncker continued to bombard Huygens via Oldenburg with requests for a new and improved watch, preferably in gold, now rendered all the more impatient because he had lent his own to the Duke of York to whom the new watch would be given. (James, Duke of York, had been High Admiral of the Navy and so head of the Navy Board, of which Brouncker was a long-standing member while Pepys, who thought little of Brouncker but much of the Duke of York as Admiral, was its Secretary). Brouncker would really have liked to order half a dozen watches in spite of the expense,[72] but to his regret Huygens never sent more than one more, this time in gold and with a minute hand like Hooke's. This watch did not arrive until December 1675 because of the difficulty of finding a trustworthy traveller by whom to transmit it; its price, which Brouncker, so Oldenburg correctly assured Huygens Brouncker would not fail to pay, was 15 louis d'or (500 livres or francs).

The chief reason for Huygens's failure to send any more watches was that in November 1675 he once more fell seriously ill.[73] His father wrote to Oldenburg in January 1675/6 telling him that his son, 'my Archimedes' as he always called him, had been 'attacked for several weeks' by an illness 'quite similar to that which he had about six years ago', that is in 1670 (see above, pp. 188–9) and was suffering from 'insomnia and other troublesome symptoms' including great weakness.[74] The illness lasted for several months and Huygens never again wrote directly to Oldenburg. Oldenburg on 7 February 1675/6 ventured to write to him to send commiserations from himself and from Boyle who, on learning that Huygens suffered from insomnia, suggested some remedies which he himself had found effective in similar circumstances, including carriage exercise. (He too had suffered from insomnia and extreme weakness after a severe illness.[75] A fortnight later[76] Oldenburg wrote a brief letter of introduction for Edward Bernard, Savilian Professor of Astronomy at Oxford and an orientalist, but Bernard never met Huygens who was presumably not well enough to receive visitors during Bernard's stay in Paris. As late as 21 April 1676 Constantijn Huygens could only report to Oldenburg[77] that his son was 'recovering his strength', sleeping better, and able to 'endure' a carriage ride 'beyond the gates of Paris' (no great distance then). But not until January 1676/7 was Huygens to be regarded as fully recovered. During all the previous ten years Oldenburg had been on increasingly friendly terms with both the younger Huygens and the elder Huygens but now, in 1677, he received no communications from either. It was a sad check to further flourishing intellectual exchange between Huygens and the Royal Society of which he had been an active Fellow; both parties had benefited greatly over the years from the exchange. It was the end also to a genuine friendship between the great Dutch scientist and the Secretary of the Royal Society.

Not until 1689 did Huygens resume relations with the Royal Society. Then, with William of Orange and his wife Mary, daughter of the Duke of York of Oldenburg's day (now the exiled James II) securely on the throne of England and many Dutchmen coming in William's wake to England, Huygens came too. He then for the first time met Newton. The two men immediately became friendly, so much so that in July 1689 Huygens accompanied Newton to Hampton Court so that Newton could present a petition (in fact unsuccessfully) to William III for securing an appointment as Provost of King's College, Cambridge.[78] No doubt Oldenburg would have been pleased by the friendship of the two men whom he had earlier helped to bring together in amiable intellectual converse, both of whom he greatly admired and benefited.

8
The encouragement of talent 1667–77

One might think that in the 1670s when Oldenburg was spending so much time and energy in organizing networks of correspondence between established scientists he would not have had any time left for any other sort of correspondence. But this was far from being the case during the last ten years of his life, for it was precisely in this very period that he found it possible to encourage a considerable number of younger men, either relative beginners in the world of international natural philosophy or living in isolation on the fringes of the European scientific scene. (It should be remembered that Newton himself appeared to belong to this class in 1670 for the important mathematical work he had already accomplished was then virtually unknown even to the Royal Society and hence hardly at all outside England. It only gradually became known after Oldenburg began to correspond with him, to encourage him, and to elicit from him accounts of his optical work for the benefit of the Royal Society and subsequently the learned world.) Oldenburg was always on the look out for new correspondents. These he encouraged to write about their work in terms suitable for presentation to meetings of the Royal Society and, if the work was sufficiently novel and interesting, to publish it in his journal for the benefit of the learned world.

Some of these correspondents involved Oldenburg in more extensive publishing work, sending manuscripts which he saw through the press, a labour which occupied him frequently in the 1670s. Some required Oldenburg to put the novice in touch with more mature natural philosophers at home or abroad, as happened with Newton, Flamsteed, and many more. Some correspondents wrote for their own self-importance, seeking contact with the now world-famous Royal Society. Then Oldenburg had to judge whether to encourage them, as he did when their interests were consonant with those of the Society and when their work seemed sufficiently reliable to merit recognition. At the same time well-established exchanges of information, in a pattern worked out during the preceding years, had to continue without cessation. All this involved Oldenburg in dealing, in several languages, sometimes requiring translation by him into English for the benefit of members of the Society, with the widest possible variety of scientific topics, often in the same letter and frequently demanding diplomatic tact. Some such correspondents were foreign, but many were English. Their scientific interests might range from mathematics (often of a minor character) through astronomy, meteorology, natural history, medicine, natural phenomena, and natural curiosities. All hoped that these would be welcomed and encouraged by the Royal Society. Obviously it is impossible here to particularize extensively or even to mention all those who wrote to Oldenburg or to whom he wrote, some hundreds in all. In what follows I have necessarily been selective. The selection has been made on the basis of my judgement of the importance or interest of the correspondence or the correspondent. (Others may disagree.) I have endeavoured to note the international as well as the local components of Oldenburg's widespread communication with natural philosophers. The international component was especially important in his eyes when it could provide material for the universal natural history of the world for which he constantly sought and with queries for which he frequently began an exchange of letters.

An important correspondence of this type opened in 1667 when Oldenburg wrote to Marcello Malpighi.[1] This, beginning like other fruitful exchanges, such as that with Hevelius, with a shot at venture was equally highly successful. Oldenburg already knew that Malpighi was an established anatomist and a distinguished medical man. Now he wrote to urge him to impart his work to the Royal Society of which, Oldenburg very reasonably believed, Malpighi must have heard. Oldenburg had already published in his *Philosophical Transactions* an account of Malpighi's microscopical study of the tongue and the brain, having taken some pains to secure a copy of the printed work.[2] Now he urged Malpighi to impart some account of his other printed work and to contribute to the universal natural history to which, he believed, Malpighi could particularly add by sending information about the mineral and natural history of Sicily, including the natural history of silkworms as bred on that island where he had been living. (He was actually by then in

Bologna, where he spent the rest of his life.) Surprisingly, the letter reached Malpighi in less than six months. It was very cordially received and Malpighi promptly[3] sent brief accounts of the work on which the leading Italian natural philosophers were then engaged. He also sent news of his own work in progress. All this was of intense interest to the English, the more so because they then found communication with Italy difficult to establish. As part of this information and, as Malpighi put it, as 'a Token of [his] release from bondage' (that is, from the constraints of medical practice) he sent a copy of his recent work on the structure of the viscera, which Oldenburg was promptly to review and later (1669) to reprint.[4] Malpighi also promised to set to work upon the natural 'history of silkworms' as Oldenburg had requested.This seemingly casual request was to produce good fruit in a manner exceedingly gratifying to Oldenburg, for what could reward his labours better than to set his correspondents to work diligently on subjects which they might not have undertaken without his urging?

Over the next year exchanges between the two men were much concerned with this work. Malpighi's manuscript of what came to be known as *De Bombyce* (on the silkworm) was sent with a dedication to the Royal Society in the form of a long letter to Oldenburg in January 1668/9.[5] When Oldenburg showed the manuscript to a meeting of the Society on 18 February he and Hooke were 'desired to peruse it' after which the Council should decide whether or not to publish it. In the event the Council did so decide on 22 February 1668/9, perhaps on Oldenburg's single recommendation. (It was three days later when Hooke 'reported, that he had perused Signor Malpighi's discourse of silk-worms and found it very curious and elaborate, well worth printing', an opinion, so the minutes read, 'seconded by Mr Oldenburg' who then read the letter of thanks which he had already prepared.[6] Perhaps, of course he had already learned of Hooke's favourable opinion.) The work was seen through the press by Oldenburg who in a preface provided an address to the reader written by him in praise of the author followed by the text of his own letter to Malpighi reporting the Royal Society's enthusiastic reception of the work and its consequent decision to elect Malpighi a Fellow. (The election was proposed to the Council by Oldenburg on 11 March and acted upon three days later.) Thus encouraged, Malpighi went on to write many of his subsequent works as letters to Oldenburg (and after his death to the Royal Society), to be seen through the press first by Oldenburg and after 1677 by subsequent Secretaries. (Seeing through the press manuscript works sent to the Society for approval was a little recognized chore undertaken voluntarily by Oldenburg and later in the century by subsequent Secretaries at the Society's request. Whether these 'publishers' received any profit from the sale of the books is not perfectly clear, but certainly they deserved to receive some reward and probably it was expected that they should do so.)

Malpighi acted at this time as a news agent for his colleagues in Bologna and possibly as a postal agent as well. Bologna was a very flourishing centre for scientific studies, the members of the university being far more active as far as natural philosophy was concerned than those of any other Italian city, in spite of the deadening effects of papal influence upon its life otherwise.[7] By this means Oldenburg and through him the Royal Society became cognizant of the work of such men as Cassini before his removal to Paris, Borelli, and Montanari, among others. Malpighi's chief importance was and remained the communication of his incredibly careful anatomical and embryological work directed towards the physiology of both animals and plants. This the Royal Society received with gratitude and praise, conveyed carefully and warmly by Oldenburg to Malpighi, who was duly grateful especially to Oldenburg for seeing all this work through the press.

Unfortunately, a potential source of conflict over English and Italian priority in work on plant anatomy arose when, in the autumn of 1671, Malpighi sent, as usual in the form of a letter to Oldenburg, what he modestly called 'a clumsy little document' ultimately published by Oldenburg (after Malpighi had supplied the necessary drawings) under the title *Anatomes plantarum idea* (1675).[8] The Royal Society was extremely approving of the work but, as Oldenburg was forced by the Society's express direction to inform Malpighi, at the same meeting at which his work was discussed Nehemiah Grew presented his own *Anatomy of Vegetables Begun*. This had been licensed by the Council in May and was to be in print by December 1675. Oldenburg told Malpighi that he would take care to send a copy of Grew's work to Bologna 'by the first convenient opportunity of a vessel sailing for Livorno, for I am confident that in your city you will easily meet with someone to interpret the English'.[9] He was soon able to add[10] that the Fellows who had read Malpighi's manuscript were 'all of the opinion that you have handled the topic you chose with the most acute judgement and extreme care, and that the essay is worthy of being published as soon as may be'. The works of Grew and of Malpighi were completely independent and different although complementary, as the Society saw and as Oldenburg was careful to tell Malpighi.

No doubt to Oldenburg's relief, after Malpighi had received Grew's book and read it he told Oldenburg that he thought well of it, as Oldenburg promised to tell Grew.[11] And since at this time Grew was equally polite about Malpighi's work, Oldenburg could write[12] happily to Malpighi that

> It is a great argument for the truth of observations when several learned and intelligent men agree upon them, as to the principal points, though separated geographically and distinct as to the manner of their research. It is a very great incentive to their pursuing their labours and researches vigorously, and thereby both building a perpetual monument of honour to their merits and affording posterity an ample guidance for the study and perfection of the sciences,

a statement reflecting his profound belief in the value of his own work in mediating between researchers in natural philosophy. Grew was also at first very carefully polite about Malpighi's work which, he said, he found very similar to his own, differing only in that Malpighi had used a microscope to detect what he himself had observed with the naked eye. But with time he came to be less generous, as Malpighi discovered when he read Grew's slightly later book, *The Comparative Anatomy of Trunks* of 1675 which Malpighi had translated into Latin in Bologna. There he found that in the dedication to Brouncker Grew's remarks about the sequence of the work of himself and Malpighi might be taken to imply that, as Malpighi put it, the latter 'had made use of several things observed by him [Grew] in his former treatises', an implication which hurt Malpighi exceedingly. This he told the English resident James Crawford, who confirmed Malpighi in his belief of Grew's meaning, all of which Crawford passed on to Oldenburg in January 1676/7.[13] Oldenburg, one imagines, must have taken steps to deal with the situation although nothing touching this survives in the correspondence, unless he felt that Crawford, who had assured Malpighi that his reputation remained as high in England as it could well be, had dealt with the matter adequately. Certainly there was no further controversy between the two men about priority in Oldenburg's lifetime and when, in 1682, Grew reviewed the past history of his botanical endeavours in the preface to his collected works, the *Anatomy of Plants*, he omitted the offensive dedication. Contact with Malpighi must have gratified Oldenburg exceedingly over the years. His tentative opening correspondence had been cordially received, and his suggestions for research and publication all bore good fruit. The correspondence produced important communications for the Society to discuss which proved worthy of publication, an opportunity which Oldenburg eagerly seized. This gave Oldenburg a personal share in the reputation accrued to Malpighi from the warm reception his books received from the world of learning. Malpighi was more than satisfied to accept a few copies and the glory of having his works printed in England and by the Royal Society, so that Oldenburg may have benefited financially. That all this was accomplished with relative ease was largely the result of assistance from English residents in Venice after 1673, first of John Dodington, secretary to the English Ambassador and then of James Crawford who had accompanied the new ambassador to Venice in 1670 in some unspecified capacity. The latter showed a decided talent both for friendship with Italian natural philosophers and for acting as postal agent for parcels of manuscript sent by Malpighi to Oldenburg.[14]

Curiously, Malpighi's rival Grew also benefited from Oldenburg's encouragement of his work without which he would not have been able to take a prominent place in its Fellowship. Grew (1641–1713),[15] according to his own later account in his *Anatomy of Plants*, had turned to botany in 1664 and four years later finished an account of his researches which he showed to his half-brother Sampson, a Fellow of the College of Physicians. Sampson Grew was impressed by what the young man had accomplished but, as he firmly told him, thought that he ought to read the work of other writers like Glisson. This Nehemiah did and went on to produce a revised version of his work in 1670, which Sampson thought so well of that he showed it to Oldenburg, obviously as a means of securing an audience for it from the Royal Society. As Sampson was never a Fellow of the Society and is not known to have belonged to any social circle intersecting with that of Oldenburg, it cannot now be said how or why he showed the work to Oldenburg; one can only surmize that Oldenburg now stood so high in the London scientific scene generally as to be an obvious point of contact with the Society for outsiders. In any case he did consult Oldenburg, who seems to have recognized the promise Grew displayed. He not unnaturally sought a second judgement, showing the manuscript to Wilkins for his opinion. Wilkins brought the work to the notice of the Society with the result already described and it was licensed by the Society and printed. (While it was in the hands of the printer in the summer of 1671 Grew dashed over to Leiden as his half-brother had done before him to secure a quick M.D. degree. Fourteen days between matriculation and the presentation of a thesis was rapid even for those days.) To complete the relations of Grew with the Royal Society and with Oldenburg at this time, suffice it to say that Grew settled in Coventry but kept in touch with the Society and Oldenburg by letter, being elected a Fellow in Novenmber 1671. The Society and Wilkins continued to encourage him; in 1672 the latter even proposed him as curator of botanical experiments for the Society, a post which he refused because his medical practice was by then very successful.[16] He only moved to London at the end of 1673 after Hooke secured him the post of deputy to Goddard as Professor of Physic at Gresham College, a post which vanished with Goddard's death in 1675. But Grew stayed in London, succeeding Oldenburg as Secretary in 1677.

Meanwhile Oldenburg tried to keep Grew and Malpighi in friendly touch but with little real success. True, when Grew published his next work, *An Idea of a Phytological History* in 1673, he was still complimentary about Malpighi, noting in the preface that from looking at Malpighi's work he 'was glad to see the Truth of [his own] Observations all along confirmed', and he also sent to Malpighi a copy of this book with a friendly covering letter (neither of which in fact ever arrived).[17] But, as already noted, he had changed his mind when he next published. This change might, one imagines, have been strengthened when in that year Oldenburg published Malpighi's *Anatomes plantarum pars prima* to which he prefixed Malpighi's 1671 sketch, sent to Oldenburg in that year, under the title assigned to it by Oldenburg of *Anatomes plantarum idea*, a title very similar to Grew's. For once in a way Oldenburg's usually sensitive tact may have deserted him, to Grew's annoyance. But possibly success in the world may have made Grew more self-assured and inclined to advance his claims to priority more strongly than he had done earlier. It was one of the difficulties of communication by correspondence, especially since transmission of Malpighi's work from Italy was necessarily slow, so that it was difficult to keep abreast of the situation.

By 1675 Grew's sensibilities might also have been irritated by a long (and essentially trivial) controversy with Lister, conducted as usual through Oldenburg. Begun in 1673, it continued even after Grew's removal to London. Here he was not the prime mover, although deeply concerned, for the dispute had been begun by Lister. Martin Lister (?1639–1712) was yet another young provincial who, patiently encouraged by Oldenburg, was to become a well-known practitioner of natural philosophy. He had travelled in France in the winter of 1665–66, meeting the eminent English naturalist John Ray with his companions Francis Willughby whom Lister had known at Cambridge and Ray's patron, Philip Skippon. At the end of 1668 Lister, then living in Yorkshire and, as he felt, cut off from the centres of scientific activity, sent to Ray a paper about the structure of snails and 'the darting of spiders', snails and spiders being his two chief interests. Ray passed this paper to Skippon who in turn handed the paper to Oldenburg, without mentioning Lister's name but suggesting that if Oldenburg thought 'fitt' he should show it to the Royal Society.[18] Oldenburg did think fit, reading the paper to the very next meeting of the Society on 18 February 1668–9. It met with approval so Oldenburg published it in the *Philosophical Transactions* during the succeeding summer, still anonymously[19] since Lister had requested that his name not appear and Skippon continued to respect his wish and did not reveal it to Oldenburg.

It was an excellent paper showing the author to be a perceptive naturalist, no credulous natural historian repeating country folks' tales as was then all too common but an author attempting exact description and beginning to search for a taxonomy. These characteristics showed more clearly in succeeding years after Lister moved to York to practise medicine in 1670. He briefly visited London in that year when, in company with Skippon, he met Oldenburg at home, still without revealing himself as the author of the interesting paper of two years earlier.[20] It was only in that summer that Willughby let slip his name in telling Oldenburg that Lister, like Willughby himself and Ray, had been making careful observations of the rise of sap ('bleeding') in trees.[21] Lister had also already compiled a catalogue of English spiders, which he had sent to Ray, and which was later published in the *Philosophical Transactions*.[22] Reading Lister's paper caused Ray to ask him whether he had observed 'the darting of spiders' (their floating through the air in autumn) which Lister had, in fact, described in an earlier paper which Ray had read, but forgotten. Lister was to be much hurt that Ray was to insist publicly that a friend of his, Dr Edward Hulse, had priority in respect to this observation over Lister, who now wrote to Oldenburg to request that, if Hulse's account was to be printed (as perhaps Ray had told him), Lister's own independent work should be acknowledged. For this, he said, he depended upon Oldenburg's 'Civilitie and prudence'.[23] Oldenburg responded favourably, three months later printing Hulse's letter together with Lister's claim for independent study of the phenomenon and probable priority.[24]

Fortunately for Oldenburg, Lister was pacified by this and, earlier, Oldenburg's soothing reception of his claim, and consequently began what was to become a steady correspondence. This included more about various interesting insects and then, urged on by Oldenburg, more about spiders and flies, all in a form suitable for printing, usually together with a covering and friendly letter.[25] Over the next few years Oldenburg was to print virtually all Lister's varied communications (on spiders, on fossils, on the rising of sap in trees, on the colours of flowers) except in those few cases when Lister requested delay so that he could engage in further researches.

And always Oldenburg responded by urging him to send more, obviously recognizing the professionalism which distinguished Lister as it did Ray. Nothing could have shown this more clearly than comparison of Lister's taxonomic accounts of spiders or his studies of the 'bleeding' of trees with the corresponding accounts sent by Israel Tonge in 1670.[26] Tonge, an Oxford graduate, friend of John Aubrey and a clergyman, was an eager reader of the *Philosophical Transactions*, but all too ready to credit the tallest of country tales. In 1670 Tonge insisted that he had been 'reliably' informed by supposed witnesses of duels between a spider and a toad, in which the toad died.[27] Willughby was much interested by this report while remaining highly sceptical.[28] Oldenburg never published Tonge's account of the duel, nor did he encourage Tonge to write any more about spiders. Tonge's account of the 'bleeding' of trees was a simple statment of their having been observed to do so, as anyone who cuts a branch or even a twig of some species of tree (like silver birch) in the late winter or early spring must have observed; there was no enquiry into circumstances or causes or conditions. Oldenburg clearly perceived the superiority of the work of Lister and Ray and gave Tonge no great encouragement to continue.

But he did encourage Lister, who was soon recognized by the members of the Royal Society as an authority on spiders, their behaviour, and their taxonomy. He worked extensively as well on fossil shells which, as he had done with spiders, he carefully classified. True, he believed them to be of purely geological origin in contrast to the then more common view that they were animal remains, since, as he accurately noted, their composition was always identical with that of the rocks in which they were found.[29] Oldenburg welcomed all Lister's communications, but delayed acknowledging those about fossils

> because I would first exhibit them to the R. Society (which I could not do sooner than yesterday, when they began their publick Assemblies again, after three months recess) who heard them read and received the contents of them with great satisfaction, which they commanded me to let you know, and to Joyne to it their hearty thanks.

He then asked Lister for more on the subject.[30] (This is very normal and characteristic example of Oldenburg at work.) Publication of Lister's accounts later stimulated others to write to Oldenburg about their views on fossils. For example John Beaumont, a Somerset surgeon, wrote describing his own large collection of fossils, accounts also published by Oldenburg.[31]

Lister's least profitable series of investigations was the only one to involve Oldenburg in prolonged although not very serious controversy. This began in 1673 when Lister sent a rather rambling letter which Oldenburg only partly printed. It was primarily concerned with a very serious subject, the existence of parasites in the gut of animals and the form of the guts.[32] Unfortunately, when the letter was read to the Society on 21 May 1673, Walter Needham, London physician and anatomist, criticised Lister's anatomy, in particular the relation of the caecum to the colon and rectum where the parasites were to be found, and wished to discuss the relation of anatomical structure to the shape of the excrements voided. This was conveyed to Lister in the acknowledgement of his letter.[33] The point of debate is of virtually no modern interest and of little real anatomical importance but it caught the fancy of various anatomists of the time, especially that of Grew. Hence controversy and discussion went on for some months in what now seems a dull exchange and which Lister in fact came to find 'very tedious'.[34] It is not possible to tell whether Oldenburg realized its triviality or not; he was merely reporting the views of two Fellows one to the other. With equal patience he handled another dispute between Lister and Grew in 1673.[35] This followed the reading by Lister of Grew's latest book, *An Idea of a Phytological History Propounded*, a study of the anatomy and physiology of 'vegetables'. Once again no very important discoveries or principles were involved, but Oldenburg was careful to relay the views of both men without comment as fully as he could, and remaining on good terms with Lister and presumably with Grew as well.

It was undoubtedly because Lister felt that he was on such good terms with Oldenburg that he was prompted to send to London the writings of his friend Francis Jessop; these were 'reflexions' on Wallis's work on tides, published as long before as 1666.[36] Jessop, who lived in Sheffield, had learned some mathematics from Ray, had been an assistant or colleague of Willughby's, and had practised chemistry with some medical friends of Ray also resident in Sheffield. In spite of his associations with Ray's circle, Jessop remained isolated from the mainstream of mathematical natural philosophy and even his knowledge of mathematics was by no means as good as he thought it was. Wallis, apprised of Jessop's comments by Oldenburg, could not resist the opportunity for instruction and entered into a two-year long exchange with Jessop, mostly indirectly. Jessop wrote to Lister who then forwarded his letters to Oldenburg with his own letters, so that Oldenburg could in turn send Jessop's letters to Wallis, who invariably replied in letters to Oldenburg, who then had to copy out the parts of Wallis's letters which related to Jessop's comments and send these to Lister for transmission to Jessop.[37] Why Wallis and Jessop did not communicate directly is puzzling; certainly Oldenburg must sometimes have wondered when the two men never seemed to tire of the indirect method of discussion which made so much work for him.

In 1673 Lister found occasion for an international controversy, this time pursued only indirectly and without greatly involving Oldenburg, although he was the recipient of Lister's complaints. This concerned the invention of a wonderful 'blood staunching' (styptic) fluid, first spoken of publicly by Jean Denis, for which many claims were vaunted and in which much hope was placed. Oldenburg first heard about it indirectly in April 1673 from Duhamel[38] and then a few days later received a full description of its supposed effects from Denis himself, a description which Oldenburg promptly published in his *Philosophical Transactions.*[39] After this he asked Denis to send him a sample, assuring him that

> If the experiment succeeds I will take care to sell a good quantity of it, without asking for a privilege and without sending a man for the purpose

evidently hoping that by this means he could 'benefit mankind' as he would have put it.[40] It is not certain that Denis ever did send Oldenburg a sample, but certainly there was one in London by June when Needham tried the fluid on a dog at the Society's meeting on 11 June, a week before Denis himself arrived in London and successfully, so it was reported, repeated the experiment at the next meeting. This obviously suggested its use on humans, particularly in warfare. Oldenburg was able to print in the October issue of the *Philosophical Transactions* a report of its successful use by naval surgeons in a naval engagement against the Dutch, probably that of 11 August 1673, acting on instructions from King Charles himself who had been much pleased by experiments made before him.[41] There were at the time other claims for the discovery of such a fluid, notably by Lister who in September told Oldenburg that he also possessed a 'blood staunching liquor' which, he insisted, was as good as that of Denis and discovered before him.[42] (No doubt many physicians and surgeons had preparations which they favoured for the purpose.) Oldenburg naturally requested a sample of Lister's fluid, offering to send him some of that of Denis in exchange, but it seemed that Lister already possessed a sample of that.[43] Lister failed to send the requested sample of his own remedy until the beginning of the new year 1673/4 and only after repeated requests by Oldenburg on behalf of the Royal Society.[44] Lister claimed that both his fluid and that of Denis were equally effective, so that there was really no ground for controversy in that respect, and their composition, possibly including nitric acid, was probably very similar. Others did not find either fluid effective, and interest in it disappeared during the course of 1674.[45]

Correspondence with Lister customarily involved Oldenburg in the discussion of an extremely wide range of subjects. In one and the same letter Oldenburg might be forced to consider mathematics, plant and animal anatomy and fossils, and to offer encouragement ranging from patient explanation to such an aspiring man as Jessop, to real, stimulating encouragement to such a clearly talented and original man as Lister was when he and Oldenburg first exchanged letters. And it must always be remembered that at the same time Oldenburg was handing and coping with controversy in optics with Newton's work, mathematics with Sluse, Huygens, Leibniz, and Newton, astronomy with Cassini, Hevelius, and Huygens, and much more. That he could keep all these topics coherent, even when several occurred in the same letter, and could deal with all of them tactfully, is some measure of his skill in handling touchy natural philosophers, as well as a tribute to his methodical handling of the vast bulk of the Royal Society's correspondence, only partially assisted by the careful indexes of names he prepared for the Letter Books at this time.

Another young provincial whom Oldenburg greatly encouraged during the 1670s was John Flamsteed (1646–1719). He was the son of a Derbyshire brewer who wanted his son to follow the family business. But the boy, educated at home, had somehow become passionately interested in astronomy by the age of sixteen and determined to devote himself to the subject in spite of the fact that early poor health seemed to militate against it. When he first made himself known to Oldenburg he was aware that he was not very cognizant of the latest astronomical advances, claiming that his poor state of health had prevented him from reading widely (although it had not, obviously, hindered him from spending his nights observing).[46] He was very confident of his skill in both observational and computational astronomy. In 1669 he compiled an almanac for 1670 calculating the date of an expected solar eclipse and of five lunar appulses (close approaches of the moon to important stars). This almanac he sent to the President of the Royal Society in a pseudonymous letter (signed J.F. and giving an anagram of his full name) in November 1669.[47] The packet containing this was naturally passed to Oldenburg who produced it at the Society's meeting of 13 January 1669/70, when the astronomers present were favourably impressed. When Oldenburg replied and conveyed[48] the Society's thanks he promised that the observations would be published quickly in the *Philosophical Transactions*, 'perhaps of this month'. As he added, it was not difficult to discover the identity of J.F. and so the tables were published with Flamsteed's full name attached.[49] Oldenburg generously offered to send Flamsteed a copy of this issue of the journal, as he did, and encouraged him to send more calculations and to supply his proper address, giving his own private address in return.

Flamsteed was so excited by this encouragement that he quickly sent off more calculations, so quickly indeed that he was forced to withdraw them before they could reach Oldenburg because he realized that they were not sufficiently complete.[50] But he promised to send more later, as indeed he did, being clearly confident that Oldenburg would continue the correspondence as, of course, he did. Although the muddle of Flamsteed's second attempt to communicate was not uncharacteristic of his methods or lack of method, Oldenburg patiently continued to encourage him, carefully printing in successive issues of the *Philosophical Transactions* the errata which Flamsteed kept finding in what he had sent and, as Flamsteed himself put it, abridging his prolixities to clarify the sense.[51] And when, three years later, Flamsteed learned that Hevelius was making comparable observations and calculations and came to feel that therefore he should present his material in Latin rather than in English, Oldenburg carefully and patiently corrected the Latin which, as Flamsteed himself realized, was as decidedly poor as his astronomy was superior, for he lacked the thorough training in Latin which a good grammar school would have given him.[52]

The stream of astronomical observations and calculations which Flamsteed produced in the next few years was mostly printed by Oldenburg in the *Philosophical Transactions*, often side by side with those of Hevelius, Cassini and others. Flamsteed soon discovered that Oldenburg could and did assist him in many other ways, for example in the acquisition of better 'glasses' (lenses) for his telescopes and in supplying information from Hevelius's printed books which Flamsteed did not possess and Oldenburg did.[53] Oldenburg continued to send him copies of the *Philosophical Transactions* as they appeared, to Flamsteed's pleasure and profit. When in June 1670 Flamsteed visited London he was delighted to meet Oldenburg in person. He was enormously grateful to him and his letters are always polite, even humble, except in the matter of his astronomical skill of which he was extremely proud, so proud as to be quite certain of the accuracy of his results. He was delighted when Oldenburg offered to communicate his predictions to foreign astronomers and also when he was sent relevant scientific news. When he heard that Newton had made a new kind of telescope he promptly turned to Oldenburg for the requisite detailed information.[54] For although he had provincial scientific friends like the Towneley family of Lancashire, long dedicated to natural philosophy and glad to encourage the young astronomer and assist him where possible, they had little connection by now with the London world of natural philosophy and now corresponded seldom with Oldenburg. Hence it was Oldenburg himself who was Flamsteed's first effective patron and Oldenburg who put him in touch with the most important active astronomers of the time.

Flamsteed's contacts through Oldenburg with G.D. Cassini (1625–1712) were of very great importance to him. Cassini had migrated from Bologna to Paris in 1669 to become the leading astronomer of the Académie royale des Sciences, Auzout having left Paris for Italy. In 1671 Cassini became the effective director of the Paris Observatoire. Oldenburg then began a steady correspondence with him, having during previous years sent messages through Malpighi and, after Cassini's arrival in Paris, through Vernon. Cassini warmly welcomed the opportunity of exchanging astronomical information with English astronomers and valued the printing of his astronomical tables and data in the *Philosophical Transactions*. With Oldenburg's encouragement he soon became a regular correspondent.[55] He learned of Flamsteed's work from its publication and by 1673 the two men were in regular touch with one another through Oldenburg (for even when they wrote directly to one another it was always by means of enclosures in letters to Oldenburg).[56] This created some problems for Oldenburg as intermediary, for the exchanges were not always entirely happy: initially Flamsteed was on the defensive, for Cassini was for some years to come to have the better telescopes and even, almost certainly, to be the better observer as regards planetary astronomy. (He was, after all, the discoverer of Saturn's first satellites and a keen and accurate observer of Jupiter and Jupiter's satellites, of which he was preparing careful ephemerides.) But he did from the first greatly respect Flamsteed's calculations of lunar motion. In turn, Flamsteed tried to recognize Cassini's merits, but he found it difficult to keep up with him and like so many other English natural philosophers resented what he and they saw as a French tendency to underrate English achievements.[57] It was Oldenburg's task to sooth and conciliate both men and to keep the interchange going as well as possible, as he did.

Flamsteed's position in the world of astronomy was in a few years to change radically as a result of his visit to London in 1670, when he met not only

Oldenburg but also Sir Jonas Moore, surveyor-general of ordnance, a friend of the Towneleys and now influential in court circles. Flamsteed made a very favourable impression, so much so that in succeeding years Moore was to join Oldenburg as his patron. Oldenburg's role was to encourage Flamsteed to send the results of his observations and calculations to the Royal Society for public appreciation, publication, and communication to foreign astronomers. Moore's was to appreciate that Flamsteed would make the perfect head of the London-based observatory which he was anxious to establish and for which he was vigorously seeking royal support.[58] Moore made sure that Flamsteed met Seth Ward, Bishop of Salisbury and Oldenburg's patron, a man of even greater influence than Moore himself in Court circles. He also introduced him to Hooke and others to ensure that he was personally known to those who mattered in astronomical affairs. Consequently when in 1675 the King became interested in the problem of longitude determination at sea (stimulated by the claims of a Frenchman to be able to do so, showing that even charlatans can have their uses) Moore was able to persuade Charles II that the real answer to the problem lay in more astronomical observations. Only these could permit the preparation of accurate tables for the use of seamen and he believed that this could best be undertaken at an official observatory. Hence the establishment of the Royal Observatory at Greenwich near the Royal palace there (now the Maritime Museum) with the director to be called the Astronomer Royal. This post was immediately given to Flamsteed. He was to have a salary which provided him with the means to leave Derby, although no instruments were provided for him, so that he was forced to continue to use those he had employed previously. Since Greenwich was then not part of London proper, being so far down river on the south bank of the Thames, Flamsteed still communicated with Oldenburg mainly by letter, only occasionally coming to meetings of the Royal Society. (He also continued to send his letters for Cassini enclosed in those to Oldenburg.) Cassini, told of Flamsteed's appointment by Oldenburg, expressed himself as being 'greatly rejoiced' and he continued to praise Flamsteed's work even when he differed from him.[59] That the correspondence between them continued for so long and was so frequent and amicable was Oldenburg's not inconsiderable achievement.

He had practised ably as patron to the young provincial and his reward was Flamsteed's success. No one else among the English natural philosophers and virtuosi rewarded him for his efforts at encouragement so greatly as did Lister and Flamsteed, but he was no less careful in his dealings with lesser men and no less zealous in encouraging them than he was with the two destined to become of lasting importance to the world of natural philosophy. That fewer of such lesser men feature in the correspondence of Oldenburg's last years may be because he then lacked an assistant to copy letters into the appropriate Letter Books and perhaps even his normally meticulous filing system for original letters suffered. But in any case there are fewer examples of unknown provincials writing in as a result of reading the *Philosophical Transactions* in 1676 and 1677 than there had been in earlier years.

It was, of course, not only in Italy and England that Oldenburg found scope for the encouragement of talent (and even in Italy there were others besides Malpighi for whom his correspondence was helpful). He had for long been in regular contact with French natural philosophers, both professionals and amateurs. True, the foundation of the Académie royale des Sciences in 1667 had lessened the need for foreign encouragement for its members received ample official recognition as well as financial support, but Oldenburg's long-term correspondents in the 1670s seemed still more than glad to be kept in touch with the Royal Society by letter. This was so even though they had some news of what the Society was doing by means of extracts from the *Philosophical Transactions* printed in the *Journal des Sçavans* with the advantages that these extracts were translated into French, for few of the French natural philosophers could read English. On the whole Oldenburg's French contacts, who were mostly in Paris, were fewer on the 1670s than they had been in the 1660s, and it was mainly long-established colleagues who wrote as occasion arose with queries or to announce new discoveries. Witness Cassini, who apparently found contact with England more rewarding than that with his French colleagues. With him Oldenburg served less to encourage than to distribute and publish the observations which Cassini was eagerly accumulating and which he was anxious to have known in England and the world.

It is worthy of note that, unusually, Jean Picard with whom Oldenburg had corresponded in the past, never wrote directly to Oldenburg about his remarkable geodetic survey described in his *Mesure de la Terre* published in Paris in 1671; Oldenburg relied for his extensive knowledge of it on his various correspondents in Paris and especially on Francis Vernon who sent him long extracts which permitted him to publish most of the book in English translation in the *Philosophical Transactions.*[60] Nor did Oldenburg have any direct contact with the Académie royale des Sciences. He had spasmodic contact only with J.B. Duhamel, titular Secretary of the Académie (who at this time seems in fact to have only rarely attended its meetings) and his correspondence with Huygens seldom touched upon its activities. Justel could supply only gossip, not altogether reliable in the case of the Paris scientific scene. In these years P.D. Huet whom Oldenburg had encouraged when he was a member of the Academy at Caen in the later 1660s was now in Paris as tutor to the Dauphin and had dropped his anatomical studies and only occasionally sought news of the English scientific scene. Oldenburg had corresponded intermittently since 1667 with the elderly Ismael Boulliaud (1605–1694) when he had secured Boulliaud's election to the Royal Society in recognition of his distinction as a mathematician and an astronomer. In the 1670s Boulliaud made use of Oldenburg as a postal agent to facilitate correspondence with Hevelius, with whom he had been in touch for many years, perhaps because the continuation of war between France and the Netherlands made direct contact more difficult. He also sent Oldenburg news of astronomical observations which Oldenburg passed on to Flamsteed, and of his plans for publication in mathematics.[61] In 1676 he suddenly demanded information about Jewish Passover ritual and textual interpretation in connection with a commentary on parts of the New Testament on which he was engaged (a reminder if it is needed of how widespread such interests still were among seventeenth-century natural philosophers).[62] His confidence that Oldenburg would certainly know to whom to turn was not misplaced: Oldenburg quickly turned to Wallis who obliged at great length, pursuing the matter as eagerly as he had such other matters as mathematics or the behaviour of tides, for he was after all a clergyman as well as Savilian Professor of Geometry.

One of the few French natural philosophers upon whose career Oldenburg had a significant influence was I.G. Pardies S.J.. He has been mentioned (Chapter 6) for his initial criticisms, always polite, of Newton's optical work, criticisms to be succeeded by a complete acceptance of Newton's doctrine of light and colours in a manner unique at the time. It had been Pardies who initiated the correspondence in 1671,[63] writing in July of that year to express his diffident admiration for the *Philosophical Transactions*, a copy of which he had been shown, correctly realizing Oldenburg's role in its production as editor. He also expressed his gratitude for Oldenburg's translation into English of a book which, published anonymously, he now acknowledged as his own. This was *Discours du mouvement local* (Paris, 1670) which Oldenburg had published as *A Discourse of Local Motion, by A.M.* [?'Anonymous Monsieur'] *Englisht out of French* (London, 1670), news of which had recently reached the author. In this book Pardies had criticized Descartes while taking favourable note of the publication in the *Philosophical Transactions* during 1669 of papers by Wallis and Wren as well as that by Huygens reprinted from the *Journal des Sçavans*.[64] No doubt the favourable attitude of Pardies to English work helps to explain why Oldenburg translated and published his own work so quickly.

Now Pardies sent an optical work just published, *Remarques sur une Lettre de M. Descartes touchant la Lumiere* and promised to send his forthcoming *Elemens de Geometrie*. Oldenburg naturally thanked him warmly for his gift,[65] (although he never reviewed the optical work in his journal) and in return sent him a copy of the English version of his work on motion together with a copy of the English edition of Leibniz's *New Physical Hypothesis* (in Latin; see below), at the same time strongly requesting a copy of Pardies's mathematical work. It was altogether a most encouraging letter as Pardies said in his reply, in which he thanked Oldenburg profusely for the two books and his letter. He declined to comment upon Leibniz's work at any length, rather quaintly remarking, 'I should not make the same difficulty with respect to the book by Mr Wallis or that by Wren if I had read them' being convinced, as he put it, of their intelligence and learning (a sufficiently slighting comment upon the somewhat juvenile performance of the young Leibniz). This letter, written in October, was accompanied by a copy of Pardies's *Elemens de Geometrie* (Paris, 1671). Sent by hand, it did not reach Oldenburg until two months later, no doubt the result of dilatoriness on the part of the person who brought it. Hence Oldenburg could only read the accompanying letter to the Society on 14 December; he was then instructed to encourage Pardies to continue his work on motion, as he did and would no doubt have done even without instruction. Pardies did not do so because his attention was now directed elsewhere, but the correspondence thus happily begun continued steadily with Pardies writing longer and longer letters, sending copies of his own books and those of others recently published in Paris (which must have pleased Oldenburg a good deal), seeking news of the work being done by Boyle and by Wallis, and soon eagerly discussing Newton's first optical paper.[68] In all this Oldenburg continued to encourage him, to make his work known in England, and to keep him informed about English achievements. It was a friendly and happy exchange which only ceased with the death of Pardies in 1673. Oldenburg must have mourned his loss, for their correspondence had an ease not altogether usual, and Pardies was clearly an attractive character as confirmed by the regrets for his death which were expressed by such disparate men as Leibniz, Duhamel, and Huygens.

So much for a hasty glance at Oldenburg's encouragement of work in Italy, England, and France. As for Germany, contacts were by no means negligible, but were largely intermittent and hence not easy to unravel. Oldenburg seems never to have maintained contact with the intellectual life of his native Bremen, although he did with that of other cities. Except for Hevelius to whom he regularly wrote, his German contacts were mainly with medical men. There were besides always German travellers who called on him when in London, wrote to him on their return home, and often put him in touch with their colleagues. There were also English travellers through Germany who, as Oldenburg himself had done on his own travels, met scientists and savants, sent back news, and made introductions. But mostly for some time these contacts were intermittent, until, through a complex series of interlocking relationships and acquaintance Oldenburg first heard from the young Leibniz in 1670.

Leibniz was then a novice mathematician aged twenty-four who had published a single work, *De arte combinatoria* four years earlier and was now earning his livelihood as councillor and legal adviser (he possessed a degree in law) to the Elector of Mainz, while aspiring to become a natural philosopher. By 1670 Leibniz had become friendly with Baron Johann Christian von Boineburg (1622–1672), a diplomat at this time in the service of the Elector. As he was a scholar, book collector, and patron of learning, he naturally knew of Oldenburg and perhaps had even met him when Oldenburg had travelled in Germany as a tutor a dozen years earlier. At any rate, in the summer of 1670 von Boineburg wrote to Oldenburg (in Latin, naturally) addressing him as both a German and as Secretary of the Royal Society encouraged, he said, by the assurance given him by an unnamed English friend that Oldenburg was a person who would willingly oblige him.[69] He had three requests to make: first, that Oldenburg would deliver a letter to Prince Rupert (then in England, but he had spent some time in Mainz in the late 1650s), sent in this way because he had never learned that earlier letters addressed otherwise had been safely delivered; second, that Oldenburg would send him news of the doings of the Royal Society; and third, that he would accept two enclosures, letters from his talented young friend Leibniz, one addressed to Oldenburg himself and one addressed to Hobbes.

Leibniz's own letter to Oldenburg[70] began with a formal, not to say flowery introduction ('Pardon the fact that I, an unknown person, write to one who is not unknown; for to what man who has heard of the Royal Society can you be unknown?') and continued by briefly stating his own interests. They were, as he put it, directed towards a universal theory of motion 'demonstrated in a geometrical way from the definitions only' (that is, in a logically rigorous way, using 'geometrical' in the sense employed by Spinoza) and were related to the work of both Huygens and Wren but encompassing far more. He also asked for news of the active scientific world of London. To this letter Oldenburg replied a month later,[71] at the same time that he assured von Boineburg that he had delivered his letter to Prince Rupert. He gave Leibniz a fairly broad summary of what the leading Fellows of the Royal Society were doing and in return asked for information about a German 'syphonic machine' for pumping water out of mines of which he had received a printed account. Leibniz ignored this request, being more anxious to send a pretty full description of his ideas about motion.[72] To this letter Oldenburg again replied warmly,[73] assuring Leibniz that his theories would be carefully and impartially considered by members of the Royal Society, even in the case of the 'perpetual motion machine' whose existence Leibniz had reported.

Rather quaintly and perhaps not fully truthfully he added

> You will readily believe, I think, that as I am a foreigner in England I shall speak out without flattering or wheedling the English. There are many men among them excelling in their experienced judgement upon questions of mathematics and mechanics. I would confidently and amicably urge you to seek their opinion of this discovery of yours, before you give it to the world and own yourself its inventor. If this advice commends itself and I could serve as a go-between in this business, I shall not refuse that responsibility and I promise all the straightforwardness proper to an honest man.

(This is particularly interesting. For one thing, it demonstrates Oldenburg's insight into men's attitudes towards their own inventions. For Leibniz had not even hinted that he personally was the inventor of whom he spoke. For another it repeats once again his earnest desire to assist the promotion of knowledge by others for the benefit and use of mankind. Finally, what is one to make of his acknowledgement of his foreign status in England at the same time as, when he wrote to von Boineburg, he referred to 'our Hobbes' as of one Englishman to another? Perhaps he was only revealing his ambiguity about his proper nationality, such as he well might feel after nearly thirty years away from his native Bremen. As much of his life shows, he mostly tended to think of himself as being on the side of the English. Hooke would have felt that for once Oldenburg was right to think of himself as a foreigner;

but there is also the probability that, writing to a German, and knowing that Leibniz knew him to be of German origin, he was exercising careful diplomacy.)

Oldenburg continued to encourage Leibniz, and when soon after this Leibniz published his *Hypothesis physica nova* (Mainz, 1671), he dedicated the first, more experimental and practical part to the Royal Society, noting that Oldenburg had encouraged him to display his conjectures about the explanation of natural phenomena for the interest of the Society's meetings.[74] Ingeniously, he dedicated the second, more theoretical part to the Académie royale des Sciences, obviously seeking its patronage as well as that of the Royal Society, but sending a copy to Oldenburg.[75] Oldenburg had both parts reprinted by the Society's printer because of the shortage of copies and because the subject seemed 'to merit consideration by scientists everywhere', as he told Pardies when he sent him a copy.[76] He published a review of the whole work in the *Philosophical Transactions* for July 1671, thus ensuring far more publicity for the work than Leibniz's private publication alone could have done and providing effectively far more patronage than Leibniz received from the Académie, although Huygens was later to take the fledgeling mathematician under his wing. The book was, in the usual way, presented to a meeting of the Society and several members (Boyle, Wallis, Wren, and Hooke) asked to give their opinions of it. Only Wallis and Hooke are known to have reported their reactions, which were not so very different from that of Pardies, already mentioned. Hooke after a rapid look merely 'said, that he was not satisfied with it', as the minutes of the Society state.[77] Wallis, kinder, wrote generally that 'I find many things expressed in it with very good reason, and to which I can fully assent since my own views are the same', but he clearly did not find Leibniz's ideas very original, while of the second part, the 'Theoria motus abstracti', he refused to give a proper opinion, only saying that he approved of much and rejected the remainder.[78] Tactfully, Oldenburg only told Leibniz what Wallis 'who seemed most of all to favour your views' had said about his points of agreement, informed him that the two parts were being reprinted in London, and promised to write more about English reactions in due course, a promise never fulfilled, and requested Leibniz to let him know what the French academicians had said about the work. During the rest of 1671 Oldenburg only answered when Leibniz wrote, but then he did so at some length with news of what the English were doing, as Leibniz had requested him to do.[79]

There seems to have been no communication at all between Leibniz and Oldenburg in 1671–2, when Leibniz was immensely busy between his revived interest in mathematics and his job at Mainz. In March of that year Leibniz left Mainz in the company of M.F. von Schönborn, the nephew of the Elector and von Boineburg's son-in-law, on a diplomatic mission to endeavour to ensure that the imminent negotiations for peace between France and the Netherlands should be held at Cologne (as in the end they were to be in 1673) and that Germany's interests should be taken into account there. Schönborn and Leibniz went first to Paris to negociate with Louis XIV where to his disappointment Leibniz found that his multiple duties and the need to learn the French language left him less time for mathematics than he would have liked. He only met Huygens six months later, but then Huygens was impressed by this self-taught mathematician, encouraged him and advised him what to read, so much so that Leibniz succeeded in doing some original work on geometrical progression by the end of the year.[80] Then, in January 1672/3, Schönborn and Leibniz travelled to London to seek the interest of Charles II in their diplomatic proposals, for the end of the war seemed further away than ever. When he arrived in London Leibniz, in spite of having neglected to write to Oldenburg for so long, promptly got in touch with him, partly because as a German, he, like von Boineburg, expected him to be particularly well disposed to a compatriot. (As noted, Germans never let Oldenburg forget that he too was a German, however closely he was otherwise associated with his adopted country in both his public and his private life.) Oldenburg responded cordially, introduced Leibniz to many of the members of the Royal Society and took him to its meeting of January 1672/3, where Leibniz produced his calculating machine (still, as he then confessed, far from perfect). This aroused interest, so much so that Sir Samuel Morland wished to show him his own for comparison.[81] At Boyle's house (to which Oldenburg presumably took him) he met Pell who discussed number theory with him, which prompted him to exhibit some further new ideas on the subject.[82]

Excited by all this, on 10 February 1672/3 Leibniz presumptuously wrote a formal letter to the Society requesting election; the request was favourably received and he was proposed by Moray, as Oldenburg told him,[83] to be elected two months later (9 April 1673), to his gratification.[84] (But Oldenburg also had to tell the young man that a formal letter of thanks to the Royal Society was required, for an informal letter such as that which he had sent to its Secretary was by no means enough).[85] About this time Leibniz must have learned something about Newton's mathematical discoveries from Oldenburg (who did not yet know a great deal about them) and from Collins and Pell, who probably did not tell him much either, for to them he still seemed a novice in mathematics. They had found him not only ignorant of what had been done in England, but, culpably, as they saw it, ignorant of much French work which had been published recently, so much so that Pell felt compelled to tell him precisely what books he must read. Leibniz then meekly did what he was told, becoming embarrassed to discover how ignorant he really was. So much was this the case that he felt obliged to write a letter of exculpation to vindicate himself from the charge of plagiarism, although he was clearly not otherwise modest about his attainments. To make matters worse, although Morland had been encouraging about his calculating machine, Hooke had rebuffed him very firmly,[86] adopting his usual tone of superiority, pointing out, correctly, that the idea of such a machine was not new, claiming, again correctly, that he had anticipated Leibniz with a similar invention of his own. He added an opinion open to dispute, that in any case mechanical calculators were inferior in their results to computation combined with tables. (Hooke here displayed to the young foreigner the very worst side of his character, the more unfairly as Leibniz's calculator was technically superior to Hooke's). On the whole in retrospect Leibniz did not feel that his London visit had been a great success, for no mathematician in London had been as kind and helpful as Huygens had been in Paris.

At the end of February 1672/3 the two diplomats left London for Paris rather hastily, after news reached them of the sudden death of the Elector of Mainz. Oldenburg and Leibniz remained on friendly terms and Oldenburg tactfully continued to ask for news of Leibniz's arithmetical machine Then the correspondence between them temporarily lapsed.[87] Leibniz broke a year-long silence, writing in the summer of 1674 to apologize for his long neglect and his delay in replying to Oldenburg's last letter, the reason for which, he said, he would reveal when the two men next met.[88] Now he could triumphantly record his successful completion of his arithmetical machine and express the hope that he could bring this second, improved machine to London one day to show it to the Royal Society. Even more triumphantly he recorded that he had 'made some discoveries in geometry' which, he said, had been praised as novel by the Parisian geometers. He briefly expounded one of them, a new method for the quadrature of a segment of a cycloid. He sent greetings to Boyle, saying explicitly how highly he esteemed him, and less enthusiastic greetings to Wallis, Haak, and Hooke. Always forgiving, and conscientious as ever, Oldenburg quickly replied. Somehow this letter failed to reach Leibniz, but nevertheless he wrote again at the beginning of October, not sending the letter by post, which perhaps he mistrusted now, but by hand, not the quickest way but the cheapest and perhaps the surest.[90]

Leibniz was still excited by his own mathematical discoveries. Later it was rightly claimed by Newton and his allies that Leibniz had been anticipated by various British mathematicians (as indeed Oldenburg noted at this time on Leibniz's letter and told him in reply, presumably deriving his information from Collins). Yet equally Leibniz's claim to independent discovery was also valid, for he knew virtually nothing of what the British had done before him. Once again he sent greetings to Wallis and to Hooke. Oldenburg's reply, while conveying the unwelcome news that English opinion still held that Leibniz had been forestalled in most of his work, particularly by Gregory and Newton, was friendly and on the whole encouraging. It should be remembered that although Leibniz could not have known any more of Newton's work than what Collins had told him, he could have read some of Gregory's which was published as far before as 1667. And that Huygens, who praised Leibniz as an innovator, also knew nothing of Newton's work and thought little of Gregory's. Perhaps a little abashed, Leibniz did not write again until he had a non-mathematical discovery to report. This was his latest invention of a design for regulating a watch movement by springs, a better account of which he afterwards drew up for publication in the *Journal des Sçavans*, later published in the *Philosophical Transactions*.[91] (The invention was ingenious but not very practical and by no means an anticipation of what Huygens was to invent later that year.) In mathematical matters Leibniz denied that what Gregory had published had anticipated his own work, politely and fairly asking for information about what Newton had done. He concluded his letter by speaking of his displeasure that Hooke, in describing his 'modification' (really an improvement) of Thevenot's level in his 'Animadversions on Hevelius' (a Cutlerian Lecture) had not mentioned Thevenot's name. Evidently Leibniz was as prejudiced against Hooke as Hooke was against him.

This letter of Leibniz initiated the difficult mathematical correspondence already outlined (Chapter 6) between, in effect, himself and Collins and later himself and Newton, sent through Oldenburg as an intermediary. Where Collins alone was concerned, he sent Oldenburg long and discursive expositions in English which Oldenburg had to sort out, translate into Latin, and then copy out in legible form, no mean task.[92] This he accomplished surprisingly well, in spite of the fact that it occupied much time, for no amanuensis could be trusted to cope with the mathematical expressions involved. That he occasionally made copying slips is hardly surprising and they were few, but when they occurred it did make comprehension all the more difficult for Leibniz. Oldenburg certainly never complained at having to deal with all this novel and difficult mathematics. It is noticeable that the young German mathematician Tschirnhaus (see above, Chapter 6) seems to have regarded Oldenburg as being well versed and competent in mathematics, having many conversations with him before Oldenburg introduced him to Wallis and then to Collins.[93] From this point on the correspondence between Leibniz and Oldenburg became almost exclusively mathematical. Oldenburg's share was largely the translation and copying of Collins's missives for transmission to Leibniz and either seeing to it that copies of Leibniz's letters reached Collins for him to copy out the mathematical portions to send to Newton, Wallis and others, or managing direct exchanges between Newton and Leibniz. It was all a massive amount of work when there was little let-up in other correspondence. He had certainly and successfully encouraged the young Leibniz to the best of his ability; his reward was much hard work and the knowledge that he had made possible the exchange of information among important contemporary mathematicians and stimulated the release of much new mathematical to the mathematical world.

Leibniz was certainly the most talented of the many Germans with whom Oldenburg had contact even among mathematicians, and the correspondence with him the most influential. Yet in the 1670s there were many, mostly spasmodic contacts with, for example, German astronomers besides Hevelius, with whom correspondence never ceased; there were also contacts with various medical men, mostly members of the Academia Naturae Curiosorum who wrote in the *Miscellanea Curiosa*, an almost purely medical journal at this time, often sending their articles to Oldenburg along with medical news. Vogel in Hamburg (see Chapter 3) continued until his death in 1675 to write on a wide variety of topics; of particular importance was his attempt to understand the nature of the drugs, especially narcotics, used in Islamic medicine.[94] He was trying to write an entire commentary on one of these, the Turkish *maslac* which had long intrigued and puzzled European botanists. To assist him in understanding its nature he turned to linguistic analysis and sought Oldenburg's help, knowing that he was in close touch with such distinguished Oriental scholars as Edward Pocock and William Seaman, as well as with English agents in the Near East like Paul Rycaut, Consul at Smyrna.[95] Through this correspondence Vogel extended his understanding of linguistics and it is more than probable that the English pharmacopoeia was enlarged by the introduction of some of the drugs he studied, a curious by-product of Oldenburg's extensive correspondence.

Of considerable interest to the Royal Society and, apparently, especially to

Oldenburg himself, was a revival of interest in the mid-1670s in the study of phosphorescent substances. This was to involve Oldenburg in a complex exchange of letters with German physicians and chemists and place demands upon his skill in persuading men to reveal their secrets. Ten years earlier in the later 1660s the Royal Society had been interested in the properties of the Bononian or Bolognian stone (mainly barium sulphate) which would shine in the dark after it had been exposed to light.[96] Now it was German alchemists who claimed to have discovered several different 'self-shining substances' or 'phosphoruses' or 'noctilucas' (substances shining in the night, without exposure to sunlight, as these materials appeared to do). In 1676 Oldenburg acquired a correspondent from Dresden, C.D. Findekeller, a curious Saxon diplomat in Paris who on his return to Dresden was to become a French spy and correspond with Justel. Findekeller wrote about an alchemist named C.A. Baldwin or Balduin (1632–82) who was anxious to be elected to the Royal Society on the strength of his alleged discovery of a new 'Hermetic Phosphorus', being already a member of the Academia Naturae Curiosorum.[97] Oldenburg probably already knew of this phosphorus since it had been described in a book by another of Oldenburg's correspondents, Christian Mentzel, chief physician to the Elector of Brandenburg and, like Balduin, a member of the Academia.[98] Mentzel's book, *Lapis Bononiensis in obscuro lucens, collatus cum Phosphoro Hermetico clariss. Christiani Adolphi Balduini* (1675), as its title indicates compared the action of the Bononian stone with Balduin's phosphorus, regarding the latter as superior. Now seemingly Balduin wished to bargain with the Royal Society, offering the 'secret' of his phosphorus in return for election as a Fellow. Oldenburg professed to be shocked at this naked proposal of possible purchase of the Fellowship,[99] telling Findekeller stiffly,

> it is not necessary to chaffer with the Royal Society; it is sufficiently led of its own accord to admit men of merit to its membership. Let him only send me his Hermetic Phosphorus and leave the rest to me, to whom he will be grateful, if he trusts himself to me.

(In what sense Oldenburg's proposal is not 'chaffering' it is difficult to see; it should be remembered that Oldenburg was now well used to acting as patron to foreigners who sought election, as so many did, as well as judging their potential merits.) Balduin did send a specimen of his phosphorus to Oldenburg in September 1676[100] enclosed in 'a silver-gilt box' which was, he said, 'to be presented to the King as founder of the Society and also to its very illustrious President and to the eminent Council and Fellows'. It is not at all obvious how the box and its contents could be 'presented' to all these persons and, in fact, Oldenburg merely produced it at the meeting of the Society on 4 January 1676/7. A trial was made of it both then and a week later. Balduin's phosphorus was calcium nitrate prepared by dissolving chalk in nitric acid, evaporating the solution to dryness, leaving the residue to deliquesce, and then evaporating the resultant fluid again and heating the second residue, after which the deposit left would shine in the dark after exposure to light like the Bononian stone. The Fellows present at the meetings at which its activity was tested seem to have been reasonably content with its performance. The exception was Hooke who wrote in his diary that 'Oldenburg produced the *Phosphorus Baldwini*. Twas affirmd it shined best. I could not perceive it' presumably meaning that he could not see that it shone better than the by now well-known Italian stone. In this he may well have been correct, for both required previous exposure to light.[101] The comparison was difficult to quantify.

There was already in existence a superior substance which shone without the benefit of exposure to light, called by Findekeller 'a true phosphorus', a denomination with which modern chemists would agree.[102] There were many claimants for its discovery as Oldenburg soon learned, all of them German physicians or alchemists.

Its true discoverer is now thought to have been most probably Hennig Brand, an alchemist-physician of Hamburg. Certainly in 1675 Johann Kunckel, at that time chemist and alchemist and gentleman of the bedchamber to the Elector of Saxony, later a respected chemist, was given a hint as to its method of preparation by Brand from which he was able to prepare a sample. Of this he boasted to Findekeller and others, claiming that he himself was the true discoverer. It was however not from Findekeller that Oldenburg first heard of the new substance, but through his old friend and correspondent J.D. Major. In the autumn of 1676 Major wrote to Oldenburg after the lapse of, as he put it, 'ten years and more' enclosing a little book by a friend of his, G.C. Kirchmeyer.[103] Kirchmeyer was Professor of Rhetoric at the University of Wittenberg and a member of the Academia Naturae Curiosorum. He was also a chemist especially interested in metals and probably Kunckel's teacher of chemistry. Either because of shyness since he was unknown in England or aware of postal difficulties, he had not sent his book direct to Oldenburg but had asked Major to send it on his behalf, apparently aware that Major and Oldenburg had corresponded in the past. The book arrived safely at the end of 1676, so that Oldenburg could read

Kirchmeyer's description of his new substance and its 'marvellous effects' at the meeting of the Royal Society on 1 February 1676/7 as he told both Major and Kirchmeyer in thanking them.[104] These 'marvellous effects' as Oldenburg told Malpighi[105] were that it not only shone in the dark without previous exposure to light but also that 'when the face is annointed with it, [this] causes the face to shine'. (A singularly dangerous experiment since true phosphorus is highly poisonous and corrosive, although of course no one could know that as yet.) Moreover, the substance 'from time to time bursts into flame and nourishes an inextinguishable fire, even under water', (White phosphorus such as this substance must have been will indeed burst into flame when exposed to air, but can be kept safely under water, with which it does not react. Perhaps Kirchmeyer and Kunckel really meant that even after a sample had been kept under water it would ignite spontaneously when exposed to the air.) It all sounded almost incredible as indeed Oldenburg told Leibniz when asking him to send any information he might have about this astounding substance.[106] Meanwhile, he was earnestly begging both Kirchmeyer and Kunckel to send a sample of this marvellous substance to England so that the Royal Society could test it for itself and assess its properties.

It was long in coming. When Kirchmeyer replied it was only to send a fuller account read to the Royal Society in mid-April (perhaps on the 19th). There was no promise forthcoming that a sample large enough for experiments would really be sent and most of the letter was devoted to a defence of Kunckel's claim to be the true discoverer of this exciting substance in the face of counter-claims by J.S. Elsholtz that J.D. Kraft or Krafft (1624–97), a physician, had purchased the secret of its preparation from Brand and communicated it to Kunckel. (This seems in fact to have been the case, as far as can now be determined.)[107] The book by Elsholtz, *De Phosphoris quatuor observatio* (Berlin, 1676) amply demonstrates the excitement with which contemporaries learned of the existence of phosphorescent substances. The first of his four phosphoruses was the Bononian or Bolognian stone; the second Balduin's phosphorus; the third an alchemical-sounding 'Phosphorus Smaragdinus' (emerald phosphorus) which was said to shine when heated on a silver or copper plate and of which little was later heard; and finally 'Phosphorus Fulgurans', Kirchmeyer's 'noctiluca constans', the modern element. Leibniz, as he later told Oldenburg, met Kraft in the spring of 1677 on his way to Holland and eventually England, who assured him that the printed descriptions of the new substance were correct.[108] Kraft naturally confirmed Elsholtz's account of the discovery but without mentioning Brand's name, insisting that Kunckel had learned the secret from him (Kraft) and was not, as he claimed, in any sense the discoverer. Kirchmeyer continued to press the claim of Kunckel, at the same time trying to persuade Oldenburg to promise to petition the King to give him (Kirchmeyer) a reward for his own discovery of 'pure Artificial Gold' as well as the discovery of phosphorus. It is by no means clear why he thought that Oldenburg would have believed him to be the discoverer of the latter.[109] Nor is it at all clear why Oldenburg should have been so credulous (or so confused by claim and counter-claim) as to believe him or to have been prepared to back Kirchmeyer's claim, as he was. There still exists Oldenburg's draft of a petition to the King on Kirchmeyer's behalf, though it was never presented.

There was no further news from either Kirchmeyer or Kunckel, but Kraft arrived in England in September 1677, too late to meet Oldenburg who had told Leibniz regretfully in July 'None of those who claim to possess phosphorus as has yet landed here'[110] much as he wished it. He might even be said to have had his own claim, although it was not one which he ever thought of pressing, for it was really he who was responsible for introducing phosphorus into England. It was after all through his network of correspondence that the Royal Society became aware of all these new and exciting discoveries. It was very probably through his urging that 'a certain German' whose name was never revealed had arrived in England by late April selling some kind of (unspecified) phosphorus. And he could certainly have claimed some responsibility for Kraft's decision finally to come to England in September, when he showed a sample of his phosphorus to Boyle. He gave the slightest of hints to Boyle of its origin, hints sufficient for that acute chemist to prepare a small quantity for himself. With this he was able to perform a notable chemical and physical investigation into its nature retailed in two books, *The Aerial Noctiluca* (1680) and *The Icy Noctiluca* (1681–2), after which any competent chemist could prepare his own samples of true phosphorus. Thus not only did Oldenburg play an important role in bringing this new chemical to England but, not for the first time, although now posthumously, he assisted in fostering the natural philosophy of his patron Boyle as he had long done for him and for so many younger men.[111]

As with Germany, so with the Netherlands, for Oldenburg never relinquished the links which he had forged when a young man and whose language he had mastered. He had made many friends there in the years of his travels and had kept in touch with many theologians, newswriters, natural philosophers, and physicians. He had been in correspondence with the Delft anatomist Regnier De Graaf since 1668, encouraging him to send news of his own anatomical discoveries (see above, p.149) and those of Swammerdam on insect anatomy. Paisen and Vogel had also sent news when travelling in the Netherlands, all of which Oldenburg reported widely, both in England and to Italy.[112] Over the years the Royal Society became very familiar with Swammerdam's work through his books as well as from De Graaf's reports: his thesis on respiration was reviewed in the *Philosophical Transactions* in 1667 and his *Historia insectorum generalis* (1669) was reviewed there in 1670. Direct correspondence between him and Oldenburg only began in 1672, to continue into 1677. The indirect and involved route which initiated it is not easy to unravel but worth exploring because it demonstrates yet again the tenacity which Oldenburg displayed in controlling the complex network of relationships in which he was so frequently involved.

Here the unlikely intermediary was Nicolaas Witsen, later burgomaster of Amsterdam, and in 1671 the author of a beautiful book on 'Naval Architecture and Conduct', as Oldenburg translated the Dutch title for a long review in the *Philosophical Transactions* for November 1671. Witsen and Oldenburg may have known one another personally. Witsen had contacts with England going back to the 1650s when he had visited the country. He had subsequently travelled extensively, visiting Russia, Switzerland, France, and Italy, and had revisited England in 1668. Moreover, he had studied law under Oldenburg's nephew Coccejus. His first surviving letter to Oldenburg dates from the end of 1671 and is in Latin, but it suggests previous correspondence; its relevance here is that in it Witsen promised to send some of Swammerdam's entomological drawings.[113] In March 1672, encouraged by Oldenburg's reponse to some unspecified 'business' brought to his attention by Witsen, Swammerdam sent a printed sheet dedicated to the Royal Society and containing a brief account of the structure of the uterus and ovaries, with plates.[114] In June he sent a copy of his important work, *Miraculum naturae*, just published;[115] and in December he sent some specimens injected and prepared according to his own method.[116] When he first wrote directly to Oldenburg (on 26 March 1672)[117] it was in French (he had lived and worked in both Saumur and Paris during the 1660s). He said that because Oldenburg had so carefully attended to Witsen's affairs he was emboldened to write to him directly to ask him to present to the Royal Society a printed sheet which he had ventured to dedicate to it. He also asked that Oldenburg inform him how best to send more anatomical specimens if they would be welcome. Oldenburg replied in Latin, telling Swammerdam how pleased the Society had been to receive the sheet dedicated to it and also that he had been 'instructed' to send the Society's thanks to the author and to inform him of the great pleasure it would give the members to receive Swammerdam's specimens. Oldenburg further gave directions for the dispatch of such material, suggesting that it might best and most safely be sent with what De Graaf was proposing to dispatch 'entrusted to a certain Ostend merchant to be brought to London in some ship from Flanders', other methods being impossible on account of 'these unfortunate times', that is, the situation created by the continuation of the war between Holland, England and France.[118] Swammerdam must have followed these instructions, for Oldenburg was able to present his 'handsome gift' to the Society on 12 June 1672 as he reported to its donor the very next day.[119] Encouraged by Oldenburg, the correspondence continued through 1674, only to cease when Swammerdam became more and more deeply immersed in religious activities, although, in response to Oldenburg's repeated queries, he did write a brief letter on 31 August 1677.[120]

This war between England and the Netherlands was really chiefly a Franco-Dutch war in which the English took part only as a result of the secret Treaty of Dover between Charles II and Louis XIV. This guaranteed the English King an annual subsidy in return for support to help further French ambitions in the Low Countries. It had not much hindered correspondence between English and Dutch natural philosophers between 1672 and early 1673/4 (when the Treaty of Westminster between England and the Netherlands was signed at Cologne). Both De Graaf and Swammerdam found it relatively easy to correspond with London, sending their letters by ordinary post. The difficulties which Swammerdam experienced in sending specimens were quite normal for parcels even in peace time. One of the few references by either Oldenburg or Swammerdam to possible effects of the war on communications between them is to be found in Oldenburg's acknowledgement of Swammerdam's 1672 gift. There he wrote that the Royal Society was particularly grateful because Swammerdam had ignored the hostilities between the two countries,[121] adding,

> the Royal Society . . . thinks the more highly and warmly of you because you so eagerly maintain a scientific correspondence and indeed do not hesitate to furnish us with a very welcome present, in this very difficult period when the weight of war lies heavy on both our peoples and tends to bring about a baleful separation between the minds of Englishmen and Dutchmen. It is indeed proper that honest

> and true philosophers should, while the princes of the world contend fiercely over questions of *mine* and *thine*, persist in the peaceful search into nature's secrets and in advancing with utmost zeal the limits of truth and knowledge.

But things did not always go so easily, either as regards the post or as regards the psychological state of mind of those faced with the stresses and strains of war. True, De Graaf writing in the summer of 1672[122] told Oldenburg that he had received his letter written three months earlier together with the *Philosophical Transactions* 'not by the hand of some friend of yours but by the post' although Oldenburg had written that the letter would go by hand. (One might guess that the traveller to whom it had been entrusted had proved dilatory or faithless or lazy and had put the letter and its accompanying parcel in the post on arrival in the Netherlands.) But on the other hand De Graaf apologized explicitly for his delay in replying because 'the disaster falling upon the whole of my country stifled all desire for correspondence'. And well he might say so, the disaster being no less than the rapid French conquest of the eastern half of the province of South Holland, including Utrecht, Amsterdam, and The Hague. Delft only escaped the French by the drastic expedient of opening the sluices and flooding so much of the countryside that the French army could not advance. Further, the result of the invasion was a revolution in the Netherlands, only the week before De Graaf finally wrote—no wonder then that he apologized for his 'disordered and hurried handwriting.'[123] Clearly it was the Franco-Dutch side of the war rather than the Anglo-Dutch side that created the real disturbance, especially in the later 1670s when communications between England and the Netherlands were often very difficult.

Writing in 1675,[123] by which time the English and the Dutch had arranged a mutual peace to which the French had not subscribed, Sluse in Liège mourned that he could not concentrate on mathematics in 'these difficult circumstances' when 'the times are such that it is scarcely possible to cast our eyes back upon the delights of our studies'. He wondered how the diploma signifying his election as a Fellow of the Royal Society could possibly be sent to him, as it could not, of course, go by post. (He had been elected in 1674 and Oldenburg had intended to send the diploma in a diplomatic bag via Williamson who was negociating the Anglo-Dutch treaty, but when this plan fell through Oldenburg found it difficult to learn of a traveller who could and would carry it; it finally reached Sluse in April 1676.)[124] Though the ordinary post continued many letters got lost *en route* even in the early 1670s, as references in the correspondence of De Graaf with Oldenburg demonstrate. Once the Anglo-Dutch peace was negotiated in 1673 correspondence between England and the Netherlands had become easier. Oldenburg thought it profitable to write an encouraging letter to De Graaf who replied cheerfully in mid-April 1673,[125] declaring that 'humane and philosophical studies are not yet banished from this place by the din of war'. To prove it he sent 'a certain little tract dedicated to the Royal Society' (his *Partium genitalium defensio*) which arrived safely in time for the accompanying letter to be read to the Society's meeting on 7 May 1673. This was gratefully acknowledged by Oldenburg a week later in what was to be his last letter to De Graaf, who died on 7 August 1673, aged only thirty-two and deep in some of his best work.[126] He had benefited the Royal Society greatly by sending results of his work for its members to study. Without the correspondence with him initiated by Oldenburg, this work might have been unknown for some years to come.

De Graaf also left another legacy to the Royal Society for in his last letter of 18 April 1673 he enclosed a paper of 'observations' which, as he put it. 'revealed what a very ingenious person named Leeuwenhoeck had achieved by means of microscopes which far excel those we have seen hitherto'. (Although he is now usually spelled 'Leeuwenhoek', De Graaf's spelling was a common one, while the 'van' was only added in 1680, as he became an important figure).[127] His observations on mould, the sting of a bee, its head and eye, and a louse (that favourite creature of seventeenth-century microscopy) are the first which the world outside his immediate circle had learned of Leeuwenhoek's remarkable work and of his equally remarkable and novel simple microscopes. (A Leeuwenhoek microscope consists of a metal holder for the lens about two inches long fitted with a device with a screw adjustment for holding the specimen, the lens being a tiny glass bead ground by hand.) Although Leeuwenhoek knew no language except Dutch he was far from being an uneducated workman. He had been a linen draper and was subsequently a civil servant who was valued in the community for his skill in mathematics. He spent most of his life in Delft, although he had travelled a little, curiously having, among other places, briefly visited the Kent coast.

Leeuwenhoek's 'Observations', presumably in Dutch like almost all his reports (the original manuscript no longer exists) were 'produced' at the meeting of the Royal Society on 7 May 1673 together with De Graaf's book. They were then translated into English by Oldenburg and printed, probably substantially in full, in the *Philosophical Transactions* for that month, showing the recognition accorded to them as important and new. Encouraged by Oldenburg's assurance to De Graaf that the Society greatly welcomed these 'observations', and instructed by De Graaf as to how to address Oldenburg, and probably also encouraged by the latter's obvious familiarity with the Dutch language, Leeuwenhoek plucked up up his courage and sent drawings of the observations he had sent previously as well as some new observations. Most of these, after editing, Oldenburg printed in the *Philosophical Transactions* for April 1674.[128] The correspondence between Oldenburg and Leeuwenhoek continued briskly thereafter. (Even after Oldenburg's death Leeuwenhoek continued to send his work to the Royal Society, now frequently having it translated into Latin in Delft, but he was never on cordial terms with successive Secretaries of the Society.) The Royal Society always valued Leeuwenhoek's work and urged its Secretary to encourage the Dutch microscopist even when, as sometimes happened, some members doubted Leeuwenhoek's observations or, more usually, the conclusions he drew from them.

It all caused a good deal of work for Oldenburg. After 'producing' the letters at meetings, which meant summarizing their contents, he had to translate these fairly lengthy accounts, usually to read at subsequent meetings, to edit them where necessary for publication and, to see to the reproduction of the accompanying drawings. Then he had to inform Leeuwenhoek of what had been done. All this he dutifully undertook without complaint. His replies to Leeuwenhoek, with thanks and with the comments and queries generated by his reading the papers to meetings, were usually prompt and full. He even remembered to tell Constantijn Huygens, who took a benevolent interest in Leeuwenhoek's work and with whom Oldenburg was in continuous correspondence at this time about the favourable reception of all of Leeuwenhoek's letters.[129] Without the Society's warm welcome and tacit endorsement of his discoveries and without Oldenburg's careful, kind, and encouraging reports of this welcome, together with his pretty full printing of the letters in English in the *Philosophical Transactions*, Leeuwenhoek would never have continued the stream of long letters in which, during the 1670s, he reported his increasingly remarkable discoveries so that they could be made known to the learned world of England and the Continent. Nothing else describing them was printed before 1684 (and then only in Dutch, a Latin version appearing only in 1715). Leeuwenhoek never wrote anything except in the form of letters, and of these nearly two hundred (some two dozen addressed to Oldenburg) which he wrote to the Royal Society were among the most important. There he announced his discovery of 'little animals' (micro-organisms) in rain-water in 1674, which he sent to the Royal Society in 1676; there he spoke of having differentiated these into various forms; there he described the discovery of 'corpuscles' in the blood (1674) and much, much more. That Oldenburg's replies to Leeuwenhoek do not survive is a pity, but their thoughtfully welcoming tone can easily be surmised from Leeuwenhoek's grateful acknowledgements and from his life-long dependence on the Royal Society for the publication of his results. Oldenburg must have reported the occasional adverse comments and even scepticism of some Fellows with his usual tact, for Leeuwenhoek was never put off. He continued his standard form of publication by letter (as did, it must be remembered, Malpighi) for the remainder of his career.

Oldenburg's linguistic skill had never stood him in better stead for pursuing with his accustomed passion the purveying of scientific information to the learned world from workers everywhere. Nothing ever came of his cherished ambition for the compiling of a universal natural history but universal communication and encouragement he did amply achieve.

PART III

Personal relations

9

From friends to enemies: Hooke and Oldenburg 1662–77

Hooke and Oldenburg had been friendly associates since Hooke's appointment by the Royal Society as its Curator of Experiments in 1662, some six months after Oldenburg's appointment as Secretary. They met on amiable terms at the weekly meetings of the Society. They met frequently at Lady Ranelagh's house when Boyle was in residence there with his sister. They met more or less weekly at Coffee Houses. They met casually at various places in London, including the shop of Martyn, official printer to the Royal Society and from 1665 printer of the *Philosophical Transactions*. And for many years they planned and managed the affairs of the Royal Society together. There seems indeed to have been remarkably little friction between these two paid officers of the Royal Society. On the contrary, Oldenburg often acted as a buffer between Hooke and some of the more domineering Fellows like Moray. That Hooke in those days thoroughly appreciated Oldenburg's activities is amply indicated by his remark to Boyle in 1667, after Oldenburg had been released from the Tower[1] 'He will not, I hope, leave off his Philosophic Intelligence'. All this was only to change in 1675 when Oldenburg defended Huygens as the true inventor of the spring-balance watch, priority in which Hooke claimed for himself. (See Chapter 7.) As a result, Hooke became Oldenburg's irreconcileable enemy, pursuing his quarrel with great vehemence even after Oldenburg's death in 1677.

The relations between the two men in the years 1662 to 1672 can be learned indirectly and from their correspondence with Boyle. From 1672 onwards we know almost too much of Hooke's feelings towards Oldenburg as well as towards his other acquaintances and friends through the medium of his cryptic but revealing diary. They must first have met in Oxford in 1656, soon after Oldenburg entered the service of the Boyle family. At that time the young

Hooke (16 years younger than Oldenburg) was assistant to Willis, working on chemistry and anatomy; by the time that Oldenburg left Oxford in 1657 Hooke had become Boyle's assistant, working on the construction of Boyle's airpump and soon helping to devise and perform experiments with it. Oldenburg presumably heard something of Hooke's activities during his own years of Continental travel, although Hooke's name does not occur in the surviving letters. Certainly on Oldenburg's return to London in 1660 he must have learned more about Hooke's work both from reading Boyle's *New Experiments Physico-Mechanicall touching the Spring of the Air, and its Effects* published that year, which mentions Hooke's role as assistant and in conversations with Boyle. In 1661 Hooke published his first book, small but with a very long title, on capillary action, derived, as its title states, from his work with Boyle. (See above, Chapter 3.) In November 1662 Hooke was appointed Curator of Experiments of the Royal Society. The two men thus became what might be termed the working members of the Society's administration, to be treated as equals, the only difference being that Hooke received a salary and Oldenburg, as yet, did not.

By November 1662 Oldenburg was securely established in the Royal Society as an early Fellow (elected in January 1660–1), Secretary of the Society (nominated in the First Charter of July 1662, when there was as yet no Provision for a Curator of Experiments), and as a member of Council. He was thus senior to Hooke in standing as a Fellow, Hooke being elected in June 1663, The two men in spite of these differences were to work harmoniously and closely together for the next ten years in the service of both the Society and of Boyle. This is amply illustrated in the regular letters from both men to their patron, then resident in Oxford, letters in which they sent news of the Society's meetings and of matters of general interest, each clearly well aware of the activities of the other. For example, on 2 July 1663 Oldenburg wrote[2] 'What was done yesterday at our meeting, I suppose Mr Hook giveth you an account of by this same carrier' while, writing on the next day, Hooke's parallel remark[3] was 'I question not, but that Mr Oldenburg has acquainted you with the news, that is extant, in this enclosed' which explains why Oldenburg knew that the two letters would travel to Oxford together, as was often the case. In 1664 Hooke frequently sent Oldenburg's 'service' to Boyle,[4] showing once again that they regularly met and discussed their postal arrangements. In the summer of that year Hooke enclosed in one of his own letters one from Oldenburg to Boyle, adding, after a brief account of the Society's meeting on the previous day, that he would not summarize a letter there read because 'I suppose Mr O. has given you a copy', once again showing the closeness with which the two men worked in serving Boyle.[5]

During the summer and autumn of 1664 Oldenburg was correcting the proofs of Boyle's *New Experiments and Observations touching Cold* (published in 1665, Oldenburg being named as the 'publisher') at the same time beginning a Latin translation of it. In November,[6] Hooke had 'procured out of Mr Oldenburg's hands some of the first sheets' so that he could draw the necessary figures of the apparatus there described, yet another instance of collaboration in Boyle's service, which continued in this year to be so close that soon Oldenburg could assure Boyle that Hooke was about to answer a query sent by Boyle to one of them.[7] Another example of collaboration occurred in the winter of 1664–5, when Hooke and Auzout were engaged in a controversy with Oldenburg acting as intermediary by means of his established correspondence. (See Chapter 5.) This, it may be recalled, involved Oldenburg in translating Auzout's French letters into English for Hooke's benefit, throwing in suggestions for strong arguments for Hooke to utilise, and then translating Hooke's replies into French. From what he said, Hooke was perfectly content with Oldenburg's handling of the somewhat long-winded exchange, and realized that Oldenburg was entirely on his side. For example on receiving the translation of Auzout's first letter he wrote[8]

> Together with my most hearty thanks for the favour you were pleased to do me, . . . I thought my self obliged, both for your satisfaction, and my own Vindication, to return you my present thoughts upon those Objections.

No one could have been more amenable to Oldenburg's role as mediator, and when the exchange died down in the first half of 1666, Oldenburg congratulated both men upon the equanimity and politeness with which the controversy had been conducted. Although he naturally did not say so, this was partly because of his tactful handling of the matter.

As already noted (Chapter 4), the outbreak of plague in London in 1665 broke up the Royal Society's meetings, sending Hooke to the country near Epsom with Petty and Wilkins, Moray with the Court to, eventually, Oxford where he joined Wallis, Boyle, and other Fellows resident there, while Oldenburg and Brouncker stayed in London. Moray, who had by now a feeling of genuine friendship for Oldenburg, was positively indiscreet in his frequent injunctions to Oldenburg to urge Hooke to complete at least some of the many scientific and technical schemes he had on hand. Oldenburg constantly defended Hooke as best he could by insisting that Hooke was 'far from idle'. This Moray only partially accepted,[9] and he continued to be highly critical of Hooke's reluctance to complete and make public his many inventions. As he wrote,[10] he thought that

> Hook concealed his invention about Watches too long: pray tell him not to do so with what other things hee hath of that kind. hee hath seen the folly & inconvenience of it.

Moray here referred to the work of a Frenchman, one De Son (or D'Essons) who was in London at this time working on spring balances for watches of very much the same sort as Hooke had earlier suggested and which, so Oldenburg told Boyle,[11] Hooke thought had been 'taken from him'. Here there is no indication that he held Oldenburg responsible for this as indeed he was not, and if anyone was it must have been Moray.[12] Curiously, Hooke did not complain directly about Moray who seems in some sense to have been De Son's patron. Fortunately for good relations, nothing seems to have come of De Son's invention.

Moray does seem to have himself told Hooke that he had communicated to Huygens news of Hooke's ideas for the regulation of watches, defending himself to Oldenburg by saying that Hooke should long ago have made his ideas public.[12]What Oldenburg in told Huygens was[13]

> that from Mr Hooke's lips I have just learned that he is convinced that the air will have less effect upon watches of his type, then on pendulum clocks; and that he can so dispose their movement that it will remain always constant, no matter what position the watch is in; moreover, that he can drive the said movement with as little force as he pleases, and that when a greater force is needed he can apply as much as is necessary

adding sceptically, 'Yet I doubt this very much.' (This can surely hardly be regarded as a 'betrayal of secrets' in any sense, for it only reveals Hooke's claim, not what he had done.) Hooke was obviously sensitive, then as later, about the possibility of anyone's revealing his work to others, even his friends. Oldenburg, although mentioning to Boyle that Hooke was currently working on watches and a new-fashioned chariot (an invention on which several other members of the Society were also working independently) when it came to making a reference to Hooke's newly-conceived 'Method for writing a Naturall History' which he thought Hooke intended to publish, he carefully added,[14] 'for the mentioning of which, I desire, that I may not be named'. Evidently, however hard he tried to be tactful, Oldenburg had not entirely succeeded in avoiding Hooke's blaming him for the comments and actions of others like Moray. On the other hand Hooke never seems to have objected to what Sprat, under the guidance of Wilkins, published in *The History of the Royal Society* in 1667.[15] There Sprat inserted a brief paper of Hooke's entitled 'History of Weather' as well as the unspecific reference to the invention of several new ways of regulating watches already referred to above (Chapter 7).

Early in 1665/6 as the force of the plague diminished, Hooke returned to London. In the following spring the Society began to function normally again and, presumably, relations between Hooke and Oldenburg also became as usual. At the beginning of September the routine of life was gravely disrupted by what the Journal Book called 'the late dreadful fire in London' which, as already described (Chapter 4), caused a break in the meetings of the Society until it could re-establish itself in Arundel House in the Strand. As a consequence of the Fire, Hooke began to be greatly occupied first with schemes for the rebuilding of London and then for many years with the actual work of designing houses and buildings in association with Wren, turning more and more later still to private architecture. Presumably he and Oldenburg continued to meet regularly. Certainly when Boyle visited his sister in Pall Mall as he often did in succeeding years, Hooke must have called on him and then it was easy to visit Oldenburg as well. Hooke continued dutifully to provide experiments at the Society's meeting which Oldenburg saw, recorded in the Journal Books, and wrote about to his correspondents. At the beginning of 1667 Hooke wrote a substantial description of the telescopes he himself used so that Oldenburg could send it to Hevelius, who had expressed the desire to learn about the instruments used by English astronomers. Oldenburg also transmitted a faithful Latin translation of a letter from Hooke which included his continuing regret, to use no stronger word, that Hevelius did not use telescopic sights on his sextants, for which Hevelius persisted in refusing to see the need.[16] At the beginning this was merely a difference of opinion to become a full-blown controversy only later; now there was no bitterness on either side. As Oldenburg was later to tell Hevelius,[17] 'Mr Hooke and I have in hand the purchase of a fine telescope for you', showing that all was then (1668) quite friendly between the three men.

There is some reason to believe that Hooke's touchiness, and not only towards Oldenburg, increased in later years until it almost amounted to paranoia. Certainly by 1668 Oldenburg was increasingly wary of reporting

Hooke's scientific work even to Boyle and even when it had been presented to a meeting of the Society. Thus in January 1667/8 when giving Boyle an account of a recent meeting of the Society he wrote,[18]

> Mr Hook brought in a contrivance for discovering the various pressure of the Air at Sea, thereby to foretell approaching storms. If he did not acquaint you with it before this, I think I may take the liberty to doe it, if you shall command it, the thing being made publick in so publick a place.

If Oldenburg, as he seems here to do, feared Hooke's possible disapproval of his communicating what had been described at a meeting of the Society and to Boyle, Hooke's dislike of publicity was evidently growing stronger.

Boyle seems to have understood Oldenburg perfectly. There was another occasion soon after this when it was Boyle who approached Oldenburg to ask for his exercise of tact and discretion in handling Hooke. In April 1668 Dr Walter Needham sent Boyle a letter in fact addressed to Oldenburg and intended for publication or at least for reading to a meeting of the Society. He obviously wished Boyle to read the letter before passing it on, a request which Boyle found disturbing,[19] for he thought

> it were not fit for me to decline, but in regard it concerns Mr Hooke and contains objections against something lately published by him [about experiments relating to the action of the lungs in breathing] I presume you will think fit, as I do, to shew it to him in the first place: and since they are both such ingenious men, and both members of the Royal Society, it were to be wished that, if it were possible, they may be brought to agree, without making their opposition éclater.

Oldenburg thoroughly understood Boyle's feelings,[20] replying that he hoped to be able to use Needham's letter in such a way 'as may rather appease than exasperate', and that he would not even read the letter to a meeting of the Society without Hooke's prior permission, and he refrained from publishing the letter in his *Philosophical Transactions*. The whole affair thus passed off amicably as far as relations between Hooke and Oldenburg were concerned. No doubt the fact that most members of the Society agreed with Hooke's interpretation of the experiments which they had witnessed made things easier for both Hooke and Oldenburg. Similarly in 1669 Oldenburg went so far as to tell Huygens[21] that Hooke at a recent meeting of the Society had spoken of 'a new way of driving a pendulum of great weight with very little force . . .

in two or more ways' but when Huygens inquired further[22] Oldenburg was more cautious, writing[23] 'When Mr Hooke is at liberty to speak of his new pendulum clock more fully you shall not be the last to have notice of it'. More and more Oldenburg was careful to praise Hooke's activities without giving details, even when those details had been recounted at the Society's meetings and even when the propective recipient was a Fellow of the Society. (He was naturally even more cautious when writing to strangers and/or foreigners.)

For most of the next two or three years we know really very little about the relations of Oldenburg and Hooke. Oldenburg regularly entered Hooke's contributions to the Society's meetings in the Journal Books, and discretely mentioned them to his correspondents, usually avoiding detail, and limiting his remarks to what he put in the minutes. Later in 1669 when Huygens, prompted by reports of Hooke's horological work by fellow Academicians and an Italian visitor who had recently been in England, asked Oldenburg for more information, the latter was carefully and cautiously laconic in his replies.[24] (These reports concerned Hooke's plans for improving pendulum clocks in several ways, particularly by the application of a magnet, the method of doing so never specified.) But on the whole Oldenburg had little to tell his correspondents about Hooke's work for the next two or three years. Certainly Hooke continued to attend the Society's meetings, to perform experiments at many of them, and usually to speak in discussion. But almost certainly he was more concerned during these years with architecture and surveying in connection with the rebuilding of the City of London than with natural philosophy.

From August 1672 onwards the situation changed and we know far more about Hooke's activities and Oldenburg's contacts with him thanks to the survival of Hooke's diary for that period.[25] This is not a totally reliable source given Hooke's temperament, but is a mine of useful information nevertheless. It clearly shows a striking degree of intimacy between Hooke and Oldenburg, who usually and for some years met several times a week, necessarily on Thursdays, the Royal Society's meeting day, but also on other days when common interests, including Royal Society business, brought them together. Then they exchanged books, talked together with Lord Brouncker or with Boyle or with 'Sarum' (the Bishop of Salisbury, Seth Ward who lived in Kensington, then a village a little to the west of London). It cannot, of course, now be determined whether their frequent meetings with Brouncker at his house near the Navy Office were in addition to Oldenburg's almost invariable weekly meetings as Secretary with Brouncker as President of the Society or not, but sometimes it must have been so. Hooke was a great frequenter of coffee houses and, as already mentioned, he and Oldenburg often met there, most usually in the 1670s at Garaways, a well-established house just off Cornhill, not very far from Gresham College in Bishopsgate, where Hooke lived as a Gresham Professor. Coffee houses had come into existence shortly after 1650 and throve immediately thereafter, providing cheap and useful places for men to meet, enjoy the new drink, converse, and conduct business of all kinds. They were used as places in which to hold auctions, as sales-rooms, for the gathering and transmission of news, as places where people could be reasonably sure of finding friends and acquaintances, in a sense foreshadowing nineteenth century clubs. Garaways was a popular and thriving house, begun probably in the 1650s, quickly re-established after the Fire, and continuing in existence for the next two hundred years. It is evident from Hooke's many references to it that it was frequented by Fellows of the Royal Society, by men of varied professions, by City men, and many others, and it was his favourite among all coffee houses. (He seldom refers to taverns as meeting places except for dinners, which coffee houses did not provide.)

As the diary reveals, from the summer of 1673 onwards Hooke and Oldenburg customarily expected to meet at Garaways on Mondays, and sometimes by arrangement on other days as well. Very frequently when they failed to meet Hooke noted 'missed Oldenburg at Garways' or 'at Garways Oldenburg not there'. Whenever Hooke called on Boyle at Pall Mall as he did with increasing frequency once Boyle took up permanent residence with his sister, he usually called on Oldenburg as well, especially in the summer when they did not meet on Thursdays, the Society being in recess. Besides this, Oldenburg sometimes called on Hooke at Gresham College. Presumably, although Hooke does not specifically say so, these meetings were usually intended to permit the two men to discuss Royal Society business together, though they also served for Hooke to learn news from Oldenburg's latest correspondence. As an example, Hooke's entry for 1 November 1673, a Saturday, reads

> At home till Lord Sarum, Sir W. Petty, Oldenburgh, etc came hither. Discoursd and concluded severall things about Society at Lord Brounckers.

It is worth noting that neither Brouncker nor Petty had been at the formal Council Meeting held two days previously, nor were they to be at the next Council Meeting on 6 November. Hence they were settling affairs behind the scenes as it were and, what is worth emphasizing, including Hooke in their discussions although he was not on the Council at this time. Evidently in 1673 Hooke was fully in the confidence of the leading members of the Society, ironically mostly the same men who were on the Council two years later when Hooke was to complain bitterly about its favouring Oldenburg over himself. Hooke's diary sheds more light upon the way in which the Society was, in practice, run, than does knowledge of the membership of its Committees or perusal of the formal and succinct minutes of Council Meetings, which naturally give only the results of discussions and not the means by which these results were reached.

Clearly, in 1673 Oldenburg and Hooke were working harmoniously together, meeting frequently, and apparently bound together not only by business but in friendship. Then on 28 January 1673–4, a Wednesday, there is an alarming and, up to this point unusual entry reading 'Oldenburg treacherous and a villain'. This is the first time that he had been so denominated, although others had been called similar names. There is no explanation of the cause. The very next day the entry says 'Met Oldenburg at Garways. Told him of New Reflex Telescope. . . . Noe meeting. Lent Oldenburg my cloke.' Evidently Oldenburg's 'villainy', whatever it consisted of, was not so black but that Hooke could forget it the next day and behave in a normally friendly fashion. Very probably Oldenburg himself was unaware of Hooke's deep anger of the previous day. A possible cause for Hooke's feelings was his evident displeasure at Oldenburg's role as intermediary between himself and Hevelius. Hooke was at this time bitterly attacking Hevelius in the Cutlerian Lecture which became 'Some Animadversions on the First Part of Hevelius his *Machina Coelestis*' published in 1674. The basis of his attack was, as already noted, Hevelius's refusal to use telescopic sights on his instruments. Hooke (rightly) insisted that open sights could never be as accurate as telescopic sights, however good the observer, proving experimentally the limitation of the ability of the human eye to discriminate between two objects very close together. No member of the Royal Society seems to have disagreed as to the matter of fact, although some, notably Wallis, said openly that Hevelius was so experienced an observer that his observations with open sights were likely to be more accurate than those of less experienced observers using telescopic sights. Wallis did not name Hooke, but the inference is obvious and Hooke so understood it. Oldenburg was involved in the disagreement because Wallis gave his opinion in a letter to Oldenburg which, as was his duty, Oldenburg presumbly read to a meeting of the Society when Hooke would almost certainly have been present.[26] Oldenburg had been cautious in writing to Hevelius, stating that he had not had time to gather the opinions of all the English astronomers on the *Machina Coelestis*. Hevelius had presented to the Society a number of copies of his book for Oldenburg to distribute, always a tactful way of avoiding dispute. As he put it, 'There are [some] among us, of whom Mr Hooke is one' who insisted upon the necessity of telescopic sights, later telling Hevelius that one of these was Flamsteed.[27] Hooke, of course, cannot have seen Oldenburg's letters to Hevelius, but he knew that his fellow members had not totally supported his denunciation of Hevelius and if not then certainly later assumed that Oldenburg himself had aggravated the case by not sufficiently taking his side, holding it as a grievance against Oldenburg. Oldenburg was in fact merely politely and properly purveying to Hevelius the views of various members of the Society both for and against him.

In May 1674 Oldenburg was able to give Hooke some better news of foreign relations when Huygens broke a ten-months' silence to thank Oldenburg for many gifts of books, including another of Hooke's Cutlerian Lectures, *An Attempt to prove the Motion of the Earth from Observations* (published that year) of which Huygens said he thought highly.[28] In reply[29] Oldenburg told Huygens that Hooke was 'very glad' of his approval, marking one of the few occasions when the two men saw eye to eye. It must have made an agreeable change for Oldenburg to be able to report praise rather than criticism to Hooke, and it must have helped to restore amiability to the two men. Relations between them continued outwardly calm, but there were hitches in June. On 4 June there was a meeting of the Society at which Hooke was 'called upon for his new astronomical quadrant' about which he had been reticent. (Curiously, Hooke did not mention this 'call' in his diary.) There was no meeting on 11 June, when Hooke recorded that Oldenburg, Hill (presumably Abraham Hill at this time Oldenburg's colleague as Secretary), and Hooke were together at Garaways and later in the day Oldenburg and Brouncker were at Gresham College although, as Hooke noted, 'Noe meeting of the Society nor noe [Cutlerian] lecture nor noe [dining] club'. Strangely, Oldenburg's rough minutes of the meeting held the next week on 18 June agree neither with what was written in the Journal Book nor what Hooke recorded in his diary. Oldenburg's rough minutes state that 'Hooke read a lecture against Hevelius his letter about telescopic sights' with Oldenburg's notes of the chief heads of the lecture. Hooke's version was that he read a paper about his quadrant (which Oldenburg said occurred the previous week) and that 'Oldenburg took notes'. This is odd, but even more curious is that there are rough minutes dated 28 June, not a Thursday, and so the Society would not normally have been meeting. Here the information about Hooke's lecture is repeated with the additional information that he also spoke about his quadrant; there is also the repetition of the information that there had been read another paper (on 18 June) on magnetic attraction, not by Hooke. This confusion is perhaps a reflection of the difficulty of their relationship at this time and it also suggests that Oldenburg and Hooke had conferred about what should appear in the minutes and that Hooke ultimately withdrew the reference to what he had done, or at least that Oldenburg supposed him to have asked for it not to be entered. (Later Hooke was to complain that Oldenburg had omitted to record Hooke's contributions fully; perhaps, at least in this case, he later forgot what had happened.)

Certainly there is no mention of any lecture by Hooke on this date in the Journal Book, which records only that the weekly meetings of the Society were to be adjourned because of poor attendance 'by reason of the season of the year, wherein many go into the Countrey' as the Council Minutes for the same day record it. Certainly meetings had been few and badly attended in recent weeks, for whatever reason. With the Society adjourned until 12 November 1674, there was still plenty of business for its Secretary and Curator to transact between them, and however prickly Hooke felt towards Oldenburg they continued to work together in apparent friendliness. They met, or sometimes expected but failed to meet, at Garaways or elsewhere at least once a week, while the most active Fellows living in London seem to have met nearly as frequently and usually at Gresham College on a Thursday, with both Oldenburg and Hooke present. Although outwardly at ease, in fact Hooke showed himself all the summer to be reluctant to trust Oldenburg with knowledge of his inventions lest he reveal what he learned to others, especially to foreigners. So on 9 July when there was an informal meeting at Gresham College (after which Hooke supplied a bottle of canary wine for the refreshment of those present) he recorded 'Shewd my quadrant to all but Oldenburg', presumably waiting until Oldenburg had left the gathering to do so. It is not clear why Hooke could not bear to have Oldenburg learn the details about his quadrant at this time except that, temporarily, he wished him not to mention it in his correspondence. But in spite of this, he called upon him on the very next day. The printing of his 'Animadversions on the First part of Hevelius his *Machina Coelestis*' was already under way. Oldenburg knew of its progress as early as the end of August. It was licensed by the Society on 9 November 1674, although it is not certain whether at this stage it was known to contain the description of Hooke's quadrant which was certainly in place by 12 December when it was printed off. It was favourably reviewed, presumably by Oldenburg, in the *Philosophical Transactions* dated 14 December. (Oldenburg mentioned the book when writing to Hevelius that autumn but said nothing about the quadrant, whether because he did not then know of its impending publication or out of tact cannot be determined. He finally sent Hevelius a copy only in April 1675, probably waiting for a means of conveyance.)[30]

After the circulation of Hooke's account, of course, even he could not complain if Oldenburg described it to his correspondents. The two men did not meet for the next couple of weeks, not until Thursday 16 July when several Fellows met at Garaways and, as Hooke noted, 'discours[ed] much about Regulating the Society'. Oldenburg was not among them, being 'not in town'. (Neither was Boyle, for two days later Hooke noted 'Boyle not returned nor Oldenburg'. One may guess that Boyle, as he often did, was visiting his sister Mary, Countess of Warwick in Essex; the reason for Oldenburg's absence cannot be now known.) By the following Thursday, 23 July, Oldenburg was clearly back in town and went to Gresham College for the Society's meeting, where Hooke was also. During the whole of the rest of July and all of August the two men continued to meet roughly once a week, either at Oldenburg's house or at Garaways. On 10 September Hooke and Oldenburg were together at Brouncker's, presumably still discussing how best to render the Society more active, and on the next day Brouncker treated them and others to a day out, going to Blackheath and, as Hooke carefully noted, paying both for the hire of the coach and for dinner.

After this Oldenburg seems to have taken his usual family holiday of, probably, a fortnight. From this he returned to find Hooke in an uncertain temper, as so often at this time varying between vehemently expresssed rage and normal friendship. He denominated Oldenburg 'a villain' in the diary entry for 25 September, with no explanation, yet the very next day called upon him at home and three days later still they went amicably together to 'Broken wharf'. And so it went on with Hooke even on one occasion borrowing Oldenburg's cloak. But it seems that Hooke was still seething inwardly, convinced that Oldenburg told his correspondents news of what went on at Society meetings which Hooke thought should remain confidential. He had complained of this to his good friend Petty, as appears from the minutes of the Council meeting on 15 October 1674. There were then present Petty in the chair, Goddard, Colwall the Treasurer, John Creed (F.R.S. 1663, a naval commissioner well known to Pepys and often on the Council) and Oldenburg (but not, of course Hooke, who was not on the Council). The discussion centred on proposals for 'restoring' the Society (many Fellows had been deplorably inactive) and possible changes to it. One change proposed was the repeal of the statute allowing admission of non-members to meetings and another the possible enactment of a statute which should enjoin secrecy upon all the Fellows about announcements by members of 'what they have discovered, invented, or contrived' whenever the presiding officer should so decree. These proposals sound very much as if Hooke had suggested them to Petty, especially as they were obviously directed against Oldenburg and his correspondence. (Although it was at this time proposed that the terms of such a statute should be drafted, this was never done, nor were non-members ever totally barred from meetings).

By December 1674 relations between Hooke and the Society had become a little strained and no doubt Oldenburg was made to feel it, although by no means all the leading Fellows sided with Hooke. On Council meeting days it was customary for the Council members to dine at Gresham College, Hooke apparently being charged with the arrangements and in return being invited to join the group. On 3 December he noted in his diary 'Councell dined here. Displeased I being absent. Colwell and Oldenburg Huffs' [self-important or arrogant]. Perhaps it was they who rebuked him, probably at the meeting of the Society later that afternoon when Hooke, as recorded in the minutes, lectured on his quadrant, presumably using the text of the version in the press. But in spite of this, relations between Oldenburg and Hooke seem smooth enough for the next couple of months. In mid-January 1674/5 Hooke introduced Oldenburg to a Mr Galoway who helped Oldenburg to chose a dish (probably majolica or Delft) at the 'Corner Shop' for which Oldenburg paid one shilling. (Hooke himself the previous week had chosen, also with Galoway's help,'12 little and 3 great' dishes (for which he did not at the time pay.) And having helped Oldenburg to buy his dish, Hooke spent a shilling treating Oldenburg at Garaways.

Sadly, trouble soon arose in late January 1674/5 when Hooke heard of Huygens's invention of a practicable way of regulating watches with spring balances (see Chapter 7). Some of the disastrous results of this for the relations of Hooke with both the Society and with Oldenburg have already been discussed insofar as they affected Huygens. Here more must be said about the effects on the relationship between Hooke and Oldenburg. To recapitulate briefly, Oldenburg's difficulties began at the Society's meeting on 18 February 1674/5 when Oldenburg read Huygens's letter of 10 February which described his invention clearly.[31] As the minutes carefully note, Hooke declared that he 'had such an invention . . . divers years ago' and that he had made such watches, appealing to the Journal Book, to Sprat's *History* and to the memory of several of the Fellows, all of which he noted in his diary. The members present were perfectly sympathetic to Hooke's claim to have made such watches, but they were also, it seems, far less certain than was Hooke that they had been successful. What Oldenburg was ordered to write to Huygens, besides sending thanks for his communication, was to tell him what had been done in London 'and what were the causes of its want of success', the latter phrase not at all confirming Hooke's belief. All this Oldenburg duly told Huygens[32], as already noted (pp.198ff).

Hooke was not unnaturally disturbed by the position the Society had taken and the fact that Oldenburg was, as it seemed to him, partisan, especially if he were to secure a patent. But there was no open rupture as yet. While all this was going on Hooke in late March 1675/6 remained on speaking terms with Oldenburg. Thus he noted in his diary that 'read paper of new propertyes of light' adding 'Explaind Light and Colours dictated to Oldenburg', no doubt so that he could be quite satisfied that Oldenburg reported him correctly in the minutes. Hooke here oversimplified, judging by Oldenburg's minutes which record that at the previous meeting Hooke had spoken at length on the subject saying that he had formerly read a paper on the same theme which he had not allowed to be registered, and as a result he was asked to read it again to the meeting. (This is fair confirmation that Hooke himself often failed to record his inventions properly, so that Oldenburg was certainly not always at fault.) What he dictated to Oldenburg must be the fairly detailed outline which can be found in the official minutes. This seems fairly amicable and for the next weeks Hooke and Oldenburg continued to meet at Garaways and Hooke even called on Oldenburg at home. There are, it is true, hints in the diary entries that Hooke thought that Oldenburg was planning something underhand. On 25 March 1675 Hooke wrote 'suspected Hill whispers with Oldenburg'. Either Oldenburg was naïvely unaware of Hooke's bitter feelings or, accustomed as he must have been to Hooke's sudden rages against individuals and usually equally sudden recoveries, discounted this episode, for on 30 March, less than a week after Hooke had entered his suspicions in his diary, he recorded that he and Oldenburg met at Martin's, the printer's, and Oldenburg not only chatted as usual about the contents of his correspondence but proudly announced the arrival of a son.

What caused the fatal rupture between the two men was the question of patent rights. In his diary entry for 3 April 1675 Hooke noted that he had learned from Sir Jonas Moore, only recently a Fellow, of what Hooke called 'Oldenburg's treachery his defeating the Society and getting a patent for Spring Watches for himself'. The source of Moore's report is naturally impossible to determine, but it was substantially correct for Oldenburg was about to draw up a petition for a patent and a draft of a Royal Warrant which should grant it.[33]Meanwhile, Oldenburg had already explained to Huygens[34] that one had to be careful not to take Hooke's hasty words too much to heart, begging Huygens 'not to be offended at some words which touch you' in the 'Animadversions' (about circular pendulums) and adding 'there are people who, not having seen much of the world, do not know how to observe that decorum which is necessary among honest folk'. Huygens was in fact considerably offended,[35] so much so that Oldenburg felt it necessary further to defend Hooke. He might have been surprized that Oldenburg practised what he preached and defended him to others, trying to deal with his explosions now as if they were temporary and soon over as they had been over the past years, apparently not recognising that this time Hooke was permanently and deeply enraged and irreconcileable.

In April 1675 Hooke was told by Jonas Moore, how reliably is doubtful, that Oldenburg would be given a patent for Huygens's design unless he could produce a working watch soon. As a result, Hooke quickly produced a watch of his own design which he gave to the King. It was found to work as well as that provided by Huygens, but no better, neither design succeeding perfectly. Hooke was at this time convinced that Oldenburg and Brouncker were jointly plotting against him, noting on 8 April that he had met Oldenburg and Brouncker at the Bear Tavern when he had 'discovered their design'. Later that day he learned from Southwell that Oldenburg's application had failed, a failure confirmed by Sir Joseph Williamson. He was greatly displeased at Williamson's advice that he should 'joyn Oldenburg' (presumably meaning that they should jointly apply for a patent) and refused to consider a reconciliation, writing 'I vented some of [my] mind against Lord Brouncker & Oldenburg. Told them of Defrauding'. After this neither Brouncker nor Oldenburg could have failed to understand Hooke's genuine animosity. This increased when Moore two days later 'told [him] the Kings message and that unless we made hast with the watch he would grant the patent', and he continued to record that he felt there had been 'treachery'.[36] During April and May while Hooke worked furiously to improve his own design all that Brouncker could do was to urge Huygens through Oldenburg to send more watches to prove the efficiency of his original design. By the early summer these watches began to arrive (to be paid for by Brouncker) while, as Oldenburg reported to Huygens, Hooke continued to improve his design and present more watches to the King. These Charles II received graciously but it was soon clear that no patent would be awarded to either party.[37]

Hooke was furious, blaming his failure to secure a patent on Oldenburg who, so he believed, must have told Huygens about what he (Hooke) had done so that Huygens's claim for invention was spurious. In his anger Hooke lashed out wildly at Oldenburg for every possible dereliction of duty, as he saw it, and especially for failing in this as in other ways to give Hooke full credit for his work. Thus, after the Society's meeting on 3 June he noted 'Oldenburg a raskell for not registering things brought into the Society, to wit that of the water engine with small pipes', about which there is indeed nothing in the surviving minutes. A week later it was 'I reproved Oldenburg for not Registering Experiments. Brouncker took his part', yet another instance, as Hooke saw it, of Brouncker's 'treacherous' partiality. Partiality on Brouncker's part there certainly was, but whether unjust or not in this instance who can now say? That there were times when Oldenburg failed to register experiments is true. An obvious example was to occur after the meeting on 18 November 1675 when, after he had read an account of experiments on various liquids in the airpump made by Denis Papin working under Huygens's direction, Oldenburg 'was desired to take care of entering these experiments in the Register' which he did not do, although he printed them in the *Philosophical Transactions* for that month. But it was hardly such an instance that Hooke meant. On 10 June 1675 after the meeting Hooke again denominated Oldenburg 'a Raskell', but without further explanation.

Hooke remained in a peculiarly irascible mood all during the summer and autumn of 1675, the period during which he was conducting war on both Huygens and Hevelius simultaneously, campaigns in which Oldenburg was unfortunately for him drawn in as an intermediary.[38] Many of Hooke's colleagues were also denominated 'rascal' and 'dog' for reasons unexplained, although one can guess that they supported Hooke's adversaries. True, as is obvious, Oldenburg was prejudiced in favour of Huygens as against Hooke, and inevitably so since he stood to profit from any patent on Huygens's design. But it does not follow that he was unfairly prejudiced, any more than it was that when he failed to register what Hooke had said and done at meetings he acted out of malice. For Hooke, as already noted, was often ambiguous about having his words recorded. Hooke now made no secret of his hatred of

Oldenburg. At the end of the summer of 1675 (10 August) when dining with Boyle he 'railed' against Oldenburg, by his own account. What position Boyle took in the quarrel is not known but it is difficult to believe that someone universally regarded as the soul of courtesy can have approved of Hooke's behaviour. It is known that when, the next summer, Hooke decided that the only solution for him was to form a 'select club' or, as he later called it, a 'decimall club' Boyle refused to join it (although Wren did so).[39] For several months in 1676 the club met, somewhat irregularly, on Saturdays, with Hooke dominating the discussion which gave him the opportunity to assert the correctness of his theories over those of others, including Newton.

In the early autumn of 1675, Hooke began a public campaign, apparently with Wren's approval, which was directed against Oldenburg as a foreigner who 'sold' secrets to other foreigners (like Huygens) and who further betrayed the English, that is Hooke, by refusing them a due share of the credit for their inventions and theories. (Hooke never regarded Huygens and Hevelius as truly Fellows of the Royal Society, although they were properly of the same status as Hooke, there then being no category of foreign member. Huygens indeed had been elected a Fellow only three weeks after Hooke himself, both in 1663.) As for 'selling' secrets, Oldenburg got nothing from most of his correspondence, the only exception being his supply of *political* news to Williamson. Having attacked Oldenburg privately, Hooke determined on a public attack on both Oldenburg and Huygens. In the autumn of 1675 he published his Cutlerian Lecture entitled 'A Description of Helioscopes and some other Instruments', offically published in 1676 but in fact printed off by early October 1675.[40] To the original text Hooke added a seemingly irrelevant postscript in which he attempted to vindicate his claim to be the true inventor of spring-balance watches. It is a little difficult to unravel the thread of Hooke's argument, this time presented in a less intemperate manner than usual. He spoke of what he merely called 'unhandsome proceeedings' by Oldenburg which consisted in his publishing in the *Philosophical Transactions* for 25 March 1675 an English translation of Huygens's description in the *Journal des Sçavans* of his then recent invention without any mention of the fact that Hooke had suggested the idea many years before on several occasions.

It is true that Oldenburg *could* easily have mentioned Hooke's ideas at this time since, as Hooke stated, they had been made public in his Cutlerian Lectures and at 'Publick meetings' of the Royal Society. That he did not does not necessarily mean that this was a deliberate reflection on Hooke, since what was in question was, as has been explained, an idea as against Huygens's reality. Hooke felt strongly that Oldenburg should have mentioned his idea, especially as he believed that Huygens *knew* that Hooke had proposed springs among other methods of regulating watches. Hooke was correct to believe that Moray had mentioned his ideas to Huygens, but whatever he now believed Oldenburg never had, and was not then corresponding with Huygens.[41] Now Hooke 'doubted not' that a full account of all that he had done about watches, including their potential use in longitude determination, had reached Huygens through Oldenburg, and that Oldenburg, and by implication the Society, had 'betrayed' him. Somewhat contradictorily he blamed Huygens for not mentioning what had been printed in Sprat's *History* of 1667, about which Oldenburg had no need write. Why it would have been wrong for anyone to have sent to Huygens a report of what Hooke had said publicly is not obvious, but Hooke in a muddled way seems to have thought that his public words were for the ears of Englishmen only. Whether he would have accepted the reality of Huygens's invention if the latter had acknowledged Hooke's priority of mental invention cannot be known. The original text which, according to Hooke's diary Oldenburg saw him correcting on 9 October and which Hooke no doubt thought he had read, seems to have been more violent and inflammatory. However, as Oldenburg told Huygens, Hooke had been persuaded (or even possibly ordered) to tone down his attack.[42] This probably happened on 15 October when Hooke recorded 'A Grubendollian Caball at Arundel House' (where the Society was still meeting), the use of Oldenburg's postal address for his name no doubt an indication of Hooke's belief that Oldenburg was constantly involved in 'betraying' English secrets to foreigners. Hooke was apparently encouraged in his attacks, then and later, by the printer Martin, whose reason was perhaps that, like many a modern journalist, he believed that virulence increased sales, or, perhaps more realistically, he had a grudge against Oldenburg over the financing of the *Philosophical Transactions*.

As already noted, Oldenburg, on Brouncker's advice, appealed to Huygens, requesting him to assure Brouncker that he had known nothing of Hooke's invention at the time of his own discovery. This Huygens cheerfully did in late October, expressing great surprise at Hooke's 'accusation', for, as he put it in a tone of superiority, 'I had noted for some time that he was vain and foolish, but I did not know that he was as malicious and insolent as I see he now is', an opinion which must have both gratified and depressed Oldenburg. With Brouncker reassured, Oldenburg felt free to use a review of 'Helioscopes' to defend his own use of the *Philosophical Transactions* to disseminate the latest scientific news.[44] There he gave an account of Huygens's various communications, pointing out that Hooke had seen them all before they appeared in the journal. So, Oldenburg claimed, probably truly, that he would also have published there Hooke's own invention and claim for priority had he been asked to do so. (It must be remembered, as no doubt Oldenburg did remember, that Hooke would have been extremely angry had Oldenburg published an account not initiated and carefully reviewed by Hooke.) Oldenburg therefore, surely quite fairly, resented having his actions called 'unhandsome proceedings' and thought Hooke's epithets quite unjust.

The next stage in the battle was set when Hooke saw what he called 'the Lying Dog Oldenburg's Transactions' on 8 November a fortnight after the date on the journal, and so presumably as soon as it was printed off. He must either have haunted Martin's shop to get early news or been alerted by Martin. His immediate reaction was 'resolved to quit all employments and to seek my health', the latter no doubt affected by his extreme agitation. In fact he did no such thing, but continued his relationship with the Royal Society in perfectly normal fashion during the rest of November 1675, except that in his diary he continually complained about Oldenburg. He complained that Oldenburg had failed to register experiments performed and discourses read by Hooke at meetings, but also that 'Oldenburg viewd [an experiment] and took notes' when Hooke read the discourse which was to be published as *Lampas*. He particularly insisted that Oldenburg's 'fals suggestions' were the cause of what Hooke called the 'quarrel' (which others must have thought of as a disagreement) between himself and others, notably Newton.[45]

It is well to remember that during these years when the prolonged dispute between Oldenburg and Hooke was at its most bitter, Oldenburg was much occupied with his usual and generally more significant business, and particularly with the complex correspondence involving Newton, a far more important occupation for both Oldenburg and the Royal Society than Hooke's hurt feelings. Oldenburg was deeply wounded by Hooke's attack on his reputation for integrity, so precious to a man of such upright morality as he was. Fortunately for him, most of the senior members of the Society seem to have taken his part. This was not for want of trying by Hooke, as shown by his frequent and sometimes somewhat devious attempts to attach to his side of the dispute numerous Fellows, including Wren, John Hoskins, Hill, Aubrey, and several more or less obscure virtuosi. Yet references in Hooke's diary show that Oldenburg's professional and social life within the Society proceeded normally. He and Hooke seem quite often to have dined or visited the coffee house together in company with others, some of them Hooke's adherents, even if the two men no longer met alone together. Thus at the end of February 1675/6 Sir Joseph Williamson entertained a company composed of a number of people, including Oldenburg, Evelyn, Wren, Petty, with all of whom Oldenburg remained on friendly terms, although both Wren and Petty were warm adherents of Hooke,[46] while Boyle kept aloof from all dissension and remained on close terms with both men.

In the early months of 1675/6 Hooke's anger merely simmered. He assuaged his wounded ego by striking at what he saw as the tyranny of a Royal Society dominated by Brouncker and Oldenburg by the formation of his 'Decimall Club' which met first at Joes Coffee house off the Strand and later at various places in the City. Hooke's most prominent supporter was Wren, with, among others, those already named above as Hooke's partisans.[47] Here Hooke, who ran the meetings, could hold forth to an audience assuredly disposed to be friendly, setting forth his own ideas and denouncing those of others. Meanwhile he planned his revenge on Oldenburg, plans rendered more urgent, as he saw it, after 6 March 1675/6 when, as he heard from Williamson, 'Councell Resolvd against me', perhaps only a friendly warning. The Council did meet on 6 March, but the minutes, perhaps out of discretion, do not mention either Hooke's name nor anything about his quarrel with Oldenburg, so that it is tempting to imagine that Williamson reported the unofficial rather than the official views of the Council members. Hooke's diary entry for that day continued 'Brouncker, Oldenburg, Colwall and Croon, busy bodys with them at the Crown' where he spoke with Williamson and Brouncker, in spite of his expressed opinion of the latter. But after this he began to regard Brouncker as his bitter enemy. Dining with him and others on 22 August 1676 he called 'Brouncker and Colwall Doggs'. He then circumvented Oldenburg's allies by securing a licence to print his Cutlerian Lecture *Lampas* not from the President or a Vice-President as normally, but from someone who could know nothing of the affairs of the Society, George Hooper, Chaplain to the Archbishop of Canterbury. He had granted it the day before Hooke dined with Brouncker. This work[48] was a description of Hooke's various inventions, principally a lamp which would burn for a very long time, but here he also entered his claim to the invention of a watch regulated by a circular pendulum before 1673 when Huygens had published his version in *Horologium Oscillatorium*, with great restraint not here accusing Huygens of plagiary.

What he did instead was to add a postscript attacking Oldenburg as the publisher of the *Philosophical Transactions* both for refusing to back Hooke's claims for priority over Huygens in the matter of spring-balance watches and other discoveries and for being as he put it 'one that made a trade of Intelligence', certainly meaning acting like a spy. At the same time he asserted that Oldenburg knew of (and should have stated) their having been made public knowledge because they were 'publickly read of' in his Cutlerian Lectures and had been shown and described to very many, both English and foreigners, and their existence made known by a passage in Sprat's *History*. The logic is far from rigorous, for if Hooke's ideas had all been made so public, how could Oldenburg have 'betrayed' Hooke by writing to his correspondents about them and publishing accounts in the *Philosophical Transactions*? At the same time he said that he had not told Oldenburg about the success of many of his inventions to prevent Oldenburg from 'betraying' them to foreigners. But if Hooke had not revealed the success of his watches and other inventions to Oldenburg, how could Oldenburg have known of this success? And if those to whom Hooke *had* made known his inventions had told Oldenburg of them, it was they who had betrayed Hooke, not Oldenburg. Angry men are seldom truly logical, and to Hooke's inflamed suspicions there was conclusive evidence that Oldenburg had betrayed him somehow.

To Hooke's shocked surprise, it was Oldenburg who received the Society's support, even from some of Hooke's friends. And speedily too, as soon as the Council could reassemble after the usual summer holiday. Under the date of 3 October 1676 Hooke's diary entry reads 'A councell against me at Lord Brounckers' (although the official minutes of the meeting say nothing about this, perhaps discreetly). But Hooke was correct in spirit, for after the imprimatur, obtained as usual from Brouncker as President, for the issue of the *Philosophical Transactions* dated 25 September (it was 'for the months of August and September') Oldenburg added an 'Advertisement' saying that he intended to reply to 'the Aspersions and Calumnies of an immoral Postscript put to a book called *Lampas*, publisht by Robert Hooke' and that he hoped that until this appeared 'the Candid Reader' would suspend judgement in the matter. When the Council considered the affair at their next meeting on 12 October those attending decided to blame the printer for accepting Hooke's postscript when it was improperly licensed. Hence Martin was told to declare in the next number of the *Philosophical Transactions* that this postscript 'was printed without the leave or knowledge of the council', as the minutes state. What Hooke thought, as his diary entry shows,[49] was that 'he was warned to Councell of Royal Society about Oldenburg. They decreed nothing against me, but against Mr Martin.' (After this he wrote, 'Resolvd to Leave Royal Society' which of course he did not do). There was no issue of the *Philosophical*

Transactions in October, but Martin did not forget his rebuke, for although he never made the required declaration, on 5 May 1677 he refused to print Hooke's *Cometa* without a licence and Hooke was forced to postpone its publication for some months.

Meanwhile during the autumn of 1676 Hooke grumbled about 'Grubendolian Councell order' and 'Great intrigues of Councell' and 'Grubendolian Councell', although at this time Henry Hunt, Hooke's private assistant, was 'chosen' to be the Operator, Hooke's official assistant as Curator.[50] Evidently Hooke was fully prepared to find the Society and its Council acting towards him as he would, given the chance, have acted towards Oldenburg. By November 1676 there was no need for Martin to make any statement for on 20 November Croone and Hill who had been asked to draft an answer to Hooke reported favourably to the Council on Oldenburg's own proposal for an answer. This was printed in the issue of the *Philosophical Transactions* dated 20 November 1676, being for October and November. (Hooke complained in his diary three days later that he had not known of the Council meeting which concerned him. Although obviously it would have been polite to inform him of it and perhaps politic as well, it was not unjust for he was not a member of Council. He naturally called it a 'packed Councell' relying on gossip, and his list of those present was not accurate.)

The official defence of Oldenburg in the *Philosophical Transactions* read,

> whereas the Publisher of the *Philosophical Transactions* hath made a Complaint to the Council of the Royal Society of some passages in a late Book of Mr. *Hooke* entituled *Lampas, etc.* and printed by the Printer of the said Society, reflecting on the integrity and faithfulness of the said Publisher in his management of the Intelligence of the said Society: This Council hath thought fit to declare in the behalf of the Publisher aforesaid, That they knew nothing of the Publication of the said Book; and further, That the said Publisher hath carried himself faithfully and honestly in the management of the Intelligence of the Royal Society, and given no just cause of such Reflections.

To this Oldenburg added his private defence, namely that 'The Council having thus justified the Publisher; he shall only add . . . part of a Letter' from Huygens, about giving the profits of any patent which might be granted for his invention to the Royal Society or to Oldenburg. This restrained answer to Hooke would, Oldenburg obviously hoped, prevent the world from taking Hooke's attack (which, it must be remembered, implied that Oldenburg had tried to steal the advantages of a patent from him) as justifiable.

Oldenburg must naturally have been pleased to have the Council support his claim to 'integrity and honesty' and fully support his conduct of his correspondence. He had intended to reply to Hooke's attack in kind and in considerable length and drafts of such a longer reply exist[51] in which he complained of Hooke's want of logic in having claimed both to have read about his inventions publicly and to have been hurt by their being communicated to Huygens (a member of the Society), of implying, as he had done, that Oldenburg's attempt at a patent involved something underhand, and of claiming that Oldenburg had publicized the work of foreigners without mentioning Hooke's own work of the same kind (not in fact the case).

In fairness to Hooke it must be said that at about the same time Flamsteed also complained that in his view Oldenburg was not as careful as he should have been to proclaim to the world the priority of the English in scientific discovery.[52] He thought that Oldenburg had published Cassini's theory of sunspots without sufficiently emphasizing that Flamsteed had anticipated him, as he believed. He insisted that Cassini had only been able to develop his theory because he had learned of Flamsteed's by means of Oldenburg's 'communication'; that is, its publication in the *Philosophical Transactions* (which he had been pleased to have done). He therefore attempted to admonish Oldenburg by saying

> Sr it will concerne you to be so studious as you have affirmed your selfe to give every man his due & vindicate good discoveryes to their proper inventors not to let us be altogether deprived of this in so far as that you may cleare yourself of their aspersions which call you more a freind to strangers then to the English . . . I hope when you print that paper of Mr Cassinis [which he never did] you will also intimate that the theory was derived from the September transaction & that it was first intimated to me by Ed. Halley then communicated to you by me whereby you will oblige me to continue
>
> Your affectionate freind & Servant

This very solemn warning from a young man to an elder. was a little ungracious considering Oldenburg's role in bringing Flamsteed and his work to public notice. (Of course he could not have known of all the many previous occasions when Oldenburg *had* defended English priority against foreign claims.)

To return to Oldenburg's defence against Hooke, it is obvious that the Society was wise to reduce Oldenburg's original longer draft to the short and dignified form in which it was published, for in that form it gave Hooke little handle for another long attack such as Oldenburg's original version would certainly have provoked. Hooke was furious in any case about what he regarded as the both unjustified and excessively partisan attitude of the Council. He decided to stand for election to the Council on 30 November, St Andrew's Day, and when he was not elected he was angrier than ever,[53] claiming 'Much fowl play used in this choice', insisting that he had in fact received two more votes than Oldenburg announced that he had (though Oldenburg cannot have been the only teller) and that the ballot must have been rigged, adding 'Resolvd to Reforme these abuses'. It would be easier to credit the truth of Hooke's aspersions were it not that several of his friends *were* elected and that his account differs from the minutes. Thus Hooke insisted 'Account not audited nor reported' whereas the minutes, more plausibly, state that a committee was appointed at the meeting, a committee which included Haak and Aubrey, both close friends of Hooke. This was to examine the accounts and report on them, as it had done before the elections. There is no reason to believe that the minutes are incorrect; it is more likely that Hooke's recollection of exactly what took place at the meeting was far from precise, possibly being written up some days after the events it describes, which was often his practice. Two days later,[54] still fuming, Hooke button-holed Seth Ward, Bishop of Salisbury in Piccadilly to complain to him about the state of the Council and the Society; the diary records of Ward that 'He promisd much' which no doubt Hooke found soothing, for Ward was always a power in the Society behind the scenes. A month later (28 December 1676) Hooke reported a gathering at the Crown Tavern (there was no meeting of the Society on that Thursday) where, although he found 'Moor a dog' (for all that Moore had been one of Hooke's partisans some months before) yet he was pleased that there was 'Much discourse about regulation of Royal Society, reading books, noting Experiments, order of Councell, &c.' Clearly, Oldenburg was not present since he was not named and there had not been a formal meeting. Probably Brouncker was also absent.

In the new year of 1676/7 Hooke began, with Hill's assistance as a former Secretary, to study the archives, reading letters exchanged between Oldenburg and Huygens, clearly seeking for more evidence of Oldenburg's 'treachery'. It is a measure of his antipathy that he omitted Oldenburg's name from the list of those present at the meeting of the Society on 11 January 1676–7, although he was certainly there. Only slowly did the hated name reappear in normal context in entries in Hooke's diary about the Society's meetings. And there were apparently no more pre-arranged meetings at Garaways and Hooke commonly refers to his enemy as Grubendol, presumably to emphasize to himself both Oldenburg's foreign nationality and his 'trade' in news. All friendship between the two men was now at an end, so much so that when Oldenburg died Hooke could only rejoice, pursuing his vendetta beyond the grave. Over the years he had often been angry with Oldenburg, but so he had with countless other men, always with an anger that flared up quickly and was soon forgotten. Two years of bitter anger against Oldenburg had now wiped out any recollection of a dozen years of amiable cooperation in the service of the Royal Society. He now could not forget his anger and continued to denounce what he saw as Oldenburg's 'unfair' treatment of himself and his claims. Since Hooke outlived Oldenburg and had many gossiping friends like Aubrey and Richard Waller who wrote of his work eulogistically, it is Hooke's side of the story that had often been uncritically accepted by historians. It is therefore only fair to give here Oldenburg's side of the story, which was that he had done nothing unfair nor underhand nor treacherous, but had endeavoured to act fairly and justly by all members of the Society he served for fifteen long years.

10

Colleagues, friends, and family

Increasingly, as the years went by, Oldenburg's life was so closely bound up with the affairs of the Royal Society and he was so ineluctably identified with it that the history of one is nearly the history of the other. Hence the history of his public and private lives are difficult to separate. His public life has been easy to explore, since he himself documented it pretty fully through his meticulous collection and preservation of the minutes of the Society's meetings in its Journal Books and Council Minutes and of the bulk of his correspondence (which was chiefly that of the Royal Society). Traces of his private life before the 1670s are far more difficult to discover except insofar as his own affairs impinged on his official life, are mentioned in correspondence, concern the publication of his *Philosophical Transactions*, or are contained in references by his colleagues. These were so often his friends as well that it is at least partly ambiguous to try to distinguish public and private relations.

Much about his character can be deduced from his official correspondence. There over the years he displayed diplomacy and tact, showing a sureness of touch in his handling of his varied correspondence as Secretary, writing with equal success and patience to country virtuosi and learned academics English and foreign, among them the greatest natural philosophers of the age. Oldenburg clearly had a talent for friendship; it is remarkable indeed how quickly his most usual correspondents became his friends, so much so that it is difficult to say who among them was merely a colleague and who a real friend. Very occasionally, as has appeared, correspondents complained to him of his failure to reply promptly to their letters but usually it was not because he failed through inattention or over-involvement in his various affairs but because the post was at fault. Occasionally they complained to him of his manner of editing their communications in the *Philosophical Transactions* or because he did or did not print what they sent to him. But that was and is a normal occupational hazard of editors. His ability, so often exercised, of stimulating or provoking those who were reluctant to make their work public, as some of the best of them were, was remarkable. He usually succeeded by diplomacy and telling one man what another was doing. It was, of course, a dangerous exercise, particularly since Englishmen were so touchy about claims for priority by foreigners (as indeed the French were too).

Great tact was required, and evidently Oldenburg had it in abundance, since he usually emerged unscathed from wrangles on this subject. Although some, Newton and Huygens among them, were in different ways sometimes annoyed or wearied by Oldenburg's persistence and importunity, on the whole the method worked and Oldenburg succeeded in extracting work from the greatest scientists of the age as well as from lesser men. In spite of their occasional protests most such correspondents came to feel genuine friendship for Oldenburg however much he nagged. The seventeenth-century world of natural philosophy owed Oldenburg a very great debt for the results of his persistence. One of the reasons for this persistence was his personal conviction that English natural philosophers were all too prone to hide their light under a bushel rather than to shine it forth to the world so that it could illuminate the study of nature and he used this conviction as a lever many times. True, he tried to extract similar information from foreigners, but as England was his adopted country his heart was particularly in the task of advancing the cause of *English* discoveries.

Oldenburg had already established friendly relations with most of his colleagues in the Royal Society within a couple of years after his appointment as Secretary, as can be seen in letters to and about him. And he was annually re-elected to the post of Secretary with no obvious demur until the troubled years of 1675–76. Even his imprisonment in the Tower in 1667 charged with spying did not openly diminish their trust in him for the most part. Brouncker as President clearly trusted him, indeed relied on him, and favoured him. In the 1670s the two met regularly (usually on Mondays) to discuss Society business,[1] and so presumably they had done in preceding years, perhaps drawn all the closer since they almost alone among the Council Members had stayed in London during the Plague. Boyle remained his patron. By 1667 they had arrived at terms of real friendship, as Oldenburg had with Moray, Wallis, and Brouncker, a relationship strengthened with all of these during the difficult months of the plague year.

Something of this appears in the signatures to letters. In 1657 Oldenburg had been Boyle's 'obliged and very humble servant'. Six years later he could sign himself merely 'your faithful humble servant'. Thus although Boyle was his employer and a very great gentleman, younger son and later younger brother of an earl, Oldenburg could address him as one gentleman to another. Equally significantly, the physical appearance of Oldenburg's letters altered. It has been noted[2] that throughout the seventeenth century the spacing of the text of a letter reflects the relationship of the writer to the recipient, and to a marked degree. So the more respect the writer wished to pay, the greater the space left between the salutation and the body of the letter and between the end of the text and the signature. This usage can be traced readily over the years in Oldenburg's surviving letters to Boyle, the later ones having both salutation and subscription written close to the text, with little of the blank space used to show formal respect. The subscriptions used by Boyle also show an increasing sense of equality over the years, so that by 1664 Boyle had come to sign his letters as Oldenburg's 'very affectionate friend, and very humble servant', a usage which persisted over the years. Clearly, in Boyle's eyes, Oldenburg had become a friend rather than merely a humble client. Although, obviously, he could never be completely Boyle's social equal, he was no longer a mere subordinate, but rather a colleague like any other Fellow of the Royal Society. Wallis, clergyman, university professor, much the same age as Oldenburg, habitually signed himself as to an equal, varying from 'Your friend to serve you' to 'your very humble servant'. Moray, as courtier and civil servant, midway on the social scale between Wallis and Boyle, in 1665 signed himself 'My worthy friend your realest servant' but almost at once began to content himself with a friendly, informal 'A Dieu' or even nothing at all. As already noted, both Moray and Boyle when describing the truncated gatherings of the Oxford-based Fellows of the Royal Society in 1665, spoke of Oldenburg's absence in the warmest terms, telling him how much he was missed, how often spoken of, and how on at least one occasion his health was drunk, a real compliment. And the pains they both took to get Oldenburg's *Transactions* published in 1665 show both their estimation of its importance and their friendly feelings towards the editor, whose private journal it properly was. With John Beale, clergyman and elder statesman living remotely in Somerset, Oldenburg had a very close epistolary relationship, painstakingly deciphering his deplorable handwriting. ('My frozen hand' Beale called it to explain why he sent a letter intended for Boyle via Oldenburg so that it could be transcribed).[3] Beale, a little patronizingly, offered counsel and advice about such matters as the prefaces to the separate volumes of Oldenburg's journal. He also urged Oldenburg to print an ever greater volume of material relating to horticulture and country curiosities.

Oldenburg certainly thought of Seth Ward (after 1667 Bishop of Salisbury and so 'Sarum') as a benevolently inclined patron to whom he could appeal for assistance when needed. True his appeal from prison to Ward, as to Lord Anglesey, was less a sign of ordinary friendship than of potential patronage. On the other hand Evelyn's visit to him in the Tower, although surely partly motivated by curiosity, showed real friendship between the two men, for it was troublesome for Evelyn to secure the necessary permission and not entirely without personal risk. From that frightening time Oldenburg was welcomed back to the Royal Society apparently by all. And in the months following he was to mention only one man who behaved unkindly to him after his release. This was his neighbour in Pall Mall, Dr Thomas Sydenham, a leading Fellow of the College of Physicians but not a Fellow of the Royal Society. This we know because of a letter written to Boyle who had apparently asked him to convey a message to Sydenham. Then Oldenburg, probably for the only time in their relationship, refused to oblige Boyle. He wrote,[4]

> I must beg your excuse for not seeing Dr Sidenham, who hath been the only man I hear off, who, when I was shut up, thought fit (God knows without cause) to raile against me, and that was such a coward, as afterwards to disowne it though undeniable. I confesse, that with so mean and un-moral a spirit I can not well associate.

By 1668 Oldenburg's official life and private affairs were both stabilized. He was constantly supplying more and more correspondents with news and endeavouring to help them in their relations with the Society and with the world of learning, bringing more strangers, both English and foreign, within the orbit of the Society. One might wonder whether all the Fellows welcomed his enthusiasm for provincial and, more, foreign correspondents. It would have been only natural if, as indeed seems to have been the case, some had felt that he was over-enthusiastic in bringing strangers to the notice of the Society and deplored his tendency to treat all his proteges as swans. But more often it seems that Oldenburg was giving an accurate sense of the approbation with which communications were received by those members when he read them at meetings. Although Birch never transcribed such things, the original minutes (Oldenburg's rough handwritten sheets) show how frequently his letters reflected the 'orders' (directions) which he had been given.[5] Oldenburg did try to be selective, especially when it came to publication in his journal, but might not at least some of those present at meetings have regarded some minor mathematicians and country natural historians whose letters he read out as being hardly worthy of attention? It cannot be ascertained now. Oldenburg's critical faculties were acute, but he was after all not a very competent natural philosopher, although there is no doubt that his knowledge of the subject increased greatly with the years; he was certainly capable of judging the quality of the work presented to him, but others might reasonably have disagreed. His great weakness was undoubtedly enthusiasm for any form of natural philosophy connected with his favourite subject, that universal natural history for which he always sought contributions. But on the whole others were eager for this Baconian project too, which lessened the chance of great differences in judgement.

Yet other duties came his way from time to time. Ever since the Fire of London the Society, meeting in Arundel House, had been developing vague plans for a purpose-built 'house' with rooms for meetings, an observatory, a chemical laboratory, and other purposes. In the autumn of 1667 Wilkins proposed that a committee be appointed to raise contributions for such a building, which was done and forms for subscriptions devised.[6] Naturally (but ironically since he himself was so relatively poor) Oldenburg was one of those appointed to this committee, which jokingly called itself a 'Committee of Beggars'.[7] Personally he seems to have disliked the task of soliciting subscriptions, telling Boyle[8]

> I confesse, I am as averse from and as much a Bungler in Begging, as any man, but I can deny myself and goe against the stream of my inclinations when the prosecution of Honor and publick Usefulnes is in the Case, as it here is; and therefore am not asham'd to begge when it is a means to accommodate and promote the good Ends of the Society's Institution. I think, since I have endeavor'd to serve them to the utmost of my power these 6. years *gratis*, and am a Beggar to boot as to my private fortune, I may extend my endeavors so much farther, as to goe a begging for the Society's establishment.

Initially the results of the Committee's begging were good. Several Fellows promised sums ranging from 40 to 100 pounds, while Henry Howard offered land adjacent to Arundel House (which he may in fact not have been legally entitled to do) and others promised generously. By May 1668 over £1000 had been promised and Oldenburg had come to believe in the new 'college' and made it an item of news to communicate to many correspondents.[9] Hooke was asked to prepare plans for a suitable building, which he had done in time to show to the meeting in June 1668;[10] at the same time Henry Howard showed his own plans to Wren,[11] and Wren was sufficiently interested to offer to produce a 'Modell' of his own design, although he probably never did so. However by August 1668 the scheme had lapsed, partly through lack of continuing enthusiasm, partly because of the difficulties involved in securing a firm lease on a portion of the strictly entailed Howard site. (In any case the Howard family began to plan to rebuild the now antiquated Arundel House.) Oldenburg's feelings must have been mixed. As a loyal Fellow and servant of the Society he was in favour of the scheme, but he clearly hated 'begging' for the promise of subscriptions to it, and it had little or nothing to do with his chief aim of serving the Society by his correspondence.

In the years 1667 and 1668 Oldenburg was himself in considerable financial straits, or as he put it, 'a Beggar as to my private fortune'. In the aftermath of his imprisonment Oldenburg was particularly despairing of his finances. This was not only because of the expected fall in the profit from the *Philosophical Transactions* but because he expected that his foreign correspondence would diminish. How he made money from this foreign correspondence is far from clear, but there are certainly tantalizing hints from his domestic correspondence that he did so. Certainly he obtained free postage from the State Paper Office by the good offices of Joseph Williamson, beginning in 1666 when he had arranged for his new postal address whereby letters bearing the words 'Monsieur Grubendol, London' went straight to Williamson, who paid the postage before forwarding the letters (unopened) to Oldenburg's house by the hand of a servant. Oldenburg then read the letters and extracted any political news and sent this back to Williamson for use in his own 'Gazette' and he must have paid Oldenburg for his work. Even so, since his foreign correspondence had aroused so much suspicion in 1667, he had felt it necessary to be cautious thereafter and to seek rather despairingly for other sources of income.

But it was not just lack of income that oppressed Oldenburg, for he was extraordinarily busy, as he had often complained bitterly to Boyle. Thus, retailing philosophical news from France to his patron and friend, and suggesting that he had ideas of how to utilize this, he wrote[12]

> Any competent Assistance would, by Gods blessing, enable me to bring from all parts of the world into England, as to a Center, whatever Ingenuities, Discoveries, Observations and Experiments lye scatter'd up and downe everywhere. But no man, that I see, does effectually consider this, which must needs overwhelme and oppresse [me].

A fortnight later he was to exclaim[13] in a postscript to another letter of news to Boyle, 'How large and usefull a Philosophical trade could I drive, had I any competent assistance'. A few weeks later still, he expostulated to Boyle in a passage worth quoting here at length,[14]

> I am sure, no man imagins, what store of papers and writings passe to and from me in a week from time to time, which I must rid myself off without any assistance. I have no lesse at present, than 30. correspondents, partly domestick, partly forrain. Many of them I not only write to, but also do busines for, which requires much time to inquire after such particulars, and dispatch such businesses, as they desire, if I mean to be gratified reciprocally in such things, as I bespeake of them. Besides my constant attendance on the meetings of the Society and Councill, and preserving what is said and done there, and giving order to have all registred, and reviewing all the Entries: solliciting also the performance of the manifold taskes, recommended and undertaken by the members of the Society; and distributing abroad such directions and Inquiries, as are thought conducive to advance our dessein, etc. I confesse I extend my patience as far as I can, but I am afraid, I have stretcht it so farr already, that it will break, unlesse some thing be done to fortify it in some measure, which I find not, that any of those, who are upon the place, and cannot but see my daily labors and toyle, are in the least sollicitous off; though some of them are forced to acknowledge even in my hearing, that no man in England would doe, what I doe, upon the terms, I goe away with.

That his correspondence often required services from him is demonstrated in existing letters. For example, in 1667 Carcavy, the French Royal Librarian, asked for lists of books recently published in England, a formidable task which led Oldenburg to appeal for help to Boyle and Wallis. Again, an unnamed correspondent asked for information about English oriental linguists, which required help from Wallis. From 1667 on both Justel and Hevelius often asked for Oldenburg to purchase books; they paid in time, but first Oldenburg had to lay out his own money after taking trouble to find the books and arrange transport. There were demands for information of all kinds from many and varied correspondents, which they expected the Secretary to supply. The seventeenth century was far from being an age of professionalism in administrative matters and public and private affairs overlapped to a degree far less common in modern times.

That a few at least of the Fellows of the Society recognised how hard

Oldenburg worked is shown by the words of Edward Chamberlayne, F.R.S. 1668, who four years later was to write[15]

> The secretary [of the Royal Society] reads all Letters and Informations, replies to all Addresses and Letters from Foreigne Parts, or from others, takes notice of the Orders and Material Passages at the Meetings, Registers all Experiments, all certain Informations, all Conclusions, &c. Publishes whatsoever is ordered and allowed by the Society.

It is noticeable that Chamberlayne refers to *the secretary*, presumably tacitly recognizing that at that time it was Oldenburg alone who undertook the secretarial duties for which two secretaries were provided in the Society's charter. This situation did not change in Oldenburg's tenure of office.

In 1668 Oldenburg began actively to seek other sources of income. He diffidently told Boyle that since Ward had expressed concern for his misfortune in 1667 and since he had great influence, he would have liked to have asked Boyle to recommend him to Ward if he learned of any suitable vacant 'places', naming particularly that of Latin Secretary. Ward himself now had a different proposal, as Oldenburg told Boyle.[16]

> The Bishop of Salisbury took occasion the other day of his owne accord to expresse his great earnestnesse to see me provided for with a recognition for my labors and concerns for the Society. He would, he said, have it urged in councill, and was pleased to add, that he for his owne part was ashamd, that I was so long neglected, who had for so many years spent all my time and all my pains upon the busines of the Society, without any consideration for it.

Oldenburg was naturally much pleased, but he felt that Boyle's influence would be needed to back that of Ward, writing,

> I confesse . . . I make very hard shift to weather it out; and I could be contented, it were insisted on in the Council, but that I shall want your presence and favor there, to represent a litle the particulars, that passe through my hands, and the difficulties, I groane under; which I cannot so well specify to others, as I know you will give me leave to do to you.

This clearly shows not only the great respect which Oldenburg felt for the Council, and his reluctance to spread abroad his private affairs, but also the easy terms on which he had arrived with Boyle.

Somehow Oldenburg felt encouraged to persevere, perhaps with direct encouragement from Boyle (now his neighbour in Pall Mall). He drew up a memorandum for presentation to the Council Meeting on 27 April 1668, when both Boyle and Ward were indeed present. The surviving draft (there is no copy in the Society's archives), reads as follows.[17]

> The business of the Secretary of the R. Soc.
> He attends constantly the Meetings both of the Society and Councill, noteth the Observables, said and done there; digesteth them in private, takes care to have them entred in the Journal- and Register-books; reads over and corrects all entrys; sollicites the performances of taskes recommended and undertaken; writes all letters abroad and answers the returns made to them, entertaining a correspondence with at least 30. persons; employes a great deal of time, and takes much pains in inquiring after and satisfying forrain demands about philosophical matters, dispenseth farr and near store of directions and inquiries for the Society's purpose, and sees them well recommended etc.
> Query. Whether such a person ought to be left unassisted?

This again is a thorough and quite fair statement of Oldenburg's work for the Society as well as a heartfelt call for aid. It obviously covers the same ground he had used to Boyle before when feeling the need of assistance to get through his work, but now he was asking for financial support rather than hoping that some one else would take on some of the secretarial burden under which he laboured.

There may have been some hostile feelings about Oldenburg's claim for financial assistance, and many may have thought that he made more from the *Philosophical Transactions* than was in fact the case. Undoubtedly the Council felt reluctant to commit itself to yearly payments. What was voted was 'That there shall be no standing salary allowed to either of the Secretaries' (somewhat unfair when Oldenburg did all the work, but, of course, the situation might perhaps change, as it did when he was gone), but 'that a present be made to Mr Oldenburg of fifty pounds.' This was not an adequate compensation for five years devoted work, and nothing was said about the future. But in time it seems that some Fellows regretted the Council's niggardliness, for without any previous hint in the minutes or in Oldenburg's correspondence, on 3 June 1669 the Council 'ordered, that a salary of forty pounds a year be allowed to Mr Oldenburg, one of the secretaries of the society, from the time, that the last present was ordered to him.'

This assured him a small but not despicable and luckily constant yearly income and as such must have been welcome. Moreover, it must have doubled his regular income, for the *Philosophical Transactions* brought him in no more than the same amount at best and was always insecure as sales necessarily varied from time to time, as did the amount he paid the printer. What we know is that in 1667 Oldenburg had told Boyle that he had not so far made more than £40 a year from the journal. He added ruefully that this was 'little more than my house rent'. (But it must be remembered that he lived in a prosperous, fairly newly developed part of London.) He had already had to make a new agreement with Martin which would reduce his annual revenue to what he variously estimated as £30 or £36, and to pay Martin at least £4 a volume. Things did improve later (in 1669 he paid Martin only £2 for the current volume)[18] but the journal was never a great source of profit, especially as he had to give away many copies certainly abroad and probably at home.

In the years 1667 to 1669 Oldenburg seems to have done little paid work for Boyle. His weekly letters ceased when Boyle moved to London, presumably to be replaced by regular conversations now that they were close neighbours. The few books which Boyle published during this period (he was not well and was soon to be seriously ill) were printed in Oxford, not London. The one Latin text which he then published, *The Origine of Forms and Qualities* was apparently not translated by Oldenburg although he made some other translations. He worked on J.J. Ferguson's *Labyrinthus algebrae* (The Hague, 1667), Latinizing the Dutch, but this was never published and so probably Oldenburg received nothing for his work. (This is the more likely as he was presumably inspired to undertake the task by Collins, who was the more impecunious of the two.[19] When in 1667 Oldenburg had succeeded in drawing Malpighi into constant correspondence and published the results (see above, Chapter 8) it was so well received that Oldenburg may well have made some profit, for Malpighi wanted only a few copies in exchange for permission to print. Certainly Oldenburg continued to publish all the long Latin letters with illustrations which Malpighi sent him, making them up into coherent texts. He was to undertake much more publishing work in the years to come.

In spite of the real uncertainties involved in his obtaining a reasonable income, Oldenburg's financial situation in the summer of 1668 must have seemed far more secure than it had been for a good many years. It was soon to be still further improved by the time-honoured method of marriage to an heiress. Oldenburg attended a Council Meeting on 10 August (a brief one to judge by the minutes) at which it was decided to adjourn the Society until the autumn. On the next day he procured a marriage licence and on 13th August he married Dora Katherina Dury at the church of St Bartholemew the Great although the bride was then resident in the parish of St Martins in the Fields and Oldenburg himself normally worshipped in that church.[20] The licence describes her as aged 'about sixteen'. Rumour which reached Collins through 'Mr Martin the Bookseller' (the Royal Society's printer) was that she was only fifteen years old and that her fortune was either 1500 or 2000 pounds. The rumour of the marriage was correct, but the data Collins retailed was not. The bride's fortune was almost certainly not so large, and as for her age, John Pell to whom Collins wrote knew better, replying that

> I know the Brides age was but 14 years and 3 months. I very well remember that she was born in May anno 1654; when her father and I were travelling together in Germany.

That, as the licence states, her father gave his consent was almost certainly correct. Dury was now very elderly and living permanently on the Continent; obviously he saw nothing improper in the marriage to a respectable man whom he had known long and liked, a friend of many years' standing to both the father and the daughter, the latter a girl whom he could not have seen for many years if at all since her babyhood, and who had been befriended by Oldenburg and his first wife. Probably that early association had rendered Oldenburg fond of the girl whom, we must suppose, he had watched grow up, perhaps in the country air of the marshes of Kent near Crayford where he had often spent holidays and now spent their honeymoon. And presumably she had amiable feelings towards him; at least she had known him well for the past four years. The great gap in age between bride and bridegroom was not usual at the time, as the comments of Collins and Pell show. Nor were child marriages common, although it must be said that thirty years or so earlier Boyle's sisters had married at equally tender ages, but to far younger (and richer) husbands. No one seems to have taken exception to Oldenburg's marriage, being apparently rather pleased than otherwise. Wallis, writing on business a fortnight later teased Oldenburg by saying 'Amongst this Moneths Transactions, I am told this is one, That you have embarked on a second marriage; I pray it may be a happy one'. And Beale wished him 'all Ioye and happiness', hoping, as he put it, that since his young wife was the daughter of so good and pious a man as Dury, special blessings would accrue to her.[21]

In worldly terms Oldenburg had done quite well for himself, for the bride had inherited from her mother a modest estate consisting of two good farms called Battens and Wansunt near Dartford Heath not far south of the Thames, the latter commemorated in the name of a road in a 1960s housing estate.

Together the farms totalled about 200 acres and were said to bring in sixty pounds a year. There was a tenant to run the farms, which also provided a home for holidays. This sum would have meant a doubling of the family income, although of course there may have been some restriction on the use of this profit as a result of Mrs Dury's will. It is pleasant to record that Oldenburg appears to have allowed himself a fortnight's honeymoon during which he wrote no official letters. He did write to Lord Brereton, with whom Pell was staying, and perhaps to Wallis. But his correspondence remained light until the Society resumed its meetings in late October. Whether any of his contemporaries thought that he might have married for money we cannot know, nor do we know what they thought or knew about his previous relations with his wife during her earlier girlhood. But in view of his well-known piety and moral outlook it would have been only malicious tongues which commented adversely. Equally obscure are the feelings of the bride. One may surmize that, while Oldenburg was genuinely fond of his bride, she respected him and was grateful for his care. Very probably, after such an unsettled childhood as she had had, having been brought up initially in Holland, then being brought to England at the age of three to be baptised and perhaps remaining there although her mother did return to Holland until the restoration, having lost her mother when she was only ten, and almost certainly never having seen her father since babyhood, she was pleased to exchange her status of orphan for the security of a home with a husband who although elderly was a familiar, kindly, and trustworthy figure whom she remembered favourably from early childhood, and who provided some continuity with her dead mother as well as the promise of a settled (instead of uncertain) future.

She apparently never knew any of her mother's grand relations, some of whom were far higher in the social scale than either her father or her husband, and which made her almost a cousin of Robert Boyle, her mother's first husband having been a brother-in-law of Boyle's much older sister Sara. Her mother, besides, came from a good Anglo-Irish family the Kings, and was a sister-in-law of Frances Moore, Viscount Ranelagh's mother. She had brought the two sons of her first marriage to Holland for education before her marriage to Dury. One may have died young; the other returned to England to marry into the Honeywood family, Oldenburg's former patrons, but he seems never to have had any contact with his half-sister. Oddly it does not seem that Lady Ranelagh, so virtuous, pious and benevolent, had ever taken any overt interest in her small orphan relation, although the child's mother had corresponded at length with Lady Ranelagh for some years and while trying to make up her mind whether to marry Dury. It is worth remarking that for all these rather grand maternal relations, Dora Katherina's paternal family connections were no higher than those of Oldenburg, for John Dury was the son and grandson of Scots Presbyterian ministers, as Oldenburg was of Calvinist teachers and divines. But the girl had other probable family connections with the wife of the baronet Sir John Wroth of Bexley, Kent, for whom Oldenburg later wrote a letter of recommendation when he and his son travelled to France;[22] it is possible that, since they lived near the farms which Dora Katherina now owned, they had, at least in some sense, looked after the little girl in the years after her mother's death and that Oldenburg had stayed with them on his visits to 'the good air of Crayford in Kent'. Certainly in some way her father's directions for her guardianship had been effective. It must be said that family connections such as those which Dora Katherina possessed which seem so distant in modern terms, counted greatly in the seventeenth century. Moreover, a wife's social standing could in at least some sense affect that of her husband. Hence Oldenburg's social standing must have been improved by a marriage which, however remotely, made him connected with the Boyle family. There is no intimation of what any of its members thought about any aspect of Oldenburg's marriage and there was certainly no apparent change in the relations between Boyle and Oldenburg now that the latter was in some degree a relation, although the normal relationship between Oldenburg and his patron became warmer over the years. Of his wife one can only say that Oldenburg had married a gentlewoman and that she must have had an upbringing suitable to her family connections to make her so. After marriage, to judge by the few references to her in Oldenburg's correspondence, she retired into the background and comported herself as any well-bred gentlewoman should, being accepted on these terms by Oldenburg's friends. One of the very few references to her in letters during the first year of their marriage occurs in a letter from Oldenburg to Evelyn, of which the postscript reads[23] 'My wife is very sensible of your favor and returns her humble service to yourself and your Lady.'

So much for the new Mrs Oldenburg's family. It seems appropriate here to say something about Oldenburg's own surviving relations. His sister, married to Heinrich Koch, had a son Heinrich, born in 1644 (d. 1719). This son dropped the German form of the family name in favour of the Latin version of Coccejus, perhaps when he moved to Leiden in 1667 as a form suited to the eminent German jurist he later became. His uncle clearly maintained some contact with the family, for when in February 1667/8 Colepresse travelled to Leiden for study, Oldenburg supplied him with several letters of introduction, one of which was to Coccejus.[24] Three years later, in March 1670/1, Coccejus travelled to London, ostensibly to learn English and to investigate the state of England, as he put it, and there he visited his uncle. He then toured France, carrying a letter from Oldenburg to Huygens, whom he presumably met. How much English he learned is uncertain but certainly not enough to write it with any ease for, with some apology, he subsequently wrote to his uncle in Latin, either expecting that the older man would have forgotten his German or, like his uncle twenty years earlier, finding epistolary German cumbersome and slow to write.[25] As he told his uncle, he had become engaged to the daughter of a man from Strasbourg high in the service of the brother of the reigning Duke of Würtemberg and was anxious that his uncle 'and your dear wife . . . your delightful wife' should attend the ceremony. (Was this a real tribute to Dora Katherina, or mere politeness? One would like to think the former.) In fact, there were difficulties and postponements and in the end the wedding only took place in November 1673. Oldenburg did not attend it but he kept up the correspondence with Coccejus on an official basis for a short time, discussing technological matters.

In 1674 Oldenburg had more news of his German relations when Constantijn Huygens, Christiaan Huygens's distinguished father, poet and diplomat, wrote from The Hague[26] that 'A good woman here, who describes herself as your sister, some time since addressed herself to me, on the basis of the acquaintance between us' and appealed to Huygens for some assistance (which he had given), adding 'I do not know, Sir, if you are aware of the state in which she is, which seems to me quite worthy of your pity.' This is puzzling: Oldenburg's brother-in-law Heinrich Koch was still alive, so Oldenburg's sister, the mother of Coccejus, can hardly have been this indigent woman. However, Heinrich Koch's brother Johannes Koch (also called Coccejus, like his nephew) had been Professor of Theology at Leiden and died in 1669, so that the petitioner might have been his widow, who in seventeenth-century parlance and custom would have been counted as Oldenburg's sister, although in modern customary speech no more than his sister-in-law. As no letters from Oldenburg to Constantijn Huygens survive, nothing further is known. It is worth noting, to indicate the degree of friendship existing between this great gentleman and Oldenburg, that the letter concludes with the words 'I finish my paper, without formality. You know my hand, and my heart.' In seventeenth-century terms this was great condescension, and Oldenburg must have taken it so.

After the Royal Society's grant of a yearly salary for Oldenburg and his marriage, he made no more pleas of indigence to his patrons, evidently aware that an income of perhaps £100 a year would be regarded as a competence, even though obviously not by any means wealth. Nevertheless it is clear that he continued to feel the need for more income to provide for his family, growing in the 1670s, and he worked hard to supplement his income. Obviously his publication of the *Philosophical Transactions* was half public (because it would have been impossible had he not been Secretary in charge of correspondence) and half private (because he alone was the publisher and any profit accrued to him). Disrupted by the Plague, the Great Fire of London, and his imprisonment during the stormy years of 1665–7, by 1668 it was once again appearing regularly. Oldenburg had now learned much about the proper content of such a journal and was taking ever greater care about the quality of the letters and papers he published in it, taking pride in its popularity and prestige in the learned world. No wonder that he was annoyed when new readers assumed, as they so often did, that it was the Royal Society's journal and Oldenburg only the editor. True, he had had assistance from others from time to time: Wallis, Boyle, and Moray during the period of November/December 1665 when it was published in Oxford because of the disruption of the London book trade and he continued to receive some advice from them and others over the years. For example, Wallis assisted him with reviews of mathematical books sometimes, John Beale who treated Oldenburg himself as a worthy protege while urging him to print more on horticulture, read and edited many of the prefaces to the yearly volumes, while others, notably Boyle and Wallis, sent him short works which they expected him to publish without question, as he naturally did. But the journal remained his responsibility, to chose what to include, in what order, and to oversee and pay for its printing, as his colleagues well knew. By 1669, probably rendered wary by his first difficulties with Hooke, Oldenburg was careful to secure express permission for publication from all his correspondents. So in 1672 he assured Martin Lister, now a constant correspondent,[27]

> Sir, whatever matters you shall please to communicate unto me, I shall manage, as to their publishing or keeping privat, as you shall direct. Whatever you shall intimate you would not have made publick, I shall faithfully keep privat; but what you forbid not, I shall presume you will not deny the knowledge of to the philosophical and curious world.

No doubt taught by the experience of all the scientific controversy of the previous years, Oldenburg was becoming more and more cautious in securing permission before publication.

Critical appreciation of the *Philosophical Transactions* was enormous, both at home and abroad. Oldenburg was virtually bombarded with requests for copies and seems to have felt obliged to distribute a surprising number of them at his own expense. Its fame on the Continent was naturally increased when, as soon happened, the *Journal des Sçavans* began reproducing extracts in French as, after its founding in 1668, did the *Giornale dei Letterati* in Italian, all based on copies supplied by Oldenburg to the editors of these journals. In return he received copies of these journals from which he could make extracts to the benefit of English readers.

Even so, there was a considerable demand for a Latin translation of the *Philosophical Transactions* especially of the earlier numbers. This Oldenburg himself failed to supply, being far too busy, so the task was taken over by foreigners with only moderate success. In 1668 Colepresse, writing from Leiden, told Oldenburg that he had heard of plans for a Latin edition of what he properly called[28] '*your* Transactions' and the very next year the first volumes were published at Amsterdam, beginning with the current volume. Oldenburg was far from pleased, especially with the title *Philosophica Regiae . . . ab Henrico Oldenburgio conscripta* (Royal Philosophical Transactions . . . compiled by Henry Oldenburg) an ambiguous title which annoyed him almost as much as the many errors he found in it, as he bluntly told the translator, one John Sterpin, forbidding him to proceed with any further translation. The next attempt was by Christopher Sand, a young theological writer, a German living in Holland. He took the trouble to correspond with Oldenburg to clarify difficult points, but called the journal *The Philosophical Transactions of the Royal Society in England, author Henry Oldenburg*. This title did not wholly please Oldenburg either, for it was *his* journal, not that of the Royal Society and the papers and letters included in it were never limited to those by members of the Society nor even always to what had been taken notice of in meetings. He continued over succeeding years to object whenever anyone failed to understand and acknowledge him as sole proprietor as well as editor and publisher of the journal, although to little effect. This was a point of pride and of accuracy, not of finance, for he continued to make very little from it.

In early September 1668 Oldenburg returned to his work as Secretary of the Royal Society with renewed vigour. Although after his marriage he was financially more secure in one way, he needed more income to support his new family. Hence, in default of any more certain sources of income, he undertook more work as editor, publisher, and translator, especially, but not exclusively, for Boyle. The first book by Boyle for which he had acted as publisher and for which, as he usually did later, he provided a preface signed H.O., had been *Experiments and Considerations touching Colours* of 1664. This was quickly followed by *New Experiments and Observations touching Cold* of 1665 and these in turn were succeeded by *Hydrostatical Paradoxes* of 1666 and several more, not quite so quickly.[28] It has never been ascertained how much or by what means Boyle paid Oldenburg for this work but it is likely that, as was to be the case when Halley published Newton's *Principia*, the publisher received the profits from the sale of the book, which in the case of Boyle's books cannot have been entirely negligible, although Oldenburg never mentioned it. Boyle also contributed numerous short works for publication in the *Philosophical Transactions* and although, of course, Oldenburg did not derive any money from these directly, their presence undoubtedly boosted the reputation and hence the sales of the journal. Probably also he continued the work begun many years earlier of making extracts from books for Boyle's perusal, for many of these notes, undated, survive still in the Boyle Papers in the Royal Society's archives. There was further translating work for which as we know Boyle paid, for from the time of his first scientific publications Boyle had been annoyed by the appearance of unauthorized Latin translations mostly published in Amsterdam. This troubled him both because he thought unsupervised translations notoriously inaccurate and because he had then no opportunity to correct errors in the initial printing, of which there were usually many, not only because of inadequate proofreading but even more resulting from his own unsystematic method of working. (His poor eyesight prevented him from supervising the work properly himself.) Hence he was delighted to have Oldenburg ready and willing to translate his books into Latin almost as fast as the English versions came off the press, with speed and accuracy.

Oldenburg's Latin translation of the *Experimental History of Colours* had been published in London in 1665 only a few months after the appearance of the English edition, successfully forestalling Dutch editions. Oldenburg translated *Experiments and Observations touching Cold* into Latin as the sheets came from the press, so that his translation was well on its way before this English edition appeared; indeed some sheets were actually printed off, although the whole text was never to appear in print, or if it did appear it was totally lost,[29] possibly in the Fire of 1666. Other works were more fortunate.

The Latin of *Hydrostatical Paradoxes* was to be published at Oxford in 1669, the Rotterdam edition of 1670 being a reprint of this text; it is not certain that this translation was by Oldenburg, although likely. There exists, scribbled on the back of a letter, notes made by Oldenburg probably in 1674 about his past translating work for Boyle and the sums due to him as translator: he was paid at the rate of ten shillings a sheet, not at all a bad rate of pay for the time.[30] (For comparison it is relevant that his Poor Rate tax was assessed at 15 shillings while Lady Ranelagh paid 18 shillings.) These notes show that in June 1671 he had delivered ready for the press a Latin translation of 'tracts of Cosmical qualities, Cosmical Suspicions etc.' of which the English appeared in 1671 and the London Latin edition in 1672, in this case having been forestalled by printings in both Amsterdam and Hamburg in 1671. In December 1671 Oldenburg delivered to Boyle Latin translations of several other tracts, including *Of the Saltnes of the Sea* published in English with other *Tracts* only in 1674, 'respiration', a paper published first in the *Philosophical Transactions* in 1670, and what Oldenburg accurately called 'Intestin Motion' (properly so because although it is now usually referred to as 'Absolute Rest', in fact the work endeavours to show that there is no such thing), first published in English in the second edition of *Certain Physiological Essays* in 1669. Some of these translations of Boyle's various 'Tracts' seem not to have been published, but Oldenburg was paid for them all the same. At the conclusion of his jottings Oldenburg calculated, apparently with some difficulty (but simple arithmetic was then not so simple and easy as it later became) that Boyle owed him £5. The last known translation made by Oldenburg for Boyle was of the curious paper which he published in the *Philosophical Transactions* in February 1675/6 of a pseudonymous text entitled 'Of the Incalescence of Quicksilver with Gold, generously imparted by B.R.' which, as Oldenburg pointed out, no doubt obediently following instructions from Boyle, was being published prematurely without a sufficient number of trials. The English text, again no doubt following instructions from Boyle, was accompanied by a Latin translation made by Oldenburg 'to gratifie the Curious amongst Strangers, as well as those of our own nation'. (This is the text which so interested Newton as already mentioned above, Chapter 6.) A further indication of Oldenburg's close involvement in Boyle's work is indicated by 'A list of such papers as were mention'd to me by the Hon'ble Robert Boyle March 26. 1677', these being both published and as yet (or never) published titles.[31]

Besides all his editorial work for Boyle Oldenburg acted as publisher for a number of other authors from 1669 onwards. The first of these was Malpighi:

Oldenburg's efforts to foster and publicize his work have already been noted (Chapters 5 and 8, above); here it should be pointed out that he acted as the publisher of most of Malpighi's work between 1669 and 1677, no sinecure in most cases. True, in the case of the first such work, the *Dissertatio Epistolica de Bombyce*, this only involved writing a (signed) preface and seeing the text through the press. Publication of later works (those on the development of the chick in the egg of 1673 and on the anatomy of plants of 1675) involved the putting together of letters written over several months and collating the illustrations sent separately, as well as writing a preface, and seeing the material through the press, all quite troublesome tasks. Presumably Oldenburg made some profit, even if a small one, from his labours.

There was more such work which Oldenburg could undertake, and in 1671 he was particularly busy with publishing ventures. He had already translated from Latin to English the *Prodromus* of Steno published in 1669, an important geological text; now the English translation was, rather curiously, both published independently in 1671 and also annexed to an issue of Boyle's *Essays of Effluviums* published in 1673 (although not to all the issues of that year). Whether Boyle paid anything for this translation or merely allowed Oldenburg to profit from the publication of Boyle's text is not known, but in any case it must have made Oldenburg something to add to his income. In 1671 there also appeared a couple of independent ventures by Oldenburg, namely a translation from French of François Bernier's *History of the Late Revolution of the Empire of the Great Mogul*, a combination of history and travel of the kind which appealed greatly to the reading public of the time, and with a very different appeal, a translation from the German of *A Genuine Explication of the visions of the Booke of Revelations* ascribed to A.B. Peganius, most probably really by Christian Knorr von Rosenroth, a known German mystic with whom Oldenburg certainly corresponded later.[32] It was about that time that he had toyed with plans, urged on him by many correspondents and colleagues, to translate his *Philosophical Transactions* into Latin. Besides all this he was also occupied in translating incoming letters for reading to the Royal Society as well as translating letters containing political news for the use of Williamson.

The years 1672 and 1673 saw the peak of Oldenburg's correspondence activity. Between the beginning of April 1672 and the end of May 1673 his correspondents numbered at least sixty-two and probably more, since not all letters to and from Oldenburg have survived or left a trace. He normally followed the practice (as he once told Lister who said that Oldenburg 'never failed' to answer his letters promptly)[33] of 'making one letter answer another';

that is, dealing as quickly as possible with all incoming letters. He often set aside at least part of one day for writing replies to all the letters he had recently received. He could not always have followed this plan strictly for he sometimes had to wait until he had read incoming letters in whole or in part to the Society meetings so that he could report the members' reactions, while on other occasions he had to write to other correspondents for information. But the number of occasions when he had to apologize for being late in replying were few. Moreover, he tried to be meticulous about keeping all incoming letters although there were inevitable exceptions. For some were sent to other correspondents to peruse and not returned, some were borrowed on the spot and never came back, while others still, because of the lack of the competent assistance of which he so often complained, were mislaid before they could be copied into the letter book. Others were mislaid after copying, while others not deemed worthy of copying were very easily mislaid. The most carefully kept tended to be those too technical for any but himself to copy, like the mathematical letters of 1676 and 1677. There were, inevitably, failures of the post, especially in the case of Continental letters, prone to be caught up in the difficulties inherent in war-time. For it must be remembered that the Third Anglo-Dutch War came to involve both France and the German Empire and dragged on for years and only after the English withdrawal in 1674 did such difficulties lessen, although they never ceased entirely. Oldenburg inevitably continued to feel excessively burdened, for success only made his task more, not less, laborious.

It is not at all obvious quite why Oldenburg should have burdened himself with so much translating and editorial work as he did in the years 1668 to 1671, unless it was an attempt to increase his income still further now that he was married and no doubt anticipating an increase in the size of his family in the next few years. In seventeenth-century terms his 'family', which then would have included his servants, must necessarily have increased with marriage. Dora Katherina must have had one or more female servants in the house to assist her in housekeeping, as Pepys's young wife did even in their most impecunious days and very possibly Oldenburg needed a 'boy' to run errands. The family in the modern sense did start to increase with the birth of the Oldenburgs' first child in the summer of 1672. In recognition of this event John Beale benevolently wrote[34] that he was sending ten shillings by the carrier on 16 July 1672 'only to buy a Sugarloafe for your Infantry'. Although it has generally been assumed that this child was a boy, since later a boy, Rupert Oldenburg, was referred to as the elder of the two children, there seems reason rather to believe that this child was in fact a girl, not least because there once existed[35] some 'excellent Admonitions and Directions for the conduct in life' of his children (presumably not dissimilar to those I have speculatively assigned to 1664 (Chapter 4, pp.89–91), that addressed to his daughter being dated 'Pallmall. 16 October 1672'. (There is, of course, the further possibility that this child was a daughter who later died, but there are no references to such an unhappy event, so that the balance of probability is that this child was the Sophia who was certainly in existence in 1677). When Hooke called on Oldenburg on 20 August 1673 he noted that he 'Saw his wife and child and Pepys', which does not help to determine the sex of the child. It does tell us that Pepys was more familiar with Oldenburg at that time then there is any other indication of his being.

In September 1672, a few months after the birth of his first child, Oldenburg was ill, sufficiently so as to be confined to the house.[36] (It was perhaps this illness which led him to begin his 'Admonitions' to his infant daughter.) He was also in some kind of difficulty, what kind cannot be ascertained, although it seems that Boyle and Moray believed that Sir Joseph Williamson had the ability to extricate him from it. Both men were very anxious that he should do so, as both wrote. Boyle, referring unspecifically to 'the Affair of our threaten'd friend [which] will grow more difficult, the longer it remains undispatch'd', begged Moray to renew the requests which they had already made to Williamson 'who is so great a friend to the Society in Generall, & has been so to the Party in particular', calling Oldenburg a 'menaced Person'. Moray obliged by sending Boyle's letter to Willliamson, adding a letter of his own in which he politely suggested that it was only 'multitude of Business' which had prevented Williamson from acceding to earlier requests for help for Oldenburg. (These letters exist by the double chance that Boyle, who had intended to call upon Moray to discuss the matter, was prevented from doing so by the arrival of a visitor, and Moray, intending to go to see Williamson at his office, was unexpectedly called out of town.) Both Boyle and Moray regarded Williamson's intervention as essential to prevent what Boyle called 'an avoidable melancholy'. But neither wrote any more specifically, and nothing survives in Oldenburg's remaining papers to permit identification of the difficulty, which one supposes must somehow have been overcome.

For the rest of the year Oldenburg seems to have been well and to have continued his life in his usual busy fashion. Indeed he became if anything busier than ever to judge by his apology[37] early in February 1672/3 to Huygens for failing to write to him with his usual polite promptness; he had, he said, been prevented from doing so by 'the state I am in, being almost overwhelmed with business at present'. He told Leibniz on the same day he had

> received at Court this morning business which must be attended to without any delay, so much so that I shall have scarcely even one minute in which to dine at home.

It is possible that this business derived from his connections with Prince Rupert but more likely that he had met Williamson at Court and been given urgent work to do. A letter written to Williamson a week later[38] at the early hour of 7 a.m. indicates something of the nature and scope of his work for Williamson. It also throws also some light upon why he had been forced to work in haste, although the business there mentioned is probably not identical with that of a week before. It seems that Williamson had sent him a (presumably intercepted) letter which Williamson thought was written in German but which Oldenburg recognized as being in Danish of which 'tongue I have no further skill, than it hath affinity with the Dutch', as he protested. He was, however, able to make out that it was to the King of Denmark from his envoy in London and discussed current English affairs.

He obviously found this demand from Williamson an unfair imposition in addition to his other work, for he added ruefully, 'I have sate up almost all night in finishing the Extracts for the Mondays gazette'. The gazette was the twice weekly *London Gazette*, edited by Williamson as a successor to his earlier *Oxford Gazette* of the plague year. The 'extracts' were presumably from foreign papers and he had provided enough, he estimated, to fill an issue and he had more in hand. He closed by assuring Williamson that 'I hope to be at Whitehall about sermon time', the day being Sunday, which suggests the urgency with which Williamson demanded the work. Moreover, it evidently shows that he and Williamson regularly met at Court rather than in the State Paper Office. At a somewhat more leisurely pace, Oldenburg sent Williamson in April 1673 a translation of several German letters to various German nobles which detailed political news of concern to the English one of which made the rather rude remark that Williamson, who was about to go to Cologne for the negotiations just beginning for a general peace (negotiations which were unsuccessful at this time) was 'a mercenary to France'. In his covering letter Oldenburg called such remarks as these abusive and 'dangerously' defamatory, adding[39]

> Sir, when anything coms from you to me, I sett all other business aside, till yours be dispatch'd. I hope it will be considered, that I have spent many a day, and sometimes a good part of the night too, in such work, as concerns the kings service; and that I am very ready to goe on so upon all occasions.

This suggests that perhaps someone had tried to oust him from the job of Williamson's translator, or that Williamson had complained that he was dilatory, or that his support of English affairs was suspected of being lax. Fortunately for him, the problem, whatever it was, seems not to have re-occurred and he was to continue this work for the future.

In the opening months of 1672/3, as the contents of the foreign letters Oldenburg was called upon to translate show, important public events were in train which, as it happened, were to have some influence upon Oldenburg's private life in the months to come. Parliament had not met between April 1671 and February 1672/3; in its absence the King had, in March 1672, declared war on the Dutch (the third Anglo-Dutch War), in order to qualify for a subsidy from the French as specified in the secret Treaty of Dover between King Charles and Louis XIV, signed nearly two years earlier, which committed the English to support the French in wartime. This was almost immediately after Charles II had issued the Declaration of Indulgence on 25 March 1672 which relieved both Dissenters and Catholics of most of their disabilities under the then existing penal laws. When Parliament met in February 1672/3, summoned to provide taxes to support the war effort, it immediately demanded the withdrawal of the Declaration of Indulgence, a demand to which the King reluctantly acceded on 8 March. Soon he was forced as well to accept Parliament's Test Act (1673) which barred from public office all who refused to take the Oaths of Allegiance and Surpremacy and to take the Sacrament according to the Anglican Church. This had a considerable effect on public life. For example, the Duke of York, a fairly recent Catholic convert as was generally known, was necessarily driven out of his office of Lord High Admiral, by virtue of which he was head of the Navy Board of which Pepys was Secretary, to the latter's great regret, for the Duke of York was a serious and effective naval administrator.

The Test Act had also an immediate effect on Oldenburg's life. In the summer of 1673, incidentally in company with Boyle, he, like many others, conformed to the recently promulgated Act, taking the Oaths of Allegiance and Supremacy and at the same time taking Communion, thereby proving himself to be a law-abiding supporter of the Government and of the Anglican Church.[40] It was not mandatory for Oldenburg to take the Oaths, as he held no public office, but not to take them incurred certain potential restrictions: non-jurors were debarred from prosecuting law suits, or acting as guardians of children or as executors of wills, putting them in a dubious legal position. Prominent persons like Boyle, especially those resident in or about London, or connected with the Court, or being members of organisations like the Royal Society, commonly and publicly hastened to take the Oaths and the Sacraments, even when, like Boyle, they could not possibly be suspected of any irregularity either in their religion or their loyalty to the Crown, and were known to be shy of taking oaths, for religious reasons. For a foreigner like Oldenburg, never naturalized and in the public eye, it was a wise, almost a necessary precaution. It should be noted that the next year Parliament, besides its successful interference in internal affairs, forced the King to withdraw from the Dutch War.

During 1674 Oldenburg continued active in publishing and translating several works by Boyle as testified by fragments extant in the Boyle Papers.[41] He was not as busy as usual on Society business in the spring of that year because the Society itself was becoming increasingly inactive. This seems to have affected Oldenburg with a similar lassitude in his official duties, perhaps feeling that they were not worth while or perhaps merely busy with other matters. In any case he never wrote up the minutes for the Society's meetings on 9 and 16 April 1674, which consequently do not exist in the Journal Book although Oldenburg's rough minutes do survive. There are no surviving minutes at all for the whole of May 1674, although Hooke certainly recorded a meeting on 7 May when presumably Oldenburg was not present, since Hooke reported that he was 'sick'. On 13 May Hooke reported that Oldenburg was 'well' but 'not at home', with no further remark. On 14 May he was not at Gresham College nor was he apparently there on 21 and 28 May when Hooke recorded that there had been a meeting. But there are no minutes, showing how completely all Secretarial business was so firmly left in Oldenburg's hands that no one else saw the need to bother about records. He had certainly recovered from his illness by the end of May, for he then resumed his normal schedule of correspondence. However, the Society had not recovered as well as Oldenburg had done and on 18 June it was decided that the weekly meetings were to be adjourned until the autumn because of the poor attendance. (This was allegedly 'by reason of the season of the year, wherein many go into the country', but June was very early for the usual holiday migration from London to the country, so perhaps the reason given was a polite fiction intended to cover up the members' negligence.) The adjournment of the Society meant less work for Oldenburg for there were no meetings to attend, but the density of his correspondence increased markedly throughout the summer and early autumn of 1674.

Rather surprisingly, the Council resumed its meetings as early as 27 August 1674 with Petty, a Vice-President, in the chair, to begin planning the autumn programme and to consider possible reforms to ensure that the Society was once again active when next it resumed its meetings. The purpose of the meeting was announced as being to consider ways and means of making the Society 'prosper'. Petty himself was anxious to have more experimental discourses read at weekly meetings than had been the case earlier in the year. It was then suggested that the Society should pay the expenses of those Fellows willing to provide such discourses, the first example of the realization that many Fellows were not well off and found the cost of experimental apparatus a burden. It was believed that the necessary money could be provided out of the weekly subscriptions of two shillings combined with a drive to persuade delinquent members to pay their arrears, never an easy task. Hooke's diary entry 'Oldenburg about arrears' is ambiguous; it is unclear whether Oldenburg had suggested a method of acquiring funds or that it was to be his task to collect the sums owed (not his normal task), or that Hooke himself was delinquent. When the Council next met there was a new proposal, namely that it should ask for an advance on the weekly subscriptions from as many Fellows as possible in order to raise a lump sum, and no more was then said about collecting from delinquents. The Council was at this time and for some time to come so preoccupied with problems of finance and framing a form which should be legally binding on all Fellows to ensure that they paid their subscriptions promptly that little else seems to have been discussed. It was agreed that weekly meetings of the Society should recommence on 12 November 1674, as they did, and that an announcement should be made to that effect.

When the meetings resumed, Oldenburg was there to attend to the taking of the minutes. Wallis acceded to the Council's request conveyed by Oldenburg in mid-October[42] and read a prepared discourse, as was to be the pattern of meetings for the rest of the year. The difficulty was to find members who would promise to read a discourse at a meeting convenient to themselves. The scheme worked reasonably well as long as Oldenburg took a hand and wrote first to persuade and then to remind those who had promised papers that the appointed time had arrived, as he conscientiously tried to do. As a consequence, throughout 1675 and 1676 meetings were occupied by prepared papers and any letters which Oldenburg thought it worthwhile to read or abstract. On the whole a sufficient number of Fellows accepted their responsibilities, meetings continued on a regular basis, Oldenburg attended regularly and the minutes were relatively easy to take and formalize. At the same time Oldenburg's correspondence during 1675 and 1676 continued very active so that he was always busy. This was, of course, not only the period of Hooke's quarrel with Huygens and hence with Oldenburg, but also the period when Oldenburg was mediating between Newton and Leibniz, a correspondence which necessitated much tedious copying for Oldenburg, while he was also dealing with the still-continuing controversy aroused by Newton's 1675 paper on light and colours. And there were many other correspondents and their interests to be accommodated. In the summer of 1676 Oldenburg felt so overwhelmed with correspondence that he was on occasion forced to apologize for his distraction,[43] saying, 'The multiplicity of letters, I am obliged to write making me sometimes forget, whether I have written such and such letters or not'. Evidently his system, whatever it was, had broken down. At the same time work went on for Boyle, especially the publication of a number of 'letters' sent directly to Oldenburg.

Although traces of Oldenburg's purely private life are few in these years, thanks to Hooke's entries in his diaries we do know some details. A little before 30 March 1675 his son was born and named Rupert, presumably for Prince Rupert, with whom Oldenburg apparently always had had some contact, although the details are now missing. But they seem to have been in particularly close touch during the next year (1676). This was in connection with a technological matter that involved Lister:[44] he claimed to possess a particularly dark and fast black pigment which could be applied to marble, a matter which also interested Prince Rupert. In urging Lister to send him a sample, Oldenburg told him[45] that he had 'waited on his Highnesse, as I have sometimes the honor to doe' and had told him of Lister's pigment. From the way in which Oldenburg expressed his connection with the Prince one may fairly assume that some of Oldenburg's many visits to Whitehall were for the purpose of 'waiting on' him, or at least included conversations with him, and that they were on quite friendly terms. They must have been if Oldenburg's son was really named after the Prince, presumably with his permission.

In the summer of 1675 Oldenburg had received a visit from a young Frenchman, Denis Papin, who arrived in England with a letter of introduction from Huygens to whom he had acted as assistant in Paris during the previous year in experiments on air.[45] When he told Huygens of Papin's arrival[46] Oldenburg promised 'to be as much use to him as possible'. This he quickly succeeded in doing, for he rapidly found him a post as tutor, with, as Papin himself told Huygens,[47] 'a young gentleman . . . who needs a tutor and whose father has a taste for experiments', a post which he had taken up by 13 September.[48] He did not in fact long hold this post, for Oldenburg had introduced him to Boyle soon after his arrival and by the end of 1675 he had become Boyle's assistant, to be highly valued and to keep his post until 1680, when he was elected F.R.S.[49] That Oldenburg was thus able to act as a patron to others indicates a step up in the social scale.

He was also at this time working harder than ever for Williamson, translating incoming letters into English and turning Williamson's official letters into Latin, in fact acting as Williamson's Latin Secretary and thus unofficially occupying the official post of Latin Secretary to the Government, which he had earlier failed to secure. He also continued to excerpt letters from France and Holland, providing the State Paper Office with a wealth of foreign news and gossip, much of which was in turn used by Williamson for his *London Gazette*.[50] Williamson was in effect Oldenburg's employer, patron and protector and, as already noted, assisted him and the Royal Society together by paying the postage on all incoming letters with the Grubendol address and forwarding them by hand to Pall Mall. He also tried to assist Oldenburg financially in February 1675/6 by procuring for him, apparently at his own request, a warrant as Licenser of books dealing with history and affairs of state, clearly a coveted appointment.[51] It is not quite clear how a Licenser made money out of his position, but presumably, in normal seventeenth-century fashion, the author paid a fee for his licence and gave a gratuity to the licenser. It is worth noting in passing that the Licence was granted to 'Henry Oldenburg Esq', a wording which confirms his status as a gentleman. But however much he longed for such an appointment, Oldenburg held it for less than three months, of his own volition.[52] He ran into trouble over his refusal to license a scurrilous French book containing an account of the relations between the King and the Duchess of Cleveland, thinly disguised as a romance under the title of *Hattige, ou les Amours du roi de Tamaran*. Possibly in retaliation, an anonymous informer denounced the bookseller, and by inference Oldenburg, for selling copies which, the informer claimed, had been printed in England. This may have been the case, but as Oldenburg had refused to license it its possible printing in England was irrelevant. The informer also hinted that the laxity which must have permitted the supposed printing showed that Oldenburg was not loyal to the Government, which was of course far from being true. It was all too much for a man who was always conscious of rectitude. Oldenburg resigned the post at the end of April on the pretext that it was too time-consuming. That this was so was true, for, conscientious as always, he had insisted upon reading through all the books which were presented to him for licensing. His real reason for resigning was the false position in which he found himself, open to unfounded attacks for disloyalty. He apologized profusely to Williamson for giving up the post which Williamson had taken the trouble to procure for him, knowing that Williamson had meant him only kindness. He could only hope that Williamson would understand that he had truly appreciated the genuine kindness, friendship, and concern for his financial situation which had prompted the appointment. Fortunately Williamson was apparently not at all offended, nor did he pay any attention to the slur on Oldenburg's loyalty, and Oldenburg was able to continue his clearly remunerative work for the State Paper Office, troublesome and time-consuming though he found it.

The spring of 1676 was, generally speaking, not a fortunate or happy time for Oldenburg. Not only had his attempt to increase his income failed, but Hooke's quarrel with him slowly but surely increased in violence, making his proper work for the Royal Society more difficult as well as hurting his professional and personal pride. Besides this, he seems to have had no proper amanuensis during this year and the next, when his load was heavy. Now he had himself to transcribe his minutes into the Journal Book, which perhaps at least partly explains why they tended to be scrappy, especially in June 1676. Nor apparently was there anyone to transcribe incoming letters into the appropriate Letter Book for there are no surviving transcripts of letters for Oldenburg's last years in office, at the very period when he had in hand the increasingly copious and very complex mathematical exchange between Newton and Leibniz. True, his correspondence with Huygens lapsed but this must have been a cause of sadness, for he and Huygens had shared a real if formal friendship. This cessation was only partly dictated by Huygens's renewed illness, which began in November 1675, just after he had written his exculpatory letter on Oldenburg's behalf, and which continued for many months. During this time Oldenburg was kept informed about his progress by Huygens's father.

Oldenburg had been acquainted with the distinguished elder Huygens for a good many years, having met him in The Hague and having known him when he and his sons were in London in the early 1660s, when they clearly all became very friendly. The surviving correspondence[53] begins in 1674 and continues until April 1676 with the older man writing as to a friend rather than as to the Secretary of the Royal Society. He often asked Oldenburg for news and for the despatch of books, especially for those of Boyle; in return he sent news both of the health and the scientific achievements of 'my Archimedes' as he called his son. (That he enjoyed being an intermediary between his son and the English is suggested by the curious fact that it was to the father that Hooke had written in 1674 about his chariot invention as well as his claim to have anticipated the younger Huygens in the idea of applying a circular pendulum to watches, a letter which the recipient, as he told Oldenburg, found difficult to understand. And it was to be through his father rather than directly that Christiaan Huygens had defended himself against Hooke in the more vehement stages of Hooke's attack upon himself.)[54] The elder Huygens was very anxious to make sure that he possessed a complete set of the *Philosophical Transactions*; he also wrote to ask Oldenburg to supply him with a large-scale map of London to put on his wall, apparently to revive his recollections of that city where he had spent so much time.[55] He sent news of the health of his relations, and sent messages to Boyle when he was ill about remedies which he might find helpful. He also sent points of interest to the Royal Society. For example in 1675 he sent Oldenburg a book about the Eastern remedy of *Moxa* which was said to be a sovereign remedy against gout; the Royal Society was much interested in the substance and the curious method of applying it. Oldenburg not only produced the book at a Society meeting but was, most probably, responsible for its translation into English and subsequent publication.[56] Constantijn Huygens also at different times sent drugs and reported on his son's showing him a large piece of Iceland spar, noted for its ability to produce double refraction (polarization). Later he described and characterized 'Hungary Water', an alcoholic infusion of rosemary, detailing its useful effects in the case of his sister, whether for private use or for the general information of the Society he did not explain.[57] That such a very great gentleman as Constantijn Huygens should treat Oldenburg with genuine friendship was undoubtedly much in Oldenburg's favour and he must have been both pleased and proud of their relationship and found in it some personal compensation for the loss of his correspondence with the younger Huygens. That it was a genuine and warm friendship which increased over the years is revealed by the fact that although the earliest surviving letter from Huygens the elder is signed 'Your very humble, obedient servant' (a form usual between equals) soon afterwards he concluded a letter written down the page to the foot with 'I finish my paper without formality. You know my hand and my heart', a very informal conclusion indeed in place of the more usual 'Your very humble, affectionate servant', in itself an egalitarian salutation.[58] So cordial a relationship with so great a man must have been some compensation to Oldenburg, however small, for his troubled relationship with Hooke at this time.

Of Oldenburg's private life during the last half of 1676 we know a little. He took his normal family holiday in Kent in September but it was not as restful as usual for, as he was to tell Boyle, he had some trouble, precisely what he did not specify, which delayed his return to London. It concerned the tenant of one or both of his Bexley farms and had to be attended to on the spot.[59] He wrote to assure Boyle that before he had gone on holiday he had sent his Latin translation of what was probably Boyle's *Tracts: containing 1. Suspicions about some Hidden Qualities of the Air*... (English, 1674) to the printer Moses Pitt but, he complained, Pitt was very slow about getting to work on it. What friendly and even familiar terms he was on with Boyle appears in this same letter, in which he enclosed a letter to his maid, asking Boyle to 'let litle Tom carry' it to his house, a request made as to near equals. And he closed,

> Allmighty God blesse you, and confirme your health for many yeares, for the publique good, and the particular concern and Ioy of Sir Your very humble and faithfull servant.

Even the last seven words are less characteristic of the formal relation between patron and client or between simple acquaintances than they are those of close friends and equals, as certainly the earlier phrases would seem to be, while the layout of the letter is similarly informal. There is little doubt that Boyle regarded Oldenburg very highly and treated him as a friend even if he could never quite be an equal.

Of the many letters which Oldenburg must have written in the performance of his duties as Secretary during the last months of 1676 few traces remain except for what is to be deduced from incoming letters and the surviving drafts and copies of Oldenburg's assistance in the exchanges between Newton and Leibniz. (These he copied out himself and preserved, showing that he entirely appreciated their importance.) At this time he sometimes noted the date and perhaps the most important parts of his replies to the letters he received, but little more; only enough remains to show that he was carrying out his usual busy correspondence at home and abroad. Strikingly, he seems to have been in closer touch with Germany and Germans than for many years, chiefly at the initative of his correspondents, many of whom visited England. Thus in mid-November 1676 at the time of his worst problems with Hooke he took the trouble to write several letters to his Parisian correspondents as letters of introduction for an unnamed 'German gentleman from Hanover'.[60] And, as already discussed (Chapter 8), he was so excited by the rumours from Germany which reached him about the discovery there of several new kinds of 'phosphorus', substances which shone in the dark, that he wrote assiduously to various correspondents in Germany to learn more about them and to try to discover the secret of the preparation of these interesting substances.

In the early spring of 1676–7 Oldenburg suddenly applied for naturalization. The reason for his doing so after nearly twenty years' continuous residence in England is nowhere revealed in his correspondence. Although he probably discussed so important a matter with Boyle, if he did so it must have been verbally, because confidence meant conversation since, as must be remembered, all letters to Boyle were read aloud by his amanuensis. Relations with Hooke remained difficult. Oldenburg must have been acutely aware of the fact that Hooke continually wrote and presumably said to others that he regarded Oldenburg as a foreigner. (No wonder that when on 26 May Hooke, dining with Boyle, encountered Oldenburg, the latter fled to avoid confrontation.) Such gibes must have hurt, since for Hooke and some others 'foreign' meant disloyal, as Oldenburg certainly was not. Besides this, it is also possible that there was a current trend towards naturalization of resident foreigners, especially those in the public eye. Certainly when on 22 March 1676/7 David's Naturalisation Bill was sent from the House of Commons to the House of Lords it contained a quite considerable list of names besides that of Oldenburg.[61] All the applicants were required to submit a certificate stating that they had received the sacraments according to the Anglican rite Oldenburg's being dated 4 March 1676/7. Hooke took note of the act of his enemy, recording in his diary under the date 8 March 'Grubendol not to be naturalised'. But he was wrong, for Oldenburg's name was certainly on the Bill which passed from the Commons to the Lords a fortnight after Hooke's diary entry and when the Bill received royal assent early in April 1677 Oldenburg's name was securely there, although as it happens at the end of the list. So he was now English at last, as were his children.

In the summer of 1677 Oldenburg once again had news of his German relations in letters from his nephew. His sister had now become a widow and

Oldenburg wrote her a letter of condolence on the recent death of her husband, Coccejus's father, the only trace of direct communication between Oldenburg and his sister for over twenty-five years, although one imagines he must at least have sent messages through Coccejus when the latter visited England in 1670. Coccejus [62] pressed Oldenburg to visit him in Heidelberg where he was prosperously settled with a growing family of three sons; he also asked for information about the grant of Oldenburg's benefice, the Vicaria of St Liborius, whether because it had somehow devolved upon him or, much more probably, because Oldenburg had asked him to inquire into its status, perhaps so that he could pass it on to his own son, as his father had passed it to him. From Oldenburg's reply to this letter,[63] it appears that he had not long before written to various prominent persons in Bremen about the state of the benefice. He declared to Coccejus that he was much occupied with business at this time being 'prevented by a thousand occupations from writing long-winded letters', and this, he said, made it 'morally impossible' for him to make a visit to his relatives in Heidelberg, as Coccejus had pressed him to do. What his 'thousand occupations' were which clearly occupied him more than usual is not obvious, for the Royal Society was in recess for the summer and his correspondence was not abnormally heavy to judge from what survives for this time. Relations between him and his Royal Society colleagues were apparently as friendly as ever, and even Hooke had begun to relax a little, recording Oldenburg's presence at meetings or at gatherings at Levet's Coffee House without gibes.

It is to Hooke's diary that we are uniquely indebted for an account of the fatal illness which seized Oldenburg in September 1677. It has always been said that Oldenburg died while on holiday in Kent, and he was certainly buried in the parish church of Bexley. However, the entries by Hooke strongly suggest that this was not the case but that he died at home in London. On 3 September 1677 when he should normally have been in Kent or about to go there, Hooke dined with Boyle and heard from him that 'Oldenburg kept his bed for ague'. (This suggests that the illness had prevented him from going to Kent.) Whatever the feverish complaint which afflicted Oldenburg was, it is very unlikely that it was a sudden attack of common malaria, for although malaria was not uncommon in seventeenth-century England, there is no record of his suffering from it earlier. And two days later (5 September) Oldenburg was dead; Hooke recorded the fact along with the statement that he was 'striken speechless and senseless'. As Hooke's information came from Boyle, it is most probable, since he had the news so quickly after the death, that Oldenburg died in Pall Mall. And this probability is borne out by Hooke's diary entry on 14 September, 'With Mr Hill at Mr Martins, Sir J. Hoskins, and Mr Boyles [*sic*], they calld on Mrs. Oldenburg', presumably to offer condolences and perhaps help, and probably to try to discover how soon they could recover the papers belonging to the Royal Society which Oldenburg of course would have kept at home. But Oldenburg's young widow was not able to be of any help, for as Hooke noted four days later, 'To Mr. Boyles, heard that Mrs. Oldenburg died the 17 in the morn', which suggests that although Oldenburg might have died from a stroke, it is possible that his illness had been a contagious one, although it is also possible that his widow had died, as so many wives then did, in childbirth. Whatever the cause of the two deaths, the two small children of the marriage were now left orphans. And the Royal Society was deprived of the most efficient and conscientious officer it had ever had or ever would have.

11
Aftermath

When Oldenburg died his family was left bereft, his friends grieved, and the learned world felt it as an extraordinary loss. Only his few enemies rejoiced, and like Hooke saw it as creating the possibility of a major upheaval in the Royal Society. Oldenburg's importance to the world is best shown by the spontaneous tributes paid by various individuals, tributes contained in letters never intended to be made public. Boyle was desolated at his loss, as the letter of condolence written to him by his sister only a week after the event amply demonstrates.[1] Not only had Oldenburg been his assistant, translator, and publisher, a close associate for twenty years with whom he had always worked harmoniously, but he was a neighbour whom he saw with great frequency and, by now, a close and valued friend. No one was ever fully to succeed him as Boyle's helper, for all that he was to have many other useful laboratory assistants, servants, and publishers. It is a touching tribute that the last work which Boyle himself prepared for the press, the *General History of Air* published posthumously in 1692 (just after Boyle's death in 1691) begins with a note addressed to Oldenburg as a response to the latter's query as to what Boyle really thought about the nature of air and what it was. This clearly shows that Oldenburg had been able to stimulate as well as assist his patron. Oldenburg's loss to the public was summed up by his long-time correspondent Justel in a letter to Leibniz written only a fortnight after Oldenburg's death, news of which obviously had travelled quickly:[2]

> We have lost Mr Oldenburg Secretary of the Royal Society who, as I have been informed, died in the country, for which I am very sorry. He was a man of curiosity and a faithful and punctual correspondent. The whole Republic of Letters loses a good, enlightened and well-disposed subject.

As recounted above (Chapter 10), Oldenburg almost certainly died in London; it is probable that the fact that he had announced his intention of going into the country before he was prevented by illness and that he was buried in the country were the causes of the confusion among those not closely connected with him. Others besides Justel mentioned his having died in the country. For example, James Crawford, resident in Venice in the suite of the English Ambassador and Oldenburg's correspondent since 1673, wrote to Malpighi[3] as early as 24 September that, to his regret, he had returned to London just as Oldenburg went into the country where he had died of a sudden apoplexy.

More than two years after Oldenburg's death Flamsteed echoed Justel's concluding words,[4] remarking that

> Our meetings at the Royal Society want Mr Oldenburg's correspondence and on that account we are not so well furnished nor so frequented as formerly, but I hope a little time will put us into order.

Both Justel and Flamsteed expressed a salient feature of Oldenburg's life, the fact that he was uniquely the point of contact between the Royal Society and the learned world, especially in his last dozen years, both through his correspondence and his publication of the *Philosophical Transactions*. The loss of both was felt very strongly. To understand why others did not immediately take over these tasks, one must reflect not only upon Oldenburg's extreme industry and devotion, to which few others could or wished to aspire, but also upon Hooke's animosity which led him to institute a revolution in the administration of the Royal Society. His hatred was so extreme that he was to pursue it beyond the grave and, for a few years, it affected all that Oldenburg stood for.

But it was not just Oldenburg's absence through death which occasioned dismay and disarray within the Society. Since he had nowhere else to work, all of Oldenburg's papers and very many of those of the Society were kept at his home and had to be retrieved from a household suddenly deprived of both master and mistress, leaving no one capable of taking charge either of his public or his private affairs. Worse than this, neither Oldenburg nor his wife had left a will, or at least none was ever found. It is, of course, not surprising that the young widow failed to make a will after her husband's sudden death and indeed she would have had virtually no time to do so before she in turn died suddenly. It is more surprising that Oldenburg, always so careful and conscientious, had never made a will. In the Plague year he had taken every care to order his papers, especially those belonging to Boyle and to the Royal Society and to keep them separate from his private papers, and he had spoken of making a will with prominent members of the Royal Society as executors.[5]

There is no firm evidence that he did so, and certainly it would not have been applicable a dozen years later when his private circumstances were so very different. The Plague years were ones during which men were peculiarly conscious of the probably imminent danger of a sudden death. The seventeenth-century was an age when everyone, especially the devout like Oldenburg, was perpetually conscious of the transitory nature of life and the ever-present threat of death. Yet at the same time it was an age when it was usual for men not to draw up their wills until they were on their deathbeds. (Seventeenth century wills commonly begin 'I . . . . . . being sicke of body but of whole and perfect remembrance, thanks be to God . . .)'. Many had a superstitious fear of wills as certain precursors of death. If this was the reason for Oldenburg's failure to provide for his young family by making a will and for neglecting the danger that the well-being of the Society he served so faithfully would be at risk, it was a curious weakness in so pious, thoughtful and conscientious a man. There is of course always the possibility that he *had* made a will, which in the confusion of his family after first his death and then that of his widow was never found, but this seems unlikely in the circumstances. This lack of a will was to cause grievous trouble to the Royal Society, to Boyle, and to the Oldenburg children, all helpless to deal with Oldenburg's effects unless or until a will could be found or some other means of dealing legally with the problem could be obtained.

Clearly, someone had to initiate immediate action. Boyle, as might be expected from his well-known generosity and his close and friendly relations with Oldenburg, came to the rescue temporarily. He paid for Oldenburg's funeral, which took place at the Bexley parish church of St Mary the Virgin on 7 September 1677,[6] the choice of the burial place presumably being that of the widow. Three days later there was an official inventory[7] of the house which listed the belongings of Oldenburg and his family, room by room. These mostly consisted of furniture and kitchen goods, with 'wearing apparel' and linen. The 'Plate and Jems' also listed were certainly those inherited by Dora Katherina Oldenburg from her mother, and worth rather more than the furnishings; curiously, as appeared years later, they were never sold.[8] That there is no mention of Oldenburg's substantial collection of books[9] nor of his papers must be because those making the inventory were never allowed into the room where they were kept which, as appears later, was locked and sealed to protect the contents. While the inventory was being made the widow fell ill and died, to be buried in the parish church of St Martins in the Fields.[10] Boyle assumed immediate financial responsibility for the orphaned children, who were so singularly lacking in close relations. He wanted his own papers back as soon as possible along with those belonging to the Royal Society. (This had presumably been the chief reason for the formal call on the widow a week after her husband's death.) But with no one in charge of the household when only servants were left it was impossible to retrieve them.

It was all exceedingly difficult, for who had the right to initiate action? Neither Boyle nor any other member of the Royal Society could properly demand the papers until the household was in a settled state, the children's rights established, and a guardian appointed who would be capable of looking after their affairs, both personal and financial. Moreover, the Royal Society almost certainly owed Oldenburg his salary as Secretary, never very promptly paid, and there was now no one to whom to pay it. It is impossible to determine who was responsible for drawing up the curious document which survives in the Boyle Papers[11] and which was the first step towards unravelling these problems. This document is undated and unsigned, in a rather illiterate hand with uncertain spelling; its author was clearly personally ignorant and very vague as to Oldenburg and his household, probably writing from dictation which may have been derived from a customary formula. (Besides spelling the family name both 'Owelden Bourge' and 'Owelden Berge', Oldenburg is said to have been 'Of Errife' (Erith lies several miles north of Bexley, where the Oldenburgs' lands actually lay), to have died 'about the 1^e^ septr 77' and his 'wido . . . about 10° dayes a Goe' (suggesting that the document was drawn up about the end of September). The document graphically describes the chaotic state in which death had left the family:

> The Lands are un sett. The Howse in London & goods are undiss-possessed of. & None impowered to serch the defunkts writing whether a will or Nott or to dispose of any thing.

Further, there was no one to take charge although, as the document carefully noted, Boyle 'for & on behalf of the Infants hath taken the oversight of them to supply them with Mony—& allso laid down Mony to bury the deads.'

The latter half of the document deals with the needs of the Royal Society and of Boyle himself; unlike the earlier part of the document this reads clearly and is succinctly stated, almost as if it had been prepared beforehand, so that the writer was able to proceed smoothly. Appended to the document and written in a more literate hand is a clear response to the final paragraphs of the main document, which presents the problem succinctly:

1 what Cours is the Justest & Lawfullest to bee taken for the good of the orphants att present & to know whether a will or Not &c.
2 How the Royal Societie & Mr Boyle shall com by their books & papers & how Mr Boyle shall bee re imbursed.

The judgment as determined by a distinguished lawyer, Sir Richard Lloyd (an Admiralty advocate and so, presumably well known to Brouncker, which would account for his advice having been sought by the Royal Society) was simplicity itself. It read

> I am of opinion that Letters of Administration may be granted to some responsible person for the use & benefitt of the orphans during their minority, if no will appeare. A Commission of Enquiry may goe out of the Prerogative Court to search for a will.

This Commission was to consist of three persons, one to represent the interests of the Royal Society, one to represent Boyle's interests, and one that of 'the minors'. If no will was found, then a guardian and administrator must be appointed (how chosen and by whom is not specified) who should deliver to Boyle and to the Royal Society their papers. The implication was that before this had been done a securing of the papers was not legal, even if made by the undoubted owners, and future events suggest that everyone recognized this and did not try to have the sealed room opened. Although the date 1678 is written at the foot of the document after Lloyd's name, in fact the Society's Council knew of it by 24 September 1677, when Michael Wicks the Clerk and Henry Hunt the Operator (Oldenburg's and Hooke's assistants, respectively) 'were named commissioners on behalf of the Society'. Of this both Boyle and 'the persons in possession' of Oldenburg's domicile were to be informed, and the latter 'desired to keep possession' of all papers, presumably to ensure that they were not dispersed, scattered, lost, or thrown away. The next step was for someone to make application for the granting of 'Letters of Administration'. The means by which this was done and by whom is not clear, nor how an appropriate person was to be found. It is only known that, very shortly, one Mrs Margaret Lowden was formally appointed 'Administratrix of the goods Chatells and debts of Henry Oldenburg . . . during the minority of Rupert and Sophia.'[12] There is no previous reference to Mrs Lowden in Oldenburg's correspondence but later documents relating to her show that she and her father must have known Dury and probably Oldenburg twenty years or more earlier than this. She was to prove a fortunate choice for she looked after the children most benevolently. (See Appendix, below.)

With the legal aspects settled, what of the effects of Oldenburg's death on the Royal Society? In many respects it was dire, although for Hooke it meant the end of his bitter enemy whose influence in the Society he both hated and deplored. But coolly considered it was a near disaster. The first evil was the temporary loss of the Society's papers, just as the autumn session was about to begin. It was nearly a month after the appointment of the Society's commissioners before Henry Hunt could spend a day rummaging in Oldenburg's house for the missing papers, perhaps as much for Hooke as for the Society.[13] It was over a fortnight after that before a formal visit from a number of the Fellows could be paid, Hooke eagerly in the forefront of the deputation. His diary tells the story:

> 7 November . . . At Mr Boyles, Pitts, Oldenburgs. The Books [the Society's Journal Books and the Council Minutes] denyd &c., and Dr Pell noe friend to the Royal Society. Dr Pell opend the seald paper from the key hole and unlocked the door. Mrs —[Lowden] and her Solicitor, Dr Pell I and H Hunt enterd, we saw the things but she denyd delivery without paying money and giving Discharge. I bet she would doe as Mr Boyle directed, but Mr Boyles sending, she would not deliver.

The money Mrs Lowden insisted upon was probably that expended on the Society's behalf and possibly an unpaid portion of Oldenburg's salary, much needed for the support of the children until the estate was settled, while the 'Discharge' was a certification that she had handed over all the Royal Society's papers, which obviously no one could be sure of on a cursory inspection. (One might wonder what Hooke, not a commissioner, was doing there, but no doubt his anxiety persuaded him to urge that he be included.) Pell, as John Dury's old friend, was probably the only person who knew him, after all the children's natural guardian, well enough to take an interest in the children's welfare, and now he was, with Mrs Lowden, clearly trying to protect the children's interests. It is probable that the necessary legal processes were not yet completed, legal business then as now being slow. When the legalities were finally completed at the very end of the year Mrs Lowden duly handed over the papers, consisting of 'one trunke of writting papers and other Instruments with some bookes . . . [and] fifteen bundles of papers all belonging to the Royal Society'.[14]

Hooke, always impatient, had not been able to wait and in irritation tried to bully Mrs Lowden to shorten matters; as he noted on 21 December she 'found her self sick' when he called 'but was abroad', no doubt sick and tired of

Hooke's importunities. Pell also refused to accommodate Hooke's impatient desire to get at Oldenburg's papers, so he was 'a Dog'. Even Boyle was 'cool', doubtless disapproving of Hooke's importunities. Finally, on 24 December, Hooke was able to note 'To Mr Boyles about trunk' and 'with much trouble retrieved the books out of Pells hands and Loudens &c'.

He promptly opened the trunk and spent much time thereafter 'examining' and 'cataloguing' the papers, hunting for anything that might indicate the betrayal, as he saw it, of his ideas either by Oldenburg or, earlier, Moray to Huygens and/or Hevelius and copying out anything which suggested that they had so 'betrayed' him. But he was still dissatisfied, certain that he (properly the Royal Society) had not received *all* Oldenburg's papers, for on the last day of 1677 he noted 'I deliver[d] to Mr Boyle memoirs about Mrs Louden', and on 2 January 1677/8 the Council authorized its Committee to ask Boyle's help in recovering more papers, which they suspected Mrs Lowden had withheld. On 5 January 1677/8 Hooke met Pell and Grew to complain about 'more delays about our books and papers', while a week later it was 'Spake with Mrs Louden &c and spoke to Mr Boyle about her' as he did again on 18 January. Not until 27 January could he write 'Received papers from Louden and Pell, 19 bundells, 1 Letterbook, 1 pocket book, 1 pack of Malpighis papers'. (Why they had not been turned over to the Society with the trunk is not clear, but probably, as Mrs Lowden could not have wished to keep them for the children, it had taken time to assemble them and sort out any private papers belonging to Oldenburg himself and not to the Society.) In the first half of 1678 Hooke continued to scrabble through the papers, to sort and list them, to sift out and copy those relating to himself, as he noted in his diary from time to time. No remains of all this activity survive, neither the transcripts nor the catalogues, which disappeared along with Hooke's other private papers, but fortunately he mostly returned the originals to the Society where they can still be found in the archives.

Hooke, all earlier friendship and collaboration forgotten, was unfeignedly glad to have Oldenburg out of the Royal Society, which he saw as a happy opportunity both for 'reform', as he saw it, of the Royal Society and for his own aggrandizement. This began as soon as he heard of Oldenburg's death after which he promptly went 'with Cox to Boyle, talked with him about Society and method'; that is, the method of reforming the Society to eliminate what he saw as deleterious. The first thing was obviously to consider whom to appoint in Oldenburg's place and, to his pleasure, the Council at their meeting on 13 September accepted what must have been his own proposal; as he recorded, 'they accepted of me for secretary pro tempore to write the Journals without reward'. This was to him a great advantage because it gave him the right to take a hand in securing the Society's papers and to go through them when once they had been released by Mrs Lowden. He was still worried lest he be ousted by Grew who was, he believed, 'canvassing for Secretary's place' until he found that Oldenburg's last colleague, Henshaw, was not to continue in office. Hooke evidently felt very proud when on 25 October 1677 he 'first sat down as Secretary, being so commanded by Mr Henshaw and Sir J. Hoskins'. As this appointment was only to last until 30 November he felt compelled to continue to lobby various Fellows, among them Wren, Hill, Hoskins, Aubrey, Grew, and Croone, both so that he might be elected Secretary and that there might be a new President to replace Brouncker, President since 1662 and hated by Hooke since the watch dispute of 1675. Hooke was inclined to favour Wren for the post (he was in fact to serve as President from 1680 to 1682) but did not greatly care who it was provided it was no longer Brouncker. Naturally there arose a counter-party, apparently led by Colwall and Croone, but Hooke's party was to prove the stronger once Hoskins, as Vice-President, insisted that the choice should be made by ballot on Election Day (St Andrews Day, 30 November). According to Hooke at this proposal 'Lord Brouncker in Great Passion, raved and went out' and hence Hooke omitted his name from the list of those present when he wrote up the minutes. (It is not far fetched to assume that Brouncker *was* angered by the opposition to his continued tenure of Presidency which, according to seventeenth-century custom, he had expected to be a lifetime post, as it was for Newton in the early eighteenth century and for Joseph Banks who held it from the late eighteenth century to the early nineteenth century, 1778 to 1820).

In November 1668 the balloting was carefully and properly conducted with the result that Sir Joseph Williamson was elected President in his absence, while Hooke and Grew were elected Secretaries, all of which, Hooke insisted, put 'Croon in a great Huff', why he did not say. Seemingly after this Hooke did not find Hoskins as friendly as he had appeared before, and at the very end of the year he reminded himself that 'Sr J. Hoskins [is] not really my friend', although without further elucidation. Once he had secured the post of Secretary Hooke does not seem to have enjoyed it. He must have complained at the drudgery of taking minutes, or else have taken them carelessly, for, as he privately recorded of the meeting on 13 December 1677 'Grew placed at table to take notes. It seemd as if they would have me still Curator, Grew Secretary' (as, of course, Grew was). A compromise was reached and on 19 December

1677 Hooke noted 'Grew to take notes also but I to draw them up'. Hooke also disliked the drudgery of correspondence and enlisted the help of his friend John Aubrey to write letters for him (as Aubrey did to Newton) at which, Hooke discovered Hoskins, still a Vice-President 'caballd'. It was Grew, not Hooke, who drew up a circular letter in Latin. This, to be sent to all regular foreign correspondents, told them of the changes in the Royal Society, encouraged them to continue to write and instructed them how to address their letters.[15] It was approved by the Council on 2 January 1677/8 but as no copies were apparently sent out until 11 April 1678, it is not surprising that all correspondence languished. Hooke later did sometimes take charge of correspondence, but usually only occasionally, sharing the task with Grew so that both Secretaries took part in the Society's affairs, as had indeed been intended from the beginning. It was no doubt with relief that Hooke ceased to be Secretary in 1682 and reverted to being Curator.

Another problem for the Council was what to do about continuing Oldenburg's highly valued *Philosophical Transactions*. Hooke was bitterly opposed to its continuation, not unnaturally considering both his hatred of Oldenburg and his belief that the journal had been too inclined [illegible] contributions by foreigners at the expense of the English, alt[illegible] majority of the members valued it and regretted its loss. The la[illegible] produced by Oldenburg had been dated 25 June 1677 and he had intended to publish the next issue in September, attributing the anticipated delay to the printers' summer holidays.[16] It is fair to deduce that the Council must have tried to persuade both new Secretaries to resurrect the journal. Hooke was firm; as he noted in his diary for 2 January 1677/8, 'I moved against *Transactions* brought them [that is, the Council] to be of my opinion'. It is fair to deduce that they did so because he offered a compromise: he was then writing up one of his Cutlerian Lectures, *Cometa*, and must have suggested that he should publish suitable papers and letters as an appendix. This he did, the papers and letters chosen being in fact very much what Oldenburg would have chosen. As a further compromise it had been proposed on 2 January 1677/8 that the most interesting papers read to the Society during each year, together with various unpublished papers from the archives, should be published annually. This was not done but suddenly the *Philosophical Transactions* came back on the scene, almost certainly because of Grew. Early in 1677/8 there was published a general index and catalogue of books 'from the beginning to July 1677' which, it is reasonable to suppose, must have been at least partially compiled by Oldenburg during his last summer. It was available for Hooke to 'take' at Martins on 9 February. After this Grew completed Oldenburg's volume XII with six issues, dated 1677/8 to 1678/9, mostly composed of previously unpublished papers which he had found in the archives (concerned with histories of trades) together with a few up-to-date letters and papers of which there could not be many, since the Society's correspondence was languishing. After this the Fellows again urged Hooke to take over the journal, but he refused yet again. He offered instead to print his own journal, which he called *Philosophical Collections*; it was very like the *Philosophical Transactions*, but appeared only irregularly between 1679 and 1682. To complete the story, after Hooke ceased to publish his journal, the Council failed to persuade its then Secretaries to take it over until in 1683 Robert Plot agreed to revive the *Philosophical Transactions* still as a private venture, provided that the Society would agree to buy sixty copies of each number, to ensure that he lost no money. After he tired of it, Halley as Clerk edited the journal for some time with intermissions after which it was edited by successive Secretaries. It remained a private venture but became more and more tied to what happened at the Society's meetings. Finally in 1752 the Council voted to adopt it as its official journal, to be entirely devoted to the 'transactions' of the Royal Society. And so Oldenburg's brave venture of recording the 'philosophical transactions' of the whole learned world staggered somewhat lamely on, eventually becoming what the learned world had long taken it to be, an organ of the Royal Society which recorded its doings exclusively.

If one asks whether Oldenburg's death mattered to the European world of letters, science and learning, the answer is clearly that it did indeed. For after he died although the English scientific scene was still in an exceedingly flourishing state, without Oldenburg's efforts full knowledge of what the English natural philosophers were doing was less easy to come by. The tact with which he had managed his wide correspondence was possessed by no one else in England, and certainly no other man was willing to expend more than a fraction of the energy that he devoted to it. His death, which interrupted the exchange of information between Newton and Leibniz, had a profound influence on both men, and had much to do with the great calculus dispute of the early decades of the next century. Oldenburg had successfully persuaded Newton in the 1670s to impart his scientific and mathematical discoveries to the world, and had extracted from him peaceful answers to criticism and queries. When this stopped in 1677, few after that, certainly not Leibniz, appreciated what Newton had already accomplished in mathematics. After

1677, the English were poorly informed of what was being done on the Continent, and Continental mathematicians and natural philosophers were unable to secure a full picture of what the English were doing. Both suffered, especially when Oldenburg's *Philosophical Transactions*, which he had founded, invented, and published, became a shadow of its former self, and a fitfully flickering shadow at that during the closing years of the seventeenth century. No Secretary after Oldenburg had so large as correspondence as he had done, not even Hans Sloane, who in the early eighteenth century kept in touch with most Continental botanists, for he did not correspond with physical scientists. Not until the late eighteenth century was the post of Foreign Secretary created, who did (and does) deal with only one aspect of Oldenburg's multifarious business. No other seventeenth-century scientific society, not even the highly organized Académie royale des Sciences, ever had such a hard-working and successful administrative secretary as Oldenburg, and consequently no other society then or since ever played such a central role in European natural philosophy as did the Royal Society during his time in office. Not until Fontenelle became Secrétaire perpetuel to the Académie royale des Sciences did any society even approach the Royal Society in its contacts with European natural philosophy outside its own country, and then it did so in a different way in a different century. Nor did other societies, not even the Royal Society after Oldenburg's death, bring in provincial and amateur natural philosophers as he had done. Certainly no other early Secretary had anything like his vision. It is fair to say that Oldenburg was unsurpassed as a scientific administrator, publisher, and scientific correspondent. In that lies his historical importance.

Appendix
The fate of Oldenburg's children
(with the assistance of P.D. Buchanan)

The impact of Oldenburg's death followed so quickly by that of his wife was little less than catastrophic for his children. Sophia was at most five years old and Rupert only two and a half, both too young to know what was happening, and left helpless in the hands of servants. Pell, as their grandfather's old friend, did what he could and kept an eye on them, but he was not only elderly but poor (he died 'in poverty' according to the *Dictionary of National Biography*). And Boyle, although he rendered financial assistance, had no legal rights to institute any proceedings. Pell seems to have waited to write about the children to their grandfather until the necessary legal steps had been taken to appoint an administratrix of Oldenburg's estate and a guardian for the children. As noted above (Chapter 11) this was Mrs Margaret Lowden.

The exact connection between Mrs Lowden and the Oldenburgs is not known. However, much can be inferred from contemporary references to her father, Sir Robert Stone Knt (d. *c.*1671) and from legal documents relating to her and her relations.[1] She was Sir Robert Stone's heir, and hence presumably his only living child at his death, and was born in Holland. He was in the service of the exiled Queen Elizabeth of Bohemia (1596–1662), mother of Prince Rupert, which caused him frequently to obtain permission to travel between England and Holland in the 1630s and 1650s. He was both her cupbearer and captain of a troop of cavalry on her behalf. (His daughter still possessed a portrait of the Queen in 1710.) In Holland he was acquainted with members of the Honeywood family, presumably the Honeywoods of Petts in Kent with whom Oldenburg stayed after the completion of his mission to Cromwell after having been tutor to Robert Honeywood and travelled with him on the Continent. (See above, Chapter 1.) Stone must have known Dury at Queen Elizabeth's Court when Dury was her chaplain, and he may also then have known Pell. All this strongly suggests that he knew Oldenburg first in Holland and after the Restoration in London, when Stone was living in Westminster (as he was in 1671). He bequeathed to his daughter considerable property in Aldgate Without (that is, lying outside the boundary of the City of London) near what is now Liverpool Street station, and more property nearby. Margaret Lowden could and very likely did know the Oldenburg family at least in the 1670s. She married one Francis Lowden who had died by the spring of 1677, when she was naturalized. (The Lowden whom Hooke called 'scotch man' and disliked when he met him at Boyle's in 1677–8 was probably her son (who predeceased her).[2]

To move from speculation to facts. After she had taken charge of the Oldenburg children she wrote to Dury to explain what had happened, as he told Pell.[3] Pell himself wrote about the situation as he saw it to Dury, who in reply expressed some concern for the children's education (as he had done a dozen years earlier when his little daughter was left in similar circumstances), but, now over eighty, he showed no inclination to come to England or to have them come to him. He probably, unrealistically by this time, feared persecution for his involvement with the Commonwealth Government and in any case had never been a family man, having his mind fixed on on higher things. As he put it to Pell,

> I should be unthanckful, if I did not acknowledge the Paines which you have been pleased to take to let me know the Concernes of my state in England which I seeke to bring to settlement; for the education of my Grand-children, which is now the Chief care I can take for them. Mrs Lowden doth much commend the favourable assistance which you have given her, for which I owe you thancks, & beseech the Lord to requite it. I have sent herewith the letter of Atturney; and desire Mr Gerard Van Heyhusen (by whence heretofore I have received the money which Mr Oldenburg made over unto me) to deliver it to Mrs Lowden . . .

(Presumably this money was derived from the Bexley estate which was now to go to Rupert Oldenburg as heir.) Pell received this letter on 11 June 1678, when he recorded[4]

> Mr Henry Oldenburghs two children and Mrs Lowden lodged, when I last saw them, at the house of Mr Jones a Cloth-worker in the row of new buildings in Goodmans fields near Whitechapell.

This shows that Pell kept in touch with the children for a little, but also that he disapproved of their being lodged in the East End of London. But Goodman's Fields[5] lying to the south of Whitechapel High Street, on land originally part of a farm belonging to the abbey of the Nuns of St Clair, was in the 1670s in the process of development (hence the 'new buildings') and the development, although not aristocratic, lodged many good merchant families and even, later, Sir Cloudesley Shovell, Admiral in command of the Channel Fleet. (There still exists Goodman's Yard in Whitechapel.) And the Lowdens owned a share in the new development. Seven years later Pell, by then out of direct touch, noted on the same sheet, 'august 19. 1685. Will Raven was told that she now lieth in Harrow Ally among the Butchers without Aldgate.)' This was not far from their earlier lodgings, and presumably in one of the several buildings in Harrow Alley which Mrs Lowden owned. Pell seems to have implied that the children were now living in less prosperous surroundings, but he was not necessarily a good judge, and the cloth-worker and butchers could easily have been in a bigger line of business than might appear. In any case, it was then not uncommon for there to be good lodgings over shops and even workshops. In the 1660s Oldenburg and therefore Lady Ranelagh had lived in Pall Mall near a stone-cutter. The area was probably quite prosperous, although not aristocratic. After Pell's death in 1685, it is obvious, the children could have had no further contact with Oldenburg's old friends or with his colleagues of the Royal Society, not even with Boyle.

As time was to show, Mrs Lowden was a good and conscientious guardian and did her best by the two orphans. She also continued to prosper, as the relevant Poll Tax Records for the Harrow Alley area in 1689 attest.[6] Besides the Oldenburg children and the necessary servants the household contained her children, her son Robert, while he was still alive, and a daughter Frances. (This daughter married a John Smith and had three sons and a daughter; two of these sons, Robert Lowden Smith and John Eusebius Smith, were to be the executors of their grandmother's will.) In this will made in 1710 about a year before her death,[7] Mrs Lowden was careful to specify provisions for 'Sophia Oldenburg daughter of Henry Oldenburg late of Bexley in the County of Kent gentleman deceased.' (It is important to realize that Oldenburg retained the status of gentleman, even so long after death, and that he was a landed proprietor by virtue of his wife's inherited 'estate'.) From Mrs Lowden she was to receive both an assured annuity of £12 a year and more from other sources. (All the family property would have devolved upon her brother.) Special directions were given to ensure that she was to be well and properly looked after 'and kept in such a manner and such a place or places House or Houses as Eleanor Burkeridge, Wife of . . . Nicholas Burkeridge shall approve' (she was then being cared for in the Burkeridge household). Further, a London attorney was to oversee her care and receive the income due to her. Evidently there was some way in which she was quite unfitted to cope with life, for, now in her late thirties, she clearly had been for some time in the care of the Reverend Nicholas Burkeridge. He was a neighbour of Mrs Lowden in West Ham and one of those appointed to oversee her own executors; moreover, he had taken in a Lowden descendent, probably another grandson, referred to as 'poor Billy' and known to Robert and John Smith. Besides an income, Mrs Lowden with her usual conscientious care (which would have delighted Sophia's father) left other directions for Sophia. Mrs Lowden 'desired and charged' her executors to see that Sophia might be given

> the Japan Cabinet which is in my Closett it having been her Mothers also a black Trunk in the said Closett with the Things that is in it they have all been the said Sophia[']s Mothers and Grandmothers which I preserved from being sold after being appraised and accounted for I may dispose of them as I please and soe doe desire They may be disposed of to the best advantage for the Use of Sophia Oldenburg by giving them unto those who is so kind as to take care of her in looking after her to see she may be kept decent and clean.

Evidently from whatever affliction Sophia Oldenburg suffered, whether mental, physical, or psychological, she was, sadly, quite incapable of appreciating her inheritance, for the heirlooms were to go to her carer. If this carer was a 'Gentlewoman' the things were to be disposed of by her, but if not, they were to be sold for Sophia's benefit.[8] (As rather an anticlimax, both she and the carer were to share Mrs Lowden's clothes.) It was a poor life for Oldenburg's daughter, but she had been and was to be greatly benefited by Mrs Lowden's care, even if she had never been capable of appreciating either it or her rightful inheritance.

The heir to the Oldenburg property was of course Sophia's brother Rupert, about whose life we know a good deal. On 13 January 1690/1, when he was not quite fifteen years old, he was apprenticed to George Bow, Apothecary, with whom of course he lived during the period of his apprenticeship.[9] On 2 August 1698 he completed his apprenticeship, although he should have served a further five months, by payment of a fine (2 shillings and 4 pence) and was then made free of the Apothecary's Company and hence became a citizen of London.[10] Whether he practised as an apothecary and if so how and where is not known, but it is possible that he chose not to do so, for the very next year he sold his lands in Bexley[11] for £1116 2s 6d to Sir Cloudesley Shovell, who had already bought the neighbouring manor of Crayford, no doubt with prize money. What Rupert Oldenburg did with this large sum is not apparent, nor why he borrowed £40 12s in the same year from Mrs Lowden, money which he had not repaid in 1711.[12] At some time after this, perhaps having somehow spent most of his money, he joined the Marines, presumably having bought an ensigncy, putatively soon after the outbreak of the War of the Spanish Succession in which England joined in 1702.[13] He was with the 20th Foot at the siege of Gibraltar in 1704 (after which Gibraltar stayed British) but as a Marine he probably never took any part in the land wars in which the Duke of Marlborough led the English armies in alliance with the Dutch and the German Emperor against France and her allies, although he could have been in Canada since the Marines took part in the warfare there. He returns to the public gaze in 1707 (the records of the Marines were not well kept in the preceding five years) when he was gazetted to be Second Lieutenant to Major Henley in Major-General Holt's Queen's Own Regiment of Marines, and then promoted First Lieutenant in 1712. But after this fairly promising start (which, since it was in wartime, may not have required the expenditure of money) his further rise was blocked by the fact that the regiment was put on half-pay in 1713 after the Treaty of Utrecht was signed. He then diappears from the regimental records for eleven years. Not surprisingly, in view of his previous financial incompetence, as it seems with hindsight, he emerges as in need of financial help, for he then appealed to the Royal Society, evidently being aware of his father's connections with it, asking for money presumptively on the grounds that his father's salary had never been paid in full. As a result, on 28 March 1717 the Council of the Society

> Ordered, That Mr Oldenburg son of Mr Oldenburg heretofore Secretary to the Society be paid Ten Guineas as a Gratuity in consideration of his father's Services and in full of all.

It was not a great deal, being a quarter of Oldenburg's yearly salary, but it was kindly, surprisingly so after so long a lapse of time and when of all the Council only Newton, then President, could have remembered Oldenburg's long and valuable services as Secretary. One would like to think that it was Newton's friendly recollection of his correspondent which was the cause of this exceptional benevolence.

Nothing further is known of Rupert Oldenburg's life after 1717, until 1724, at which time he was a Lieutenant of Invalids on garrison duty in Plymouth, clearly as events showed a singularly unfortunate and inefficient middle-aged officer. For on 25 September 1724 he was dismissed the service 'for having neglected his Duty; and being absent when General Wills came there to review the Garrison', according to the report in *The Political State of Great Britain* for September 1724.[14] He then, it seems, gave a party for his friends 'and all the while behaved with his usual Mirth and Chearfulness' but the next morning 'shot himself with a Pistol through the Head', evidently this time being for once all too efficient. He left a touching and pathetic letter behind him, addressed to his Colonel, which reads as follows:

> Sir,
> Believing I shall meet with more Difficulty than the Thing is worth, and considering my Friends, as I imagine, have forsaken me, and that dying is much more preferable than a Miserable Life, I have taken this Method of putting an End to every thing. I left in my Pocket three Moydores, one Guinea and thirteen Shillings and Six pence. If Mr D-s Bill, is not paid, I desire it may be Satisfy'd out of what I leave; There is Lodging and some small Matter due to the House, and six pence to the Coffee-House. I wish You and every Body well, and am
> Sir
> Your humble Serv.
>
> Rupert Oldenburg
>
> The Barber is not paid.

Only the final conscientious attention to financial detail sounds at all like his father; and only his sad and obviously unusual end explains his appearance in the public eye. Evidently, quite unlike his father and in spite of Mrs Lowden's earlier care, he was one of life's failures. His letter suggests charm and conviviality but clearly he had inherited none of his father's ability to thrive.

Abbreviated titles

For full bibliographical details, consult the Bibliography.

Birch, Boyle	Birch, *The Life and Works of the Honourable Robert Boyle*
Birch, *History*	Birch, *The History of the Royal Society*
BL	British Library
Boyle Papers	In the archives of the Royal Society; they have been sorted and catalogued by M. Hunter and there is a microfilmed catalogue available in reproduction.
Correspondence	Hall and Hall, eds., *The Correspondence of Henry Oldenburg*
Early Science in Oxford	Gunther, ed. *Early Science in Oxford*
Evelyn's diary	Bray, William, ed., *The Diary of John Evelyn*
Hall & Hall, *Notes & Records* 18	Hall, A.R. and M.B., 'Some hitherto unknown facts about Henry Oldenburg'
ibid. 23	'Further Notes on Henry Oldenburg'
Hooke, diary	Robinson & Adams, *The Diary of Robert Hooke*
Newton, *Correspondence*	Turnbull, Scott, Hall and Tilling, *The Correspondence of Isaac Newton*
Notes & Records	*Notes and Records of the Royal Society*
Oeuvres Complétes	Huygens, *Oeuvres Complètes*
Pepys's diary	Latham and Matthews, eds., *The Diary of Samuel Pepys*
Philosophical Transactions	Oldenburg, ed., *Philosophical Transactions*
P.R.O.	Public Record Office
Rigaud	*Correspondence of Scientific Men of the Seventeenth Century*
Turnbull,	*Hartlib, Dury and Comenius* Turnbull, G.H. *sic.*

Notes

Chapter 1

1. Oldenburg's ancestry and early years were documented by Dr Friedrich Althaus in the Munich *Beitrage zur Allgemeinen Zeitung*, 1888–9, especially volume 212 of 2 August 1889, pp. 1–3. This article was used by Herbert Rix, then Assistant Secretary of the Royal Society, in his article on Oldenburg in the first edition of the *Dictionary of Scientific Biography* and also by A,R. and M.B. Hall in 'Some hitherto unknown Facts about the private Career of Henry Oldenburg', *Notes & Records*, 18 and in *Correspondence* I, Introduction. Details of his education may be found in *Festscrift zur Vierhundertjahrfeier des Alten Gymnasium zu Bremen*, 1528–1928 (Bremen, n.d.), p. 32.
2. See *Correspondence* I 281; Levinus Warnerus was then Dutch Resident at Constantinople.
3. For the letter from the Rector of the Gymnasium to Vossius, see *Gerardi Joan. Vosii et Clarorum Virorum ad eum Epistolae*, ed. Paul Colomesius (London, 1690), p. 210; for that from Balthasar Wille, Professor of Theology in Bremen, see *Correspondence* XIII, 379–81.
4. *Correspondence* I, 3–6.
5. The *Liber Epistolaris* (Letter Book), now in the archives of the Royal Society, MS. 1. These letters are published in *Correspondence* I and XIII.
6. Milton to Oldenburg, 6 July 1654, *Correspondence* I, 34.
7. *Correspondence* I, 124 and XIII, 385.
8. For the military basis of Cromwell's power see Timothy Venning, *Cromwell's Foreign Policy* (London, 1995, 1996) Chapter 1, and for the Swiss mission of Pell and Dury see Chapter 2 and esp. p.36. Pell and Dury were expected to visit Bremen in the course of the mission; see *Correspondence* I, 29. For the close friendship, *ibid.*, p.23.
9. *Correspondence* XIII, 381–83.
10. *Correspondence* I, 63–5 *et seqq.*
11. *Correspondence* I, 23.
12. *Correspondence* I, 7–10.
13. *Correspondence* I, 69–72.
14. *Correspondence* I, 10–14. Letters between Oldenburg and the Senate of Bremen and the Senator Albrecht Bake and also Oldenburg's formal letter to Cromwell are to be found in *Correspondence* I, 15–20, 25–30, 38–62; these give in considerable detail his diplomatic actions during 1653 and 1654, which are confirmed by the House of Commons Journal (Parliamentary Papers) for the period. The letters to Bremen are preserved in the Bremen Archives, MS. W.9.b.1.b., entitled 'Korrespondenz mit Cromwell, seinen Ministern un Heinrich Oldenburg in London wegen Neutralität der

bremischen Schiffrat im englisch-hollandischen Kriege sowie Gewährung einer englischen Anleihe an Bremen und sonstiger Hilfe gegen Schweden 1653–1656'.
15. *Correspondence* I, 32–4 *et seqq.*
16. *Correspondence* I, 40–47; for the result, *ibid.*, 48–62.
17. For the Boyle family genealogy, see Dorothea Townshend, *The Life and Letters of the Great Earl of Cork* (London, 1904). For Lady Ranelagh see esp. the entry in the *New Dictionary of National Biography* by Sarah Hutton, which utilizes the Hartlib Papers. The most valuable source for Robert Boyle's life is still the straightforward biography prefixed to *The Works of the Honourable Robert Boyle* (London, 1744 and 1772, the latter edition available in modern facsimile), hereafter Birch, *Boyle*. For a review of modern Boyle studies with a full bibliography of works about Boyle 1940–94, see Michael Hunter, ed. *Robert Boyle Reconsidered* (Cambridge University Press, 1994). For my own brief interpretation, see Marie Boas, *Robert Boyle and XVIIth Century Chemistry* (Cambridge University Press, 1958), esp. pp. 5–47 and M.B. Hall, *Robert Boyle on Natural Philosophy* (Indiana University Press,1965), pp. 3–33.
18. See Turnbull, *Hartlib, Dury and Comenius* which is, as its subtitle states, composed of 'Gleaning from Hartlib's papers', factual but unorganized, and, for a more complex and interpretative study, based on the same and wider sources, Charles Webster, *The Great Instauration* (London, 1975), *passim*. The Hartlib Papers are now available in the Sheffield University Library and on CD Rom.
19. *Correspondence* I, 137, note 5, Turnbull, *Hartlib, Dury and Comenius*, pp. 54–6 and Webster, *Great Instauration*, esp. pp. 67–77.
20. *Correspondence* I, 102.
21. Oldenburg addressed him as 'neer Charing-Cross' until until June 1659. King Street was obliterated in the nineteenth century.

Chapter 2

1. *Correspondence* I, 73.
2. *Correspondence* I, 63, 65, 86.
3. *Correspondence* I, 74–5.
4. *Correspondence* I, 78–9.
5. *Correspondence* I, 81.
6. *Correspondence* I, 92–5.
7. *Correspondence* I, 96–8.
8. *Correspondence* I, 82, letters to Lord and Lady Cork.
9. *Correspondence* I, 85.
10. *Correspondence* I, 94.
11. *Athenae Oxoniensis* (Oxford, 1721), II, 114.
12. Seth Ward, Savilian Professor of Astronomy at Oxford and later Bishop of Salisbury (Sarum); the quotation is from his letter to Sir Justinian Isham, 27 February 1651–2, in *Notes & Records*, 7 (1949), 70.
13. Cf. Thomas Smith, *De Republica Anglorum* (1583), quoted in S. Shapin, *A Social History of Truth* (Chicago and London, 1994), p. 57.

14. *Correspondence* I, 89–92. Boreel was acquainted with Hartlib and Dury; he was a learned Hebraist, a radical in religion who supported Socinian views, and a great controversialist.
15. Cf. Charles Webster, *The Great Instauration*, p. 160ff.
16. As described by John Wallis in two accounts. The older was printed in his *A Defence of the Royal Society, an Answer to the Cavils of Dr William Holder* (Oxford, 1678), p.8. (Holder had ascribed its origin to 'divers ingenious persons in Oxford'). The longer account was published in the Preface to Thomas Hearne's edition of *Peter Langtoft's Chronicle* (London, 1725), pp. clxi–clxiv, and was written in January 1696–7. Confirmation of the London meetings is contained in a letter by an instrument maker, Anthony Thompson, to the mathematician John Pell dated 2 November 1658, conveniently quoted in A. Rupert and Marie Boas Hall, 'The intellectual Origins of the Royal Society—London and Oxford', *Notes & Records*, 23 (1968), 157–68.
17. Ward, op.cit., note 15, p.69; a note by Wallis (*Philosophical Transactions*, 8 (1673), no. 98, 6146–49) referring to 1657 is reprinted in Douglas McKie, 'The Origins and Foundation of the Royal Society of London' *Notes & Records* 15 (1960), 25–6. The regulations of the Oxford Club, preserved in its minute book, were first printed by C.R. Weld, *A History of the Royal Society* (London, 1848), I, 33–4. For the spirit of the Oxford Club and its intellectual milieu, see Robert G. Frank Jr, *Harvey and the Oxford Physiologists* (Berkeley, Los Angeles and London, 1980), Chapter 3, esp. pp. 51–7 and Charles Webster, op. cit. note 17, pp. 153–78.
18. Anonymous letter of 14 September 1655, Boyle Papers, XXVII, ff. 194–5, partly quoted by Webster, op.cit. p. 162. Although it has been claimed that the writer was Boyle, internal evidence does not support this.
19. *Correspondence* I, 133.
20. Royal Society MS. l.
21. See R.E.W. Maddison, *The Life of the Honourable Robert Boyle F.R.S.* (Taylor & Francis, London, 1969), p. 91.
22. Letters to Lady Cork, 3 September 1656 (which notes the 'full recovery' of the boys) and to Boyle, 15 April 1657, *Correspondence* I, 107–8 and 117.
23. *Correspondence* I, 116, 130.
24. Letter to Boyle, 24 June 1657, *Correspondence* I, 118–21.
25. For what follows, see Samuel Mours, *Le Protestantisme en France au XVIIe Siècle (1598–1685)* (Paris, 1967).
26. As indicated in Oldenburg's letter to Dury of 14 November 1657, *Correspondence* I, 146.
27. See his letter to Lady Ranelagh, 22 August 1657, *Correspondence* I, 130.
28. *Correspondence* I, 121–3, 127, 140 (Milton), 123–6 (Manasseh ben Israel), 142–4 (Boreel), 145–7 (Dury).
29. *Correspondence* I, 136–8; the recipe is appended to a letter of 22 September 1657.
30. *Correspondence* I, 148 *et seqq.* For later relevant letters see *ibid.* XIII, 385 *et seqq.*
31. *Correspondence* I, 137, to Boyle; cf Webster, op. cit. *passim.*
32. Letter to Hartlib, 20 March 1657/8 *Correspondence* I, 157. They left in time to arrive in Geneva six weeks later, when Oldenburg again wrote to Hartlib, *Correspondence* I, 158.

33. For letters to Hartlib, *Correspondence* I, 160–73. None of the letters which he must have written to others from Frankfurt has survived. For Cromwell's interest in the outcome see T. Venn, *Cromwellian Foreign Policy*, 133–6.
34. *Correspondence* I, 163 *et seqq*, 365–66.
35. *Correspondence* I, 170, 186–7; Oldenburg later corresponded with Becher, cf. letter of 12 March 1658–9, *Correspondence* I, 208–10.
36. *Correspondence* I, 173, 174–5.
37. *Correspondence* I, 176–80.
38. *Correspondence* I, 180 to Hartlib, 235 to Baron von Friesen.
39. *Correspondence* I, 182–6.
40. *Correspondence* I, 177, 235, 243–5, 265–7.
41. In September 1658; *Correspondence* I, 177.
42. *Correspondence* I, 177–8, 239–42.
43. *Correspondence* I, 233–7.
44. *Correspondence* I, 285.
45. *Correspondence* I, 288. For Wiesel, see Inge Keil, 'Technology transfer and scientific specialization: Johann Wiesel, optician of Augsburg, and the Hartlib circle', Mark Greenglass, *et al. Samuel Hartlib and Universal Reformation* (Cambridge University Press, 1994), pp. 268–78.
46. *Correspondence* I, 198–200, a joint letter of thanks for hospitality received from Jones and Oldenburg to the Moderator of the Academy of Montpellier, certainly drafted by Oldenburg.
47. Notably Saporta, Pierre Borel (both of Castres) and Pradelleis of Montpellier.
48. *Correspondence* I, 323–5.
49. *Correspondence* I, 192.
50. *Correspondence* I, 241.
51. *Correspondence* I, 217, 227.
52. *Correspondence* I, 248–51, letter to Tolle, chemist and physician of La Rochelle.
53. *Correspondence* I, 287, letter of 23 July 1659.
54. See Harcourt Brown, *Scientific Organizations in Seventeenth Century France (*1620–1680*)* (Baltimore, 1934) for all these academies.
55. *Correspondence* I, 263.
56. Quoted by Brown (note 54, above, pp. 75–6.)
57. As told to Hartlib, *Correspondence* I, 291.
58. *Correspondence* I, 227.
59. *Correspondence* I, 291.
60. *Correspondence* I, 331.
61. *Correspondence* I, 330.
62. e.g. *Correspondence* I, 373 to the librarian of the Hotel de Thou, and 378–9, to Montmor.
63. *Correspondence* I, 291.
64. *Correspondence* I, 257–8.
65. *Correspondence* I, 277.
66. *Correspondence* I, 280 and 308.

67. e.g. to Saporta, *Correspondence* I, 293–300.
68. *Correspondence* I, 270, 327, 329.
69. *Correspondence* I, 258, 284, 288.
70. *Correspondence* I, 303.
71. *Correspondence* I, 337–9 and 340.
72. *Correspondence* I, 218–19, 314–20.
73. *Correspondence* I, 271, 308–9, 332.
74. Cf Evelyn's diary, s.v. 29 May 1660.
75. *Correspondence* I, 370, 371–4; the latter letter, to one of his French friends and dated 11 June 1660, indicates that Oldenburg thought that life in London was returning to normal.

Chapter 3

1. Sorbière, Samuel, *Relation d'un Voyage en Angleterre* (Paris, 1664). I quote from the English edition, *A Voyage to England* (London, 1709), p. 30. Sorbière noted that he had seen Oldenburg at his own house in Paris and at the Montmor Academy 'where he constantly attended'. For some account of the views held by foreign visitors to the Royal Society see Rob Iliffe, 'Foreign Bodies Part II' in *Canadian Journal of History of Science*, XXIV, 1999, 23–50.
2. He gave this address to Montmor in a letter of 28 June 1660, *Correspondence* I, 379.
3. As revealed by his letter to Lady Frances Jones, older sister of his pupil Richard Jones, *Correspondence* I, 383.
4. See his letter to Beale of 7 September 1660, *Correspondence* I, 384–6.
5. Letters to Jacques de la Rivière, Pierre Petit, *Correspondence* I, 371–4, 389, 395–7.
6. Mid-September 1660, *Correspondence* I, 390–2.
7. Letter to Boyle, 18 September 1666, *Correspondence* III, 230.
8. *Correspondence* I, 397.
9. See note 20, Chapter 2.
10. Letter from Anthony Thompson to John Pell (British Library MS. Birch 4279, f. 259) printed in J.O. Halliwell, *A Collection of Letters Illustrative of the Progress of Science in England* (London, 1841), pp. 95–6; cf. note 19, Chapter 2.
11. The Royal Society Journal Books begin with this entry. The minutes of this and all subsequent meetings in the 1660s and 1670s are readily available with only one or two exceptions in Birch, *History*.
12. Lord Edmond Fitzmaurice, *The Life of Sir William Petty* 1623–87 (London, 1845), pp. 103–4 , from a letter to his brother dated 4 February 1660–61. He was knighted the following April.
13. *Correspondence* I, 406.
14. See *Oeuvres Complètes*, XXII, 566–76, the journal of his visit to England during March, April and May 1661.
15. *Correspondence* I, 410.
16. Details of his journey are derived from the correspondence of Huygens and Moray (he carried a letter from Moray dated 17 June), for which see *Oeuvres Complètes*, III, 282,

297, and, esp. 307, and his own correspondence with Huygens, *Correspondence* I, 411–12, 420–3. A letter from Huygens to Moray, 1 August 1661 N.S., gives details of his journey, *Oeuvres Complètes*, III, 307.
17. See Oldenburg's letter to Borri of 7 September 1661, *Correspondence* I, 417–19.
18. Moray acknowledged a letter from Huygens which Oldenburg had brought back with him. Oldenburg's first letter to Spinoza is dated 16 August, *Correspondence* I, 413–15.
19. The last letter dealing with this topic is of 10 November 1665.
20. *Correspondence* I, 417–19.
21. Described by Pecquet (see Chapter 2, p. 44); cf. *Oeuvres Complètes*, III, 268–9, letter of 7 May O.S.
22. See *ibid.*
23. *Correspondence* I, 433–4.
24. The word alkahest was invented by Paracelsus who applied it to the universal solvent much sought by alchemists.
25. *Correspondence* I, 443–4. The book was to be sent via Sir John Finch, who left England for Italy on 22 October 1662. It is not certain that it was, nor that Oldenburg's letter was ever sent.
26. The first letter addressed to him 'at Mr Herbert's house' is from Beale, dated 31 January 1662/3; cf. his own letter to Hevelius of 18 February 1662/3. Southwell, writing on 14 February 1662/3, still addressed him 'at Lady Ranalaughs' as did other domestic correspondents for some time to come. See *Correspondence* II, esp. 19 and 22 and, for his telling foreign correspondents to address him at Mr Herbert's *ibid.*, 27.
27. Both the 1662 and 1663 Charters are printed in *The Record of the Royal Society*, 4th ed. (London 1940), pp. 213–17 and 237–63, respectively in both Latin and English. *The Record* also contains historical lists of the Officers and Fellows, 1663–1940.
28. 'An Attempt for the Explication Of the Phaenomena Observable in an Experiment Published by the Honourable Robert Boyle, Esq.' (London, 1661); the dedication to Boyle is signed by Hooke with his full name.
29. Beginning in the summer of 1672 such encounters are documented in Hooke's diary and there is no reason to suppose that they were then a novelty. Earlier meetings are sometimes deducible from their letters to Boyle; a few of those by Hooke are printed in Birch, *Boyle*, VI, 481–509.
30. For the letters dated 1663–65, see *Correspondence* II.
31. *Correspondence* I, 25–8.
32. *Correspondence* II, 136–39, 4 January 1665 N.S.
33. Beginning in 1662–3; *Correspondence* II, 21.
34. Beginning in 1662; *Correspondence* II, 59.
35. The earliest extant letter of this exchange is from Quintinye, 17 March 1662/3 O.S. in reply to one from Oldenburg, now not extant, acknowledging receipt of melon seeds. There was an active interchange during 1663 and an intermittent one thereafter.
36. Sorbière's *Relation* (see note 1) went through many editions; Balthasar de Monconys's *Voyages*, 3 vols. (Lyon, 1665–6 and in 5 vols. 1694 reprinted Paris, 1887) include accounts of his travels on the Continent as well as his 1663 visit to England. For

Sorbière's behaviour, cf *Correspondence* II, 128–9.
37. See *Correspondence* II, 45–7, addressed to the Royal Society and Oldenburg's reply of 23 September, *ibid.*, 110–11. Eccard Leichner (1612–90), Professor of Medicine at Erfurt from 1646, was a staunch conservative, opposed to all new medical, chemical, and philosophical ideas.
38. See *Correspondence* II, 50, 57, 72–3.
39. See *Correspondence* II, 112–14 *et seqq.* There are gaps in the Justel correspondence for some years from time to time; either Oldenburg destroyed the letters as unsafe or they passed into official hands because of their political news.
40. Serrarius was already a regular correspondent in the summer of 1663 when Oldenburg told Spinoza to send any of his works via Serrarius who would forward them to London; see *Correspondence* II, 99. There are few surviving letters between Oldenburg and Serrarius in the Royal Society but many references to them. There are more in the British Library, for which see M. Hunter, *Establishing the New Science*, pp. 257–60.
41. Lambeth Palace Archives: Faculty Office (Canterbury) Allegations of Marriage Licenses 1663–64, under date 20 October 1663. See further Hall & Hall, *Notes & Records* 18, 94–102.
42. 18 September 1666, *Correspondence* III, 230.
43. *Correspondence* II, 255.
44. *Correspondence* II, 59.
45. For Mrs Dury's death see Turnbull, *Hartlib, Dury, and Comenius*, pp. 297–8, where is quoted Dury's letter of 11 June 1664 to the ministers and elders of the Dutch Congregation at Austin Friars Church in London asking them to oversee his daughter's 'education' (upbringing). From their reply six weeks later (25 July) it appears that she was already living with the Oldenburgs, but they promised to look after her should Mrs Oldenburg die. It is, of course, possible that Dorothy West had been known to Mrs Dury but more likely that it was her father's friendship with Oldenburg which caused her to be placed in his household. Of course if Dorothy West had been associated with Mrs Dury it would explain how Oldenburg came to know her.
46. Postscript to a letter dated 25 August 1664, *Correspondence* II, 209–10.
47. On the 24th; *Correspondence* II, 319–20, also the first reference in the correspondence to the projected *Journal des Sçavans*.
48. See *Correspondence* II, 341 *et seqq.*
49. *Correspondence* II, 504, letter of 9 September 1665.
50. As Oldenburg reported to Boyle; see note 47.
51. Printed in Birch, *History*, II, 18; see also David Kronick, 'Notes on the Printing History of the Early *Philosophical Transactions*', *Libraries and Culture*, 25, 1990, 244–67.
52. See *Correspondence* II, 315, 316 note. For Captain Holmes see Chapter 4, p. 92. He had with him one or more pendulum watches made after the design of Christiaan Huygens in order to test whether they were accurate enough to be used to determine longitude at sea. (They were not.) It was presumably Moray who had arranged this.
53. Cf. Peter Dear, *Discipline and Experience*, Chicago and N.Y., 1995, p. 209.
54. Cf. *Correspondence* III, 563.

55. The first publication of a translation was to be in 1671; see below, Chapter 10.
56. *Correspondence* II, 211–13; thus the English edition dated 1664 was printed off at the end of 1663, and the Latin translation was dated 1665.
57. See M.B. Hall, 'What happened to the Latin Edition of Boyle's *History of Cold?*', *Notes & Records* 17 (1962), 32–35.

Chapter 4

1. Burial Register of St Martins in the Fields, now in Westminster archives and *Correspondence* III, 230, to Boyle, 18 September 1666.
2. Now British Library Add. MS. 4458, folios 108–16. Initially my husband and I ascribed this document to the 1670s, when Oldenburg had both a son and a daughter; going further, Michael Hunter ('On Oldenburg and Milleniarism', *Establishing the New Science* (Woodbridge, 1989, pp. 257–60) assumed without question that it was addressed to his son Rupert. Both these assumptions now seem to me less likely than that it was written in 1664 in anticipation of the birth of a child.
3. See note 45, Chapter 3.
4. *Correspondence* II, 358, printed from J.H. Hessel, *Ecclesiae Londino-Batavae* (Cambridge, 1897), III, pt 2, p. 3663.
5. *Correspondence* II, 459.
6. *Correspondence* III, 206.
7. For Robert Holmes, see Chapter 3, p. 85 and its note 53. For French feelings see letter to Boyle, 21 August 1664, *Correspondence* II, 206 and *ibid* (from Paris) p. 207, 'Your carryings on show clearly that you mean war.'
8. *Correspondence* II, 222. for the relative size of the two navies, see J.R. Jones, *The Anglo-Dutch Wars of the 17th century, passim.*
9. *Correspondence* II, 241, letter of 29 September 1664.
10. *Correspondence* II, 249, letter of 6 October 1664, to Boyle.
11. See, e.g. *Correspondence* II, 282–3.
12. See, e.g., *Correspondence* II, 297–8, letter of 10 November to Boyle.
13. *Correspondence* II, 324, 24 November and 326, 26 November, sending a copy of the speech to Boyle, and 332, 10 December.
14. He had probably been visiting his sister Mary Rich, Countess of Warwick, in Essex; he returned to Oxford via London in mid-June.
15. As it is now usually called; the English fleet had been in Sole (now Southwold) Bay and so called it the Battle of Sole Bay.
16. *Correspondence* II, 433, 4 July, to Boyle. But by the end of September (*Correspondence* II, 533) he was telling Boyle that he had heard from Amsterdam that the Dutch were becoming pessimistic about the chances of success.
17. *Correspondence* II, 449.
18. *Correspondence* II, 481, 24 August.
19. *Correspondence* II, 489–90, Moray's letter of 31 August in reply to one from Oldenburg of 26 August 1665.
20. *Correspondence* II, 582.

21. So Boyle (14 June) told Oldenburg that he had hunted out works on the plague by the Italian Ludovico Settala (1542–1633) published a generation earlier.
22. To Boyle, 4 July 1665, *Correspondence* II, 430.
23. See above note 22.
24. *ibid.*
25. *ibid.*
26. Letter of 10 August, *Correspondence* II, 459.
27. *Correspondence* II, 575, letter from Moray 19 October 1665.
28. See the memorandum for a letter to Boyle of late September, *Correspondence* II, 523 and to Moray, 28 September, *ibid.* 526–7.
29. Moray to Oldenburg, 16 September 1665, *Correspondence* II, 506 and Oldenburg to Moray (note 28).
30. 23 August, *Correspondence* II, 479.
31. Oldenburg to Boyle, 24 August, *Correspondence* II, 481 and 18 September, 511–12.
32. Pepys's diary under date 7 June 1665 and *Correspondence* II, 543, letters to Moray, 4 or 5 October and Boyle, 5 October.
33. *Correspondence* II, 582, letter from Moray 29 October.
34. *Correspondence* II, 446, letter from Moray 23 July.
35. *Correspondence* II, 524, letter from Moray 28 September and 563 of 11 October.
36. *Correspondence* II, 501, 509, 522, 536.
37. *Correspondence* II, 537, letter from Boyle 30 September.
38. *Correspondence* II, 563, letter from Moray of 11 October which also mentions his lodging with Wallis.
39. *Correspondence* II, 571, 17 October and 578, 24 October.
40. *Correspondence* II, 573, from Boyle *c.* 18 October.
41. *Correspondence* II, 580. This and other letters from Boyle and from Moray mention Wallis's work for the *Philosophical Transactions*, esp. II, 582–3 from Moray.
42. For Oldenburg's own reaction, see *Correspondence II*, 584; for Boyle's comment, 588.
43. For Moray's comments, *Correspondence* II, 582, 590.
44. *Correspondence* II, 591, 606.
45. *Correspondence* II, 642–7, 649–50, 652–3.
46. *Correspondence* III, 69.
47. *Correspondence* II, 650 and note 45 above.
48. *Correspondence* III, 8.
49. *Correspondence* III, 69, 24 March 1665/6.
50. *Correspondence* III, 17, 16 January 1665/6 and, to Lord Brereton, *ibid.* 22. The Bills of Mortality were issued weekly, detailing all deaths in the City classified by cause.
51. The Fire of London (see below) probably ended plague in the City, although not, of course, elsewhere. But for some reason, still not wholly understood, the years 1665 and 1666 saw the last great flare-up of plague anywhere in Britain, although it had been endemic there since the fourteenth century. The last recorded case was in 1679 before it was re-imported in 1900 from China and India, where it remains endemic. See Christopher Morris, 'Plague', Vol. X 'Companion', Latham and Matthews, ed. *The Diary of Samuel Pepys*, p. 336.

52. *Correspondence* III, 11.
53. *Correspondence* III, 33, 27 January 1665/6. The information about Louis XIV was derived from Justel's letter cited in note 52. The Queen Mother of England, widow of Charles I, was living in Paris.
54. Moray to Oldenburg, 26 September 1665, *Correspondence* II, 524.
55. *Correspondence* III, 8, 8 January 1665/6. Sir William Morice was a Privy Councillor and Secretary of State from 1660 to 1668; as Moray says, he and Boyle were very friendly. In September 1666 Oldenburg hoped to use Boyle's influence with Morice to help him to obtain a post: *Correspondence* III, 226. For Williamson, see below.
56. To Boyle, 24 February 1665/6, *Correspondence* III, 46.
57. Oldenburg to Hevelius, 24 August 1666, *Correspondence* III, 221, telling him of the Grubendol address for letters (but not parcels). It is curious that there is no earlier surviving example of such instructions in Oldenburg's extant foreign correspondence, but this perhaps results from the fact that almost no letters from him to his Dutch and French correspondents survive. Certainly from at least May 1666 he was sending French news to Williamson, extracted from the letters which Williamson presumbly forwarded to him. There is nothing about the Grubendol address in Oldenburg's first letter to Huygens after the latter moved to Paris as a pensionary of Louis XIV (*Correspondence* III, 128, May 1666), but he did then ask him to send his letters via Justel, who presumably added them to his own. No letters to Justel survive from this period.
58. *Correspondence* III, 39 from Auzout, 2 February 1665/6 and to Boyle, 46 of late February.
59. *Correspondence* III, 129–30, 15 May 1666.
60. *Correspondence* III, 16 May 1666. 'We hope that God will perform a miracle on our behalf and ruin your fleet by a gale for pride and presumption are sometimes punished.'
61. *Correspondence* III, 155, 8 June 1666.
62. *Correspondence* III, 202, *c.* 31 July 1666.
63. *Correspondence* III, 213, Wallis's letter of 18 August 1666.
64. *Correspondence* III, 231, 18 September 1666.
65. Off the town of Margate; *Correspondence* III, 234 and, below, letter to Boyle, *ibid.* 238, 2 October 1666.
66. The best contemporary account is to be found in Pepys's diary. For a good brief modern account, see 'The Great Fire' in Vol X, 'Companion', in the Latham and Matthews edition, pp. 138–40.
67. *Correspondence* III, 226, 10 September 1666, to Boyle.
68. As he told Boyle, *Correspondence* III, 230–1.
69. On 10 September, cf. note 37.
70. See his letter to Boyle, note 68.
71. *Correspondence* III, 244, 16 October 1666.
72. *Correspondence* III, 227, to Boyle. For his hearth tax (he paid 8 shillings compared with Lady Ranalagh's 14), see P.R.O. E179/152/31.
73. *Correspondence* III, 229–30.

74. The Earl of Bedford paid the tutor to his two sons £100 per annum in 1660; see G.S. Thomson, *Life in a noble Household* (Ann Arbor paperback, 1959), p. 97.
75. *Correspondence* III, 273, letter to Boyle of 23 October; for the cause, see above, p. 94. The date of the declaration of war was 19 September 1666, the royal proclamation being published as *A True Deduction of all Transactions between his Majesty of Great Britain and the King of Denmark*. Cf Pepys's diary for 20 October.
76. *Correspondence* III, 273, to Boyle 23 October. The supplies cut off were those normally coming from Denmark.
77. *Correspondence* III, 282, to Boyle, 15 November 1666.
78. Quoted to Williamson *Correspondence* III, 323, 26 January 1666/7; see also Justel's letter of 20 March 1666–7 *ibid.*, 369.
79. For the catastrophic state of the Navy which permitted the Dutch raid, see Pepys's diary and, for a modern summary, see Richard Ollard, *Pepys a Biography* (London, 1974), ch. XII, pp. 169–82. The Medway flows into the Thames estuary.
80. See Ollard *Pepys* (note 79), pp. 200ff.
81. Peter Pett was strictly examined in the autumn, as was Pepys, but although there was still some talk of impeaching Pett, in the end all charges were dropped, no doubt because so many were at fault. He then disappeared into obscurity. Pepys, obviously, weathered the storm.
82. *Correspondence* III, 444–5. One such letter dated 5 July from Serrarius in Amsterdam, survives in the Public Record Office *ibid.* 446–7. It is concerned with news of both naval matters and Jewish milleniarists.
83. *Correspondence* III, 448, letter to Seth Ward dated 15 July 1667 from the Tower; it was written at Oldenburg's dictation by an unnamed friend who obviously did succeed in visiting him there. For his having written to Boyle, see note 87 below.
84. *Correspondence* III, 452–3, enclosed in a letter to Arlington, *ibid.*, 450–1.
85. *Correspondence* III, 457, letter from Colepresse who lived near Plymouth, 26 July 1667.
86. It was signed on 21/31 July 1667 to be followed by a separate peace with France. The Dutch secured a revision of the Navigation Acts in their favour and in an exchange of overseas territories they ceded New Amsterdam to the English.
87. *Correspondence* III, 471–2, to Boyle, 3 September 1667, which describes events immediately after his release.
88. Birch, *Boyle*, VI, 509, 5 September 1667.
89. Cf. note 87.
90. *Correspondence* III, 473–4, 12 September, to Boyle.
91. Who produced the letter by Denis at the meeting of the Royal Society on 4 July and who was, after all, also Secretary.
92. *Correspondence* III, 480, to Boyle, 24 September 1667. Cf. A.R. and M.B. Hall, 'The first Human Blood Transfusion: Priority Disputes', *Medical History*, 24, (1980), 461–65.

Chapter 5

1. *Plus Ultra* (London, 1668), p. 103. Glanvill (1636–80) was at this time Rector of the Abbey Church at Bath. He was F.R.S. 1664 and wrote several works in defense of the 'new philosophy' and the Royal Society.
2. *Correspondence* IV, 58, 17 December 1667.
3. *A Journey to Paris in the Year* 1698, 3rd. ed. (London, 1699), p.81.
4. *Corrrespondence* IV, 58, 17 December 1667.
5. For later examples, see Chapter 6, below.
6. Fairfax (1637–90) was a Cambridge graduate who had been presented to a perpetual curacy in Suffolk at the time of the Commonwealth from which he was ejected in 1662 for refusing to conform; this seems to be the reason why he turned to medicine. He was a friend of Sir Thomas Browne and in 1674 the author of *A Treatise of the Bulk and Selvedge of the World*. He first corresponded with Oldenburg in 1666/7, *Correspondence* III, 315 *et seqq*; for later letters see vols. IV,V, and VI.
7. Colepresse (d. 1669) lived about five miles from Plymouth and had been befriended by a local magnate. A paper on tides by him was read before a Royal Society meeting on 19 December 1666. His earliest known letter to Oldenburg is dated 8 January 1666/7; he wrote to Boyle the next month. For further letters see *Correspondence* III through VI.
8. Childrey (1623–70) is usually denominated an antiquary but in fact he was keenly interested in miscellaneous natural history. He was an Oxford graduate, Rector of Upwey in Dorset when he first wrote to Oldenburg in February 1668/9, apparently encouraged to write by Walter Pope (for whom see Chapter 2, p. 30), at this time Gresham Professor of Astronomy. Childrey had published *Britannia Baconica* in 1660.
9. *Correspondence* IV, esp. p.232.
10. His last letter is dated 16 August; *Correspondence* VI, 194.
11. *Correspondence* V, 48–50; his account was partially printed in *Philosophical Transactions* no. 40 (19 October 1668), 803–5; cf. *Correspondence* VI, 67.
12. *Correspondence* VI, 68–70.
13. *Correspondence* III, 316–19, 358–60, 623–4.
14. *Correspondence* VII, 23–8, 6 June 1670. Tonge, a London clergyman, was later to become notorious as an associate of Titus Oates in working up the crisis of the Popish Plot of 1678–9, said to have aimed to put the Catholic James, Duke of York, on the throne after the plotters had assasinated Charles II: there was in fact no such Catholic plot of this kind. Oldenburg published some of Tonge's work on the rising of sap and read a paper of his on grafting trees to a meeting of the Society on 12 March 1672/3.
15. For Willughby, see below, p. 134.
16. For Childrey's letters see *Correspondence* V and VI. In 1669 he wrote what he called 'Animadversions' on Wallis's hypothesis of tides printed in the *Philosophical Transactions*, no. 16 (6 August 1666), 264–81; Childrey's comments were ultimately published in *ibid.*, no. 64 (10 October 1670), 2061–8 (he was slow in completing them), and Wallis's reply to them in *ibid.*, no. 64, immediately following Childrey's essay.

17. Born in 1608. See *Correspondence*, XII, 273–7 (*Philosophical Transactions*, no. 125 (22 May 1676), 599–603) and *Correspondence* IV, 475–6, 505–6 (printed in the *Philosophical Transactions*, no. 37 (17 July 1668), 727–9 and 729–31, and V, 9–10, VIII, 140, IX, 404–7, XI, 381–3, 384–5, for examples. See further Mayling Stubbs, 'John Beale, Philosophical Gardener of Herefordshire, *Annals of Science*, 39 (1988) 463–89 and 46 (1989) 323–63.
18. John Ray (1627–1705) lived with Willughby until the latter's death, after which he became his executor and completed his work. Skippon (1641–91), F.R.S. 1667, knighted 1674, was a protege of both Wilkins and Ray; he began corresponding with Oldenburg, whom he already knew, in February 1668/9, *Correspondence* V, 409–10. Willughby (1635–72) first wrote to Oldenburg in May 1669, *Correspondence* V, 571–2, sending him observations made with Ray on the rising of sap (*Philosophical Transactions*, no. 48 (21 June 1669), 963–5) and on the laws of motion (see below, p. 137).
19. See Wallis's letter of 24 July 1666, *Philosophical Transactions* no. 16 (6 August 1666), 289–94, criticizing a recently published book by Hobbes.
20. François Du Laurens, *Specimina Mathematica* ((Paris, 1667); he had learned from Justel that a problem which he had proposed in the then customary manner had been solved 'by one of your Fellows' and sought more information. (*Correspondence* III, 335–6.) The parcel and probably the letter arrived only slowly: Oldenburg told Wallis of the book in December 1667 (*Correspondence* IV, 23) and sent him a copy in March 1667/8 (*ibid.*, 285). Wallis's first reaction followed a few days after its reception and he continued to discuss the work in letters to Oldenburg throughout 1668 in 'animadversions' which Oldenburg published in *Philosophical Transactions* no. 39 (21 September 1668), 775–79 and no. 41 (16 November 1668), 825–32.
21. *Correspondence* IV, 398–401. He published his defence as *Responsio Francisci Dulaurens ad epistolam D. Wallisii ad clarissimum Oldenburgium scriptam* [Paris, 1668?].
22. *Correspondence* V, 104, 16 October 1668. For Huygens's reply, see *ibid.*, 126–8.
23. On 11 January 1668/9, *Correspondence* V, 331–3.
24. William Neile (1637–70), F.R.S. 1663, son of Sir Paul Neile, a pupil of Wilkins, had in 1657 devised a method of rectifying the cycloid, published by Wallis in *De Cycloide*, 1659. In 1666/7 he sent some ideas about impact motion to Oldenburg, which the latter sent to Wallis for comment; see *Correspondence* III, 372, Wallis's letter of 21 March 1666–7. He took the matter up in mid-December 1668; *Correspondence* V, 263 *et seqq.*
25. *Correspondence* V, 361–2, 27 January 1668/9.
26. *Correspondence* V, 450–3, 20 March 1668/9; it was published in the *Journal des Sçavans* for 8/18 March 1668/9.
27. *Correspondence* V, 462–6, 29 March 1669.
28. *Correspondence* V, 554–7, 19 May 1669.
29. For an authoritative discussion, see E.J. Dijksterhuis, 'James Gregory and Christiaan Huygens' H.W. Turnball, *James Gregory Tercentenary Volume* (London, 1939), pp. 478–86.
30. *Correspondence* V, 128 and 178.
31. For Gregory, see *Correspondence* V, 250–9, published in *Philosophical Transactions* no. 44 (15 February 1668), 883–6 for Wallis, *ibid.* 135, 193–4, 204, 273, 336–7; and for Moray, *ibid.*, 383.

32. *Correspondence* V, 466 (29 March 1669) and the Society's minutes in Birch, *History*.
33. Auzout's first announcement of his views was in his *L'Ephéméride du Comte . . . fait le 2 Janvier* 1665 [N.S.] sent to Oldenburg later that same month; see *Correspondence* II, 341. Auzout also sent to Oldenburg for the Royal Society a letter of Cassini's which supported his view, *ibid.* 359–62, 363–67. Auzout's first reaction to Hevelius is *ibid.*, 404–6, sent to Oldenburg in a letter now no longer extant, but of which an extract was sent to Boyle in June 1665. The controversy continued into 1666. For the differences in the assumed paths of the comets see *Correspondence* III, 5; for Hevelius's first letter (6 January 1665/6) *ibid.* 3–7.
34. See *Correspondence* III, *passim* for comment; for Hevelius's use of plain sights, *ibid.*, 313 (Wallis's opinion) and 347–48 (Hooke's opinion).
35. For Hevelius's initial request see *Correspondence* III, 256, 19 October 1666; for Hooke, *ibid.*, 349 and IV, 447.
36. *Correspondence* III, 354 and 519 and IV, 447. For further negotiations, IV, 581–2, Richard Reeve to Hevelius. It was in the end made by Christopher Cock and finally dispatched in August 1669: *Correspondence* IV, 168 and 173.
37. This is one reason why the controversy between Hevelius and Auzout died down. Cf. Auzout, *Lettre à M. l'Abbé Charles* (Paris, 1665) which Auzout sent to Oldenburg who received it in the spring; he sent two more copies through Justel on 22 June 1665: *Correspondence* II, 245. He later sent, more promptly, his second pamphlet, *Lettre du* 17 *Juin (*1665*) à Monsieur Petit* (*Correspondence* II, 428).
38. Oldenburg presumably made the translation for the benefit of Hooke who did not read French. For its text and that of Hooke's reply see *Correspondence* II, 383–8.
39. *Correspondence* II, 441–2, 23 July 1665.
40. For Auzout's letter in Oldenburg's translation see see *Correspondence* II, 461–74; extracts were printed in a slightly revised form in *Philosophical Transactions*. For Oldenburg's notes and comments see *Correspondence* II, 474–5, notes.
41. See *Correspondence* II, 605, 610.
42. For Major's reaction to the news from his fellow countryman, a lawyer named Theodore Jacob, see his letter to Oldenburg of 13 December 1664 O.S., *Correspondence* II, 334–7. His own work, *Prodromus invertae a se chiurgia infusoriae* had been published in Leipzig that same year.
43. For Oldenburg's letter of 11 March 1664/5—it is not clear why he waited two months to reply, contrary to his usual custom—see *Correspondence* II, 379–80.
44. No. 35 (18 May 1668), 678–9.
45. Letter of 27 August 1665, *Correspondence* II, 484.
46. No. 7 (4 December 1665), 128–30. That the account was written by Oldenburg is evidenced by his letters to Boyle of 31 October and 21 November 1665, *Correspondence* II, 585 and 616. In the latter he asked Boyle to edit it for details of dates etc.; see also Moray's note that he had explained to Wallis what Oldenburg wanted by way of editing, 27 November 1665, *Correspondence* II, 625.
47. Letters to Boyle of 17 September and 1 October 1667, *Correspondence* III, 476, 486, 504 and *Philosophical Transactions* no. 27 (23 September 1667), 490–1, all relating to C. Fracassati and M. Malpighi, *Tetras anatomicarum epistolorum* (Bologna, 1665).

48. Letter of 17 October 1667, *Correspondence* III, 532–5, partly from *Philosophical Transactions* no. 29 (11 November 1667), 551–2.
49. Richard Lower (1631–90/91) of Christ Church, Oxford, M.D. 1665, F.R.S. 1667. He had worked with Willis in preparing the latter's *Anatome Cerebri* (1664). His own *De Corde* (London, 1669) contains accounts of these experiments and records Boyle's share in devising some of them. He later practised as a physician in London.
50. *Philosophical Transactions* no. 20 (17 December 1666), 353–58.
51. *Philosophical Transactions* no. 22 (11 February 1666–7), 385–88 and no.25 (6 May 1667), 449–52.
52. *ibid.*, p. 453.
53. 'Touchant une nouvelle maniere de guerir plusieurs maladies par la tranfusion du sang, confirmées par deux experiences faites sur les hommes', addressed to Montmor; it was reprinted in the *Journal des Sçavans* XI (28 June 1667). J.B. Denis (d.1704) was a surgeon, not a physician; he began lecturing in Paris in 1664 and became a member of the Montmor Academy on whose behalf he performed his transfusion experiments. He edited two periodicals in the early 1670s and having, as he believed, discovered an effective 'styptic fluid' he came to England in June 1673, when he demonstrated it to the Royal Society.
54. Cf. *Correspondence* III, 480 to Boyle, 24 September 1667.
55. No. 28 (21 October 1667) 517–25.
56. See the minutes of the meetings of 21 September *et seqq*. The subject, named Arthur Coga, gave an account of the effects in a Latin paper on 28 November. Since he was willing, the experiment was repeated, as reported in the minutes for 12 December and described by Coga a week later; he said he had experienced a slight fever. Dr King wrote a report which he presented to the meeting on 9 January 1667/8, which was registered. Cf. *Correspondence* III, 617 and *Philosophical Transactions* no. 30 (9 December 1667) 557–59.
57. Printed accounts by Denis in the form of letters to Oldenburg were published in *Philosophical Transactions* no. 32 (10 February 1667/8), 617–24 and no. 36 (15 June 1668), 710–15 and in *Correspondence* IV, 40–53 and 372–87, the latter containing an account of the death of the subject and the official condemnation banning any further such experiments in France.
58. *Correspondence* IV, 350–67, as published in *Philosophical Transactions* no. 35 (18 May 1668), 672–82.
59. See *Philosophical Transactions* no. 34 (13 April 1668), 663 for the review of De Graaf's *Epistola, De nonnullis circa Partes Genitales Inventis Novis* (Leiden, 1668).
60. Regnier De Graaf (1641–73) practised medicine in Delft after 1666 when he returned to Holland from a tour of France; he first wrote to Oldenburg in July 1668, sending him a copy of his latest work, *De Virorum organis generationi inservientibus* (Leiden and Rotterdam, 1668); *Correspondence* IV, 523–4. Oldenburg in reply told him that his first book had been discussed at the Royal Society on 19 March 1668/9 and presumably sent him the appropriate issue of the *Philosophical Transactions*. His first defence of his own priority against Clarke's claim is in his letter of 25 September 1668, *Correspondence* IV, 67–71. Exchanges of letters between himself and Oldenburg continued until his death; Oldenburg's letters to him do not survive.

61. See the opening of Clarke's letter of 20 December 1669, *Correspondence* VI, 385.
62. His interest in universal natural history is recorded throughout his correspondence, especially after 1662.
63. See *Correspondence* III, 343–4 and 602 (letters of Paul Rycaut, consul of the Levant Company in Smyrna; *ibid.* 340–1, 461–2 from Benjamin Lannoy, English consul at Aleppo;and *ibid.*, 384–5 to Sir George Oxenden, Governor of Bombay and President of the East India Company at Surat.) So the next year the Royal Society took advantage of the opportunity of a group of ships going to the East Indies to send with them 'Philosophical Commissions', as Oldenburg told Boyle, *Correspondence* IV 207, 25 February 1667–8.
64. *Correspondence* IV, 316–17, letter to Jeronimo Lobo, 13 April 1668.
65. See e.g. *Correspondence* IV, 451–2, letter of 6 June 1668 to Richard Kemp asking about Mexico and *ibid.* VIII, 220–48 sent to Thomas Hill in Lisbon for an unknown Jesuit in Brazil.
66. Begun in 1663–4; *Correspondence* II, 146.
67. Begun in 1663 and continued until 1672.
68. *Correspondence* IV, 528; Paisen's first letter was of July 1668 and his last shortly before his death in October 1670 for which see VII, 455.
69. See *Correspondence* IV to XI.
70. *Correspondence* III, 179–81. *Theatrum Cometicum* was published in Amsterdam.
71. 6 February 1666–7, *Correspondence* III, 337–39.
72. *Mesolabum* (Liège, 1659, 1668); the word signifies a mathematical instrument used to determine mean proportionals; the book deals with mean proportionals in relation to circles, ellipses, and hyperbolas. See *Correspondence* III, 594–8, 616–17, 629.
73. 10 May 1667, *Correspondence* III, 412–13.
74. See Oldenburg's reply of 6 June 1667, *Correspondence* III, 430–2, which enclosed 'Rules for Reducing Biquadratic Equations' (*ibid.*, 433–4), sent from Paris with, as Oldenburg noted, Collins's comment that this work had been published in J.H. Rahn, *Teutsche Algebra* (Zurich 1659, English translation with notes by Pell, 1668).
75. For Collins, see *Correspondence* VI, 227, *c.* 12 September 1669, sent to Sluse by Oldenburg in his letter of 14 September 1669, *ibid.*, 233. See further, Chapter 6 below.
76. For Peter Serrarius and his first correspondence with Oldenburg, see *Correspondence* II. For his reports on milleniarism see *ibid.*, III, 447, XIII, 407–8 and 408 note 2. Michael Hunter (*Establishing the New Science* (Woodbridge, 1989), p. 259 has found other letters copied by Oldenburg in BL MSS. Add. 4299, ff. 50–1 and hints of Oldenburg's interest in the subject in or after 1673 in MSS. Add. 4458, ff. 146–7. These are manuscripts extracted by Thomas Birch from the Royal Society archives and bequeathed by him to the British Museum in the later eighteenth century.
77. *Correspondence* V, 461–2 and notes and *ibid.*, VI–IX; also below, Chapter 7.
78. *Correspondence* III, 235, 22 September 1669. Cf. Hunter, *Establishing the New Science*, Chapter 3, for more detail concerning committees of the Royal Society.

Chapter 6

In *Correspondence* vols. VIII through XIII all letters exchanged between Newton and Oldenburg are summarized with references to the complete text as printed in Newton, *Correspondence*. vols. I and II, except for some short memoranda given in full. Letters to and from Leibniz are printed in full in both Latin and English except for those few which, at the time of editing, had been printed in the monumental German edition of Leibniz's *Sämtliche Schriften*, still now in course of editing; these last are given in English translation only.

1. For the relation between Collins and Barrow, and then Collins and Newton, see especially A. Rupert Hall, *Isaac Newton. Adventurer in Thought* (Blackwell, Oxford, 1992 and Cambridge University Press, 1996). For Collins's letters, see *Correspondence* IV and XIII and Newton, *Correspondence* I and II. Letters of Collins to others besides Oldenburg and Newton are printed in Rigaud.
2. See e.g. *Correspondence* IV, 408–9, 558 and *ibid.*, V, 211–13, 345–6.
3. See especially *Correspondence* V, 468–71 and notes 471–2. That Oldenburg then already knew of Barrow's optical work is shown by his postscript to his letter to Huygens of 8 March 1668/9, *ibid.*, 436, 437.
4. For Collins's letter of September 1669, see *Correspondence* VI, 226–9 and notes 230–1; it was sent to Sluse on 14 September (*ibid.*, 232–6). For Sluse's indication of its loss see *ibid.*, 23 and 24 December 1669, p.395. Oldenburg sent its replacement on 26 January 1669–70, *ibid.*, 520–5. Newton's *De Analysi* was not to be printed until 1711.
5. In a letter dated 24 September 1670, *Correspondence* VII, 188–9.
6. Collins (c. 20 February 1670/1) told Oldenburg to tell Sluse that he expected that a mathematical work by Newton would soon be 'put... into the Presse'. (*Correspondence* VIII, 545) which Oldenburg did on 4 March 1671/2, *ibid.*, 576.
7. For Sluse, 17 December 1671, see *Correspondence* VIII, 411.
8. *Correspondence* VIII, 484–9, especially 489 which deals at length with Pardies.
9. No. 81 (15 March 1672) 4010–16; Collins told Gregory that this resulted from his hearing that Sluse intended to publish on the subject: H.W. Turnbull, *James Gregory Tercentenary Memorial Volume* (London, 1939), 224. Huygens (17 September 1672, *Correspondence* IX, 259) thought that Wallis's 'rules for tangents' were very like those of Fermat and others.
10. See Sluse, 7 January 1672/3, *Correspondence* IX, 386–96 and for Oldenburg's reply with Newton's 'method', 29 January, *ibid.*, 429–30 and 430 note 2. Newton's 'method' was part of his development of fluxions (his version of the calculus).
11. For Sluse's acknowledgement of the similarity of their methods see *Correspondence* IX, 618, 23 April 1673.
12. *Correspondence* X, 44, 23 June 1673, sent to Sluse on 10 July, *ibid.*, 79. For Sluse's acceptance of this, *ibid.*, 99.
13. See, *inter alia*, A. Rupert Hall and A.D.C. Simpson, 'An Account of the Royal

Society's Newton Telescope', *Notes & Records* 50 (1996), 1–12; I. Bernard Cohen, 'Newton's Description of the Reflecting Telescope', *ibid*, 47 (1993), 1–9; and A. Rupert Hall, 'John Collins on Newton's Telescope', *ibid.*,49 (1995), 71–78. That Leibniz's suggestion might have been influential in encouraging Newton to develop his ideas further was first conjectured by Hall.

14. *Correspondence* VIII, 443; for Wallis's reply, *ibid.*, 466–7. Oldenburg's letter is lost.

15. *Correspondence* VIII, 443–6,468–73.

16. *Correspondence* VIII, 447, summarizing Newton, *Correspondence*, I, 73.

17. *Correspondence* VIII, 523, 4 February 1671/2. Hooke's 'method' presumably relates to his machine for grinding telescope lenses which, he said, would grind elliptical instead of simple spherical lenses. (Cf. Chapter 5).

18. Cf. note 16.

19. *Correspondence* VIII, 468–73 with drawing; it was to be printed in the *Journal des Sçavans*.

20. The original does not survive. There is a transcript by Newton's usual copyist which was used in Newton, *Correspondence*, I, 92–102 to supplement what is printed in *Philosophical Transactions* no. 80 (19 February 1671/2), 3075–87; summary in *Correspondence* VIII,328.

21. Summary in *Correspondence* VIII, 528, 6 February 1671/2, text in Newton, *Correspondence*, I, 107–8.

22. *Correspondence* VIII, 533, 10 February 1671/2 and Newton, *Correspondence*, I, 108–9.

23. See Newton, *Correspondence*, I, 105, note 19.

24. Others had thought of using mirrors in place of, or in conjunction with, lenses: Gregory had suggested this and had produced a design as early as 1663, but the model built was not successful. After the publication of Newton's work, Cassegrain in 1672 published a scheme in the *Journal des Sçavans* claiming that he had invented an improved modification of the Gregorian design. Both of these versions are, in modern times, more used than the Newtonian version, but, of course, the mirrors are now made of glass, not metal.

25. *Correspondence* VIII, 568, 28 February 1671/2, also the letter to Fermat, 25 January 1671/2, *ibid.*, 507. Oldenburg had told Cassini (*ibid.*, 476, 477, 15 January 1671/2) that he could obtain detailed knowledge of the instrument from Huygens.

26. e.g. to Cornelio, 9 February 1671/2, *Correspondence* VIII, 531; to Sluse, 4 March 1671/2, *ibid.*, 575; to Hevelius, 18 March 1671/2, *ibid.*, 593, and to Vogel, 23 March 1671/2, *ibid.*, 613.

27. Hooke's paper was read of 15 February 1671–2 and is printed in Birch, *History*, III, under that date. It is also printed in Newton, *Correspondence*, I, 110–14 from the holograph manuscript which Newton saw, and in *Early Science in Oxford* from Birch.

28. His notes (Cambridge University MS. Add. 3958, ff. 1–2) are printed in A. Rupert Hall and Marie Boas Hall, *Unpublished Scientific Papers of Isaac Newton* (Cambridge, 1962), pp. 400–13.

29. Newton's original letter occupied most of *Philosophical Transactions* no. 80 (cf. note 21) and does not mention Hooke's name; the original (which did give Hooke's name)

is printed in Newton, *Correspondence*, 171–88. To this paper Hooke never replied publicly, although there exists a partial draft of a reply in the form of a letter addressed not to Newton but to an unnamed nobleman, plausibly Brouncker as President. This Newton certainly and Oldenburg most probably never saw. In the remainder of 1672, Hooke performed many optical experiments at meetings, some a repetition of Newton's work and some striking out on different lines, all reported in the minutes. Ironically, none of these ever disproved Newton's 'doctrine' in the slightest degree, although Hooke continued to believe that his own theory was correct and Newton's 'doctrine' erroneous.

30. See *Correspondence* IX, 119 (21 June 1672) and letters throughout 1672 and 1673.

31. See especially *Correspondence* IX, 7–10, 3 April 1672, with Newton's reply, *ibid.*, 13 April 1672, 25–8 and, for Pardies's acceptance of Newton's comments, *ibid.*, 21 May 1672, *ibid.*, 59 and 29 June 1672, 134.

32. *Correspondence* IX, 111, 30 May 1672.

33. *Correspondence* IX, 52–3, 2 May 1672, Newton, *Correspondence*, I, 150–1 in reply to Newton's letter of 13 April, *Correspondence* IX, 25, Newton, *Correspondence*, I, 136–9. For Newton's comments on this proposition, see *Correspondence* IX, 53, Newton, *Correspondence*, I, 153–5.

34. See above, note 28.

35. *Correspondence* IX, 255, 21 September 1672; Newton, *Correspondence*, I, 237–8. For a more detailed discussion of Oldenburg's role and Hooke's attitude, see A. Rupert Hall and Marie Boas Hall, 'Why Blame Oldenburg?' *Isis*, 3 (1962), 482–91.

36. Cf note 35.

37. *Correspondence* IX, 256, Newton, *Correspondence*, I, 241–3.

38. *Correspondence* IX, 421, 18 January 1672/3; Newton, *Correspondence* I, 255–6 shows that Oldenburg had written to Newton earlier and received an answer (both now missing).

39. *Correspondence* IX, 383, 4 January 1672/3.

40. *ibid.*, 3 April 1673, Newton, *Correspondence*, I, 264–6; it was sent to Huygens on 7 April, *Correspondence* IX, 570–1.

41. Newton, *Correspondence*, I, 674–6, 31 May 1673.

42. For all the above, see *Correspondence* X, 43–4 and Newton, *Correspondence*, I, 290–5; for Oldenburg's action, Newton, *Correspondence*, I, 284. Cf. *Correspondence* XI, 165, late January 1674–5.

43. *Correspondence* X, 212, Newton, *Correspondence*, I, 305–6. The first of Saturn's satellites was discovered by Huygens in 1655; two more were discovered by Cassini in 1671 and 1672.

44. The first letter was dated 26 September 1674 and was to be printed in *Philosophical Transactions* no. 110 (25 January 1674–5), 217–19; Newton, *Correspondence*, I, 317–19. The pupils were to write after Line's death in 1675.

45. Boyle had at first never used the Torricellian tube, but rather a glass vessel from which the air was evacuated by means of an airpump. Line's attack led him to devise an elegant experiment with a J tube, by means of which the relation between pressure (from mercury poured into the long arm of the J tube) and volume (of air trapped in the

short, closed end of the tube) could be measured. But Line insisted that the mercury in the closed end of the tube was held in position by invisible little threads or 'funiculi'.
46. Cf Line's letter of 14 November 1674, *Correspondence* XI, 123–4.
47. *Correspondence* XI, 137, 5 December 1674, Newton, *Correspondence*, 328–9. Oldenburg replied very briefly to Line on 17 December 1674, *Correspondence* XI, 146–7.
48. *Correspondence* I, 137 and *Philosophical Transactions* no. 110 (15 January 1674/5), 219.
49. Line wrote at length on 15 February 1674/5, *Correspondence* XI, 191–3 and briefly, complaining, on 1 September 1675, *ibid.*, 481.
50. *Correspondence* XII, 53, Newton, *Correspondence*, I, 356–8, 13 November 1675. Oldenburg's own letter is lost.
51. All this is in the letter cited in note 50.
52. See *Correspondence* XII, 87 and Newton, *Correspondence*, I, 362–86; it was first printed in Birch, *History*, III, 247–60, 262–9, 272–8, 280–95, 296–305. Reading it occupied several meetings.
53. This letter survives only as an endorsement, for which see *Correspondence* XII, 88 and Newton, *Correspondence*, I, 362 note 6. The account of the unsuccessful attempt to repeat Newton's experiment, described in a letter from Oldenburg, is inferred from Newton's reply.
54. *Correspondence* XII, 91 and Newton, *Correspondence*, I, 392–3.
55. *Correspondence* XII, 102, 21 December 1675 and Newton, *Correspondence*, I, 404–6.
56. See *Correspondence* XII, 84–6 and 195–301 respectively.
57. *Correspondence* XII, 133–4, *c.* 10 January 1675/6, Newton, *Correspondence*, I, 407–11.
58. Newton, *Correspondence*, I, 412–13.
59. *ibid.*, 416–17, 5 February 1675/6.
60. *Correspondence* XII, 149, 25 January 1675/6, Newton, *Correspondence*, I, 413–15.
61. John Gascoines was an English Catholic in the Jesuit College of Liége who had replied to Oldenburg's last letter to Line.
62. For Anthony Lucas see *Correspondence* XII, 84, 5 December 1675 and for Oldenburg's reply *ibid.*, 144, 18 January 1675/6; see also *ibid.*, 295–301 of 17 and 18 May 1676.
63. See *Correspondence* XIII, 51 and 51–2, Newton, *Correspondence*, II, 76–81 and 83 for Newton's letters conveyed to Lucas in August 1676, *Correspondence* XIII, 60 and for Lucas's answer *ibid.*, 99–104. This last was sent by Oldenburg to Newton *c.* 25 November 1676, *Correspondence* XIII, 146. To the earlier Newton replied on 28 November 1676, *ibid.* and Newton, *Correspondence*, II, 183–5. Lucas in turn replied on 23 January 1676/7, *Correspondence* XIII, 191–4, sent to Newton by 19 February when he acknowledged it briefly, *Correspondence* XIII, 217 and Newton, *Correspondence*, II, 193–94.
64. Newton, *Correspondence*, II, 239 to Hooke, 18 December 1677.
65. *Ibid.*, 253 to Hooke, 5 March 1677/8 enclosing two letters to Lucas of the same date, *ibid.*, 254–63.

66. *Ibid.*, 266–8, 269, undated but 1678.
67. No. 122, 515–33 in English and Latin, printed in parallel columns.
68. *ibid.*, p. 524.
69. *Correspondence* XII, 258–9, 26 April 1676.
70. See A. Rupert Hall and Marie Boas Hall, 'Newton's chemical Experiments', *Archives int. d'histoire des sciences*, 11 (1958) 113–52, reprinted in A. Rupert Hall, *Newton, His Friends and his Foes* (Variorum. Aldershot, Hants, 1993), no. IV, for the chemical notebook in use 1678–82 and, more generally, B.J.T. Dobbs, *The Janus Faces of Genius, the Role of Alchemy in Newton's Thought* (Cambridge, 1991).
71. On 26 April 1676, *Correspondence* XII, 258–9, and Newton, *Correspondence*, II, 1–2.
72. The copy is now in the Royal Society, MS. OB, no. 94; the original went back to Newton.
73. *Correspondence* XII, 288, Newton, *Correspondence*, II, 6, 11 May 1676.
74. No. 131, 29 January 1676/7, 775–87; a continuation was published in no. 132, 26 February 1676/7. It was Boyle's Experiment 13 which interested Newton.
75. On 19 February 1676/7, *Correspondence* XIII, 217–18, Newton, *Correspondence* II, 193–4.
76. For what follows see, besides the letters themselves, A. Rupert Hall, *Philosophers at War* (Cambridge, 1980), esp. Chapters 1 (for Newton) and 4 (for Leibniz). See also, for Leibniz's development, J.E. Hofmann, *Leibniz in Paris*, 1672–76 (Cambridge, 1974).
77. *Correspondence* XI, 399–400, Tschirnhaus's first letter to Oldenburg; cf. *ibid.*, 409–11. He obviously wrote out his mathematical ideas and these Oldenburg sent to Collins.
78. *Correspondence* XI, 390–1 and, for Tschirnhaus's meeting with Collins, *ibid.*, p. 324 note, 323–4, 325–6, and 326 note. Tschirnhaus's last surviving letter to Oldenburg is dated 22 August 1676, *Correspondence* XIII, 52–8. He returned home to Saxony and later was instrumental in developing the first European hardpaste porcelain (Meissen).
79. Leibniz resumed correspondence with Oldenburg on 5 July 1674, *Correspondence* XI, 42–6, revealing his progress in mathematics, and continued to write regularly for the next three years. Oldenburg replied as coached by Collins, translating Collins's somewhat disorganized English into correct and logical Latin. Cf. e.g., *Correspondence* XII, 95–9, 338–9 and *Correspondence* XIII, 1–13. The news had been retailed by a Dane named Mohr travelling from London after meeting Collins; see *ibid.*, 268–70.
80. *Correspondence* XII, 288 and Newton, *Correspondence*, II, 7, 15 May 1676.
81. *Correspondence* XII, 335–36, Newton, *Correspondence*, II, 20–32, 13 June 1676.
82. *Correspondence* XIII, 1–13, 26 July 1676, Oldenburg to Leibniz.
83. On 17 August 1676, *ibid.*, 40–9.
84. *Correspondence* XIII, 89 note.
85. Sent to Oldenburg on 24 October 1676, *Correspondence* XIII, 119–21, Newton, *Correspondence* II, 110–29.
86. *Correspondence* XIII, 303–13 and 316–18.
87. *Correspondence* XIII, 336–9.
88. Newton, *Correspondence* II, 237, 30 August 1677.

Chapter 7

1. *Correspondence* X, 42, 526.
2. *Correspondence* VI, 436, 19 January 1669/70.
3. *Correspondence* I, especially pp. 445–7.
4. For 2 July 1668 N.S.; see also *Correspondence* IV, 560 and V, 138–9.
5. *Correspondence* V, *passim*, e.g. 3–4 (3 August 1668), 128 (4 November 1668); for Wallis's view, 138–9 (4 November 1668), 161–3 (14 November 1668), etc., and for Gregory's defence, 250–9 (11 December 1668) also published in *Philosophical Transactions* no. 14 (15 February 1668/9).
6. For Oldenburg's letter to Huygens, *Correspondence* V, 331–3 and 383 and for that to Gregory *ibid.*, 340–1 (19 January 1668/9). For Collins, *ibid.*, 374 and 375 note 7. Cf. *ibid.*, 466 to Huygens, 29 March 1669.
7. Cf. *Correspondence* V and VI, *passim*. Oldenburg wrote some dozen letters to Huygens in 1669 and received eight in return. For astronomy see *Correspondence* V, 427 and for longitude experiments see e.g. *Correspondence* IV, 175, 228, V, 503, 557, 582, VI, 44–5 etc.
8. Huygens sent his solution on 16 June 1669 in an example of his newly invented method of printing in facsimile (*Correspondence* IV, 556 and 558 note 2, 19 May 1669); the solution to the problem was printed in *Philosophical Transactions* no. 97 (6 October 1669), see *Correspondence* V, 45 and 46 note 5. Oldenburg told Wallis of it on 29 June 1669, *ibid.*, 67.
9. 15–16 July 1670, *Correspondence* VII, 79. Oldenburg sent Huygens's solution to Sluse on 24 September 1670, *ibid.*, 177–8, 185–6. For Sluse's own solution see *ibid.*, 246ff.
10. Oldenburg to Sluse, 12 January 1670/71, *Correspondence* VII, 372; this letter also contains Gregory's comments on Barrow's method of tangents, taken from a letter from Gregory to Collins. Sluse gave permission to quote his method on 27 February 1670/71, *ibid.*, 483. The letter is also concerned with problems of complex equations. Oldenburg sent Sluse's Alhazen letter (note 9 above) on 22 July 1671, *Correspondence* VIII, 170.
11. *Correspondence* VIII, 316, 28 October 1671.
12. *Correspondence* VIII, 370, 21 November 1671 and for Sluse's reply of 17 December, 407–10, reported by Oldenburg to Huygens on 1 January 1671/2, *ibid.*, 446; a copy was sent later as it was 'too long to be copied in the small time available to me' as Oldenburg said.
13. *Correspondence* VIII, 3 February 1671/2, 521 in a letter mainly concerned with Newton's new telescope.
14. *Correspondence* VIII, 537, 12 February 1671/2, a prompt reply. He sent another long piece from Sluse in September 1672, *Correspondence* IX, 235.
15. *Correspondence* IX, 436, 31 January 1672/3.
16. Huygens wrote on 30 March 1672, *Correspondence* VIII, 637; Oldenburg sent this to Sluse on 11 April along with Newton's first paper on light and colours *Correspondence* IX, 17. Sluse gave permission for printing his work on 29 May, *ibid.*, 79, asking

Oldenburg to tell Collins of his latest idea, then correcting and extending his work two days later, *ibid.*, 81–2, and again on 12 June 1672, 100–1, and yet again on 16 November, 330–6. Huygens sent his latest thoughts on 21 June 1672 with permission to print them, *ibid.*, 119, 120–2, 249, 17 September 1672. Sluse's letter of 31 May was sent to Huygens on 13 January 1672/3, *ibid.*, 410.

17. For Vernon's letters from Paris March 1669 to April 1672 (after which he left Paris) see *Correspondence* V–IX. He set off on a Mediterranean journey in September 1674 writing from Smyrna which he reached via Italy, the Dalmatian Coast, and Greece on 10 January 1675/6, *Philosophical Transactions* no. 124 (24 April 1676) 575–82. He arrived in Persia in 1677 only to be killed in a quarrel over a penknife.

18. For Huygens's letter of 12 January 1669/70, see *Correspondence* VI, 425; for Vernon's letter of 19 January 1669/70, *ibid.*, 435.

19. *Correspondence* VI, 502–6, 15 February 1669/70.

20. *Philosophical Transactions* no. 60 (20 June 1670) 1065bis–1074bis. For the anagram on motion sent to the Royal Society 25 August 1669, *Correspondence* VI, 213–18.

21. What remains reads, '& you Mr Oldenburg hee applauded for a most . . . and all the whole Society in general'.

22. *Correspondence* VII, 5 October 1670.

23. On 21 October 1670, *ibid.*, 217–20.

24. Oldenburg sent news, copies of the *Philosophical Transactions* by means of travellers) and tried to interest Huygens in Leibniz's comments on the laws of motion framed by Huygens and Wren; *Correspondence* VII, 239–42, 8 November 1670, and 317–18, *c.* 12 December 1670, 532, late March 1670/71, 537–9, 28 March 1671, *Correspondence* VIII, 167–71.

25. *Correspondence* VIII, 517–22.

26. *Correspondence* IX, 247–50.

27. *Ibid.*, 380–84, 4 January 1672/3 and 433–6, 31 January 1672/3.

28. Mentioned in his letter of 31 May 1673, *Correspondence* IX, 575; they had arrived by the end of May, cf. *Correspondence* X, 2,3 and, for the list, 58, 66.

29. *Correspondence* X, 61–2, 27 June 1673.

30. *Ibid.*, 2.

31. *Ibid.*, 30, 14 June 1673. Huygens's own list of intended recipients may be found in his *Oeuvres Complètes*, VIII, 321 note 2.

32. *Correspondence* X, 39.

33. *Ibid.*, 3–4, extended 23 June, *ibid.*, 40–3, in Latin for sending to Huygens as was done on 27 June, *ibid.*, 68.

34. No. 93 (17 November 1673) 6146–9, *Correspondence* X, 276–82.

35. *Correspondence* X, 42.

36. *Ibid.*,525–26, 20 March 1673/4.

37. Note 30, above.

38. Letter of 30 June 1673, *Correspondence* X, 73; cf. his letter of 14 June, 30–1.

39. *Correspondence* XI, 235–6, sent to Oldenburg in March 1674/5, *ibid.*, 233, because the original never arrived.

40. Summary in *Correspondence* X, 74, 30 June 1673, Latin text in *Oeuvres Complètes*, VII, 339–40.
41. Note 39 above.
42. They were all published in *Philosophical Transactions* no. 98 (17 November 1673) as letters to Oldenburg, the first in Latin (cf. note 33 above) and the last two in English.
43. 4 August 1673, *Correspondence* X, 113 and *Philosophical Transactions* no. 97 (6 October 1673) and 98 (17 November 1673).
44. 8 December 1673, *Correspondence* X, 373.
45. 2 March 1673/4, *ibid.*,490–91.
46. *Ibid.*, 525–6, 20 March 1673/4.
47. *Ibid.*,547–8, 30 March 1674.
48. *Correspondence* XI, 1–3, 5 May 1674.
49.*Ibid.*, 24–5, 25 May 1674.
50. *Ibid.*, 32, 49, 144, 11 June, 23 October, and 9 December 1674.
51. *Ibid.*, 20 January 1674/5.
52. That is, a list of the letters contained in a brief message with their frequency, here decyphered as 'Axis circuli mobilis affixus in centro ferreae' i.e., 'The axis of the moving circle is fixed in the centre of an iron spiral'.
53. *Correspondence* XI, 176–8, 2 February 1674/5.
54. Pp. 105–6, in the middle of descriptions of various instruments 'contrived' (his word) by Hooke. For a facsimile, see *Early Science in Oxford*, VIII (1931) 'The Cutlerian Lectures of Robert Hooke'.
55. *Correspondece* XI, 184–87, 10 February 1674/5. The sentence about the possibility of a patent in England is printed in English in *Philosophical Transactions* no. 112 (25 March 1675), 273.
56. Thomas Sprat, *The History of the Royal Society* (1667), p.247. The printed text of Hooke's diary reads 'when' but 'where' seems a better reading.
57. *Correspondence* XI, 220–2, 11 March 1674/5.
58. *Ibid.*,224–6; cf. 225 note.
59. *Ibid.*, 245, 27 March 1675, 281–2, 19 April 1675, and 300 *et seqq.*
60. *Ibid.*, 495, Oldenburg to Huygens, 13 September 1675.
61. *Ibid.*, 326–8, 29 May 1675; for regulation, 342, 11 June 1675, 406, 15 July 1675, and 442, 31 July 1675.
62. It was sent by Huygens a month earlier. *Correspondence* XII, 50, 11 November 1675, the price being 27 crowns (ecus) or 80 pounds (livres). For its arrival, 89, 13 December 1675.
63. *Correspondence* XI, 328, 29 May 1675.
64. *Ibid.*, 335, 7 June 1675.
65. *Ibid.*, 379–80, 1 July 1675.
66. *Correspondence* XII, 3, 2 October 1675.
67. *Correspondence* XI, 380, 1 July 1675.
68. *Correspondence* XII, 11–14, 11 October 1675.
69. *Ibid.*, 14–15, 12 October 1675, and 16–17, 15 October 1675.
70. *Ibid.*, 21–4.

71. *Ibid.*, 41, ?8 November 1675.
72. *Ibid.*, 90, 13 December 1675.
73. *Ibid.*, 115, letter from D'Alencé, 25 December 1675.
74. *Ibid.*, 148, 21 January 1675–6.
75. *Ibid.*, 185–6.
76. On 22 February 1675–6, *ibid.*, 200–1.
77. *Ibid.*, 256–7.
78. *Oeuvres Complètes*, IX, 333 note and XXII, 744, cited by A. Rupert Hall, 'Huygens and Newton' in *The Anglo-Dutch Contribution to the Civilization of Early Modern Society* (OUP for the British Academy, 1976), p. 45.

Chapter 8

1. *Correspondence* IV, 28 December 1667.
2. *Philosophical Transactions* no. 27 (23 September 1667), 491–3. This is not a review as such, for no specific work is mentioned, but must be based upon *Tetras Epistolarum De Lingua et Cerebro Clarissimi Domini Malpighi Phil. ac Med. Bononien. nunc Messanensis Primarij* . . . (Bologna, 1665), published with works by C. Fracassati, which Oldenburg described. For Oldenburg's knowledge of the book, see *Correspondence* III, 476, 17 September 1667, letter to Boyle.
3. *Correspondence* IV, 269–72, 22 March 1667/8.
4. *De viscerum structura exercitatio anatomica Marcello Malpighii* (Bologna, 1666, London, 1669), reviewed in *Philosophical Transactions* no. 44 (15 February 1668/9) 888–91.
5. *Dissertatio epistolica de bombyce* (London, 1669); see *Correspondence* V, 323.
6. *Correspondence* V, 457–60, 25 March 1669. It is worth recounting the action in some detail to illustrate the care with which Oldenburg and the Society handled such matters.
7. *Correspondence* V–IX, *passim*. For background, see Marta Cavazza,'Bologna and the Royal Society in the Seventeenth Century', *Notes & Records*, 35 (1980),105–23.
8. *Correspondence* VIII, 308–9, 22 October 1671.
9. *Ibid.*, 398–401, 14 December 1671; the meeting was on 7 December 1671. Oldenburg sent off Grew's work on 26 April 1672, *Correspondence* IX, 50. Malpighi did have it translated into Latin; see *Correspondence* IX, 263–4.
10. 18 January 1671/2, *Correspondence* VIII, 491–2.
11. For Malpighi, *Correspondence* IX, 263–4.
12. *Ibid.*, IX, 473, 18 February 1672/3.
13. For Malpighi's reaction, *Correspondence* XII, 178–9, with Crawford'sletter of 14 January 1676/7. For Crawford, see note 14 below.
14. John Dodington (d. 1673), rather curiously known in Venice as Dorrington, had accompanied the new English Ambassador, Viscount Fauconberg, to Venice as his Secretary in 1670; he was dismissed from his post for rowdiness on the journey but was shortly appointed English resident in Venice until 1672. For his correspondence with Oldenburg and his activity in sending him news, see *Correspondence* VI–IX, *passim*, esp.

IX, 157 note 5. A little over a year after he left Venice for London he died of a carouse in the Bear Tavern on Leadenhall Street. James Crawford accompanied the English legate to Venice in April 1674; he himself had no official post but, like Dodington before him, acted as a postal agent for Oldenburg, and sent him news of Italian natural philosophers. See *Correspondence* X, 297 note 1 and XI–XIII, *passim*.

15. For this information see Grew, *The Anatomy of Plants* (1682), Preface, also used by Michael Hunter for 'Early Problems in Professionalizing Scientific Research' in *Establishing the New Science*, 261–78.

16. *Correspondence* XIII, 431–3.

17. There is a copy in Royal Society MS. G 1, no. 34; for the failure of the letter to arrive, see Howard B. Adelmann, *Marcello Malpighi and the Evolution of Embryology* (5 Vols., Ithaca, N.Y., 1966), I, 699.

18. *Correspondence* V, 409–10; Skippon did not mention Lister's name.

19. No. 50 (16 August 1669), 1011–16; the original is in Royal Society MS. Classified Papers XV (i), no. 23.

20. *Correspondence* VII, 104–5, 9 August 1670.

21. *Ibid.*, 52–4, 4 July 1670.

22. *Ibid.*, 355–9, printed in *Philosophical Transactions* no. 72 (19 June 1671), 2175. It was sent to Oldenburg on 10 January 1670/71.

23. Lister's first direct communication with Oldenburg is dated 9 August 1670; perhaps surprisingly his address for Oldenburg was 'to be left with Mr Martin Bookseller at the Bell neer Temple-Barr London' at the same time requesting Oldenburg 'to send me how I may direct a Letter to you'. *Correspondence* VII, 104–5.

24. For Hulse's letter and Lister's claim, see *Philosophical Transactions* no. 65 (14 November 1670) 2103–4.

25. See *Correspondence* VII, esp. 351–9.

26. See *ibid.*, 388–90; it was published in *Philosophical Transactions* no. 68 (20 February 1670/1), 2072–4. In 1676 Tonge was to be associated with the notorious Titus Oates whose infamous insistence upon the existence of a 'Popish Plot' to replace Charles II by a Catholic monarch (i.e. the future James II) Tonge helped to foment.

27. *Correspondence* VII, 23–8, 6 June 1670; it appears from Oldenburg's letters to Willughby that he had written about this earlier, *ibid.*, 3, 3 May 1670.

28. *Correspondence* VII, 30, 7 June 1670.

29. His first discussion of fossils, in which he openly dissented from Steno's biological view, was in 1671: *Correspondence* VIII, 214–15, 25 August 1671. His first 'Table' (classification) of fossil snails was published in *Philosophical Transactions* no. 105 (10 July 1674) 96–9 in Latin, sent with a letter of 23 October 1672, *Correspondence* IX, 292–3. The printed version, somewhat emended, was sent on 12 March 1673/4; for this see *Correspondence* X, 498–502.

30. *Correspondence* IX, 310, 31 October 1672.

31. For Beaumont (d. 1730/1) see *Correspondence* XII, 238–45, 341–50, both published in 1676.

32. *Correspondence* IX, 658–63, 21 May 1673, mostly printed in *Philosophical Transactions* no. 95 (23 June 1673), 6060–5. For the reference to the 'figure of the excrements' see *Correspondence* IX, 660.

33. On 12 June 1673, *Correspondence* X, 26.
34. Cf. *Correspondence* X, 56–7 (Lister 24 June 1673), 82–3; for Needham, sent by Oldenburg 15 July 1673, 95; for Lister's opinion of Grew, 301 ff., 25 October 1673.
35. It began with a letter from Lister on 26 July 1673 (above, note 34) which Oldenburg sent to Grew on 30 August. Grew replied at length on 13 September, *Correspondence* X, 201–10. The controversy continued into the spring of 1674.
36. Lister introduced Jessup by sending with his own letter of 28 June 1673 one from Jessup himself of 25 June, *Correspondence* X, 69 amd 70.
37. The exchange lasted for about a year before Jessup gave up, *ibid.* X, *passim*, esp. XI, 54.
38. Duhamel had been appointed Secretary to the Académie royale des Sciences at its foundation in 1666 but did not exercise that function regularly until 1673. He later wrote a history of the Académie.
39. *Correspondence* IX, 612–14, 21 April 1673, *Philosophical Transactions* no. 94 (19 May 1673), 6039.
40. *Correspondence* X, 642.
41. *Philosophical Transactions* no. 97 (6 October 1673), 6115, *Correspondence* X, 151, August 1673; see also *ibid.*, 137, Oldenburg's report of the King's action, letter to Lord Herbert of 11 August 1673. The third Anglo-Dutch conflict had begun in March 1672 and was to last for two years.
42. *Correspondence* X, 177, 2 September 1673.
43. For the request, *ibid.*, 220, 18 September 1673; for Lister's having a sample of his rival's fluid, *ibid.*, 221, also 18 September.
44. *Ibid.*, 302, 23 October 1673, 364, 368, 18 November 1673, and 437, 17 January 1673/4.
45. For example, *Correspondence* XI, 12, Sir George Croke to Oldenburg, 19 May 1674. Croke was High Sheriff of Oxfordshire, F.R.S. 1676, with an interest in medicine and much else. See *Correspondence* X, 250 and XI, 17.
46. Flamsteed's 'History of his own Life' was published by Francis Baily in *An Account of the Revd. John Flamsteed* (London, 1835, Dawson reprint, 1966), pp. 7–105, the relevant portion (to 1677) being pp. 7–43; for his first relations with Oldenburg, see pp. 23–24, 28.
47. The letter, without the calculations, was printed in Rigaud, II, 76–90, where the date is given as 24 November; however Flamsteed in 1707 remembered it as dated 4 November and that is the date ascribed to it by Oldenburg (below, note 48).
48. *Correspondence* VI, 427–8, 14 January 1669/70.
49. *Philosophical Transactions* no. 55 (17 January 1669/70) 1098–1112.
50. *Correspondence* VI, 468–70, 7 February 1669/70.
51. *Ibid.*, 513–15, 26 February 1669/70.
52. *Correspondence* VII, 265, 16 November 1670 and IX, 158–9, 15 July 1672.
53. *Correspondence* VI, 515–16, and VII, 21, 30 May 1670, and 266, 16 November 1670, etc.
54. *Correspondence* VIII, 525–6, 5 February 1671/2.
55. *Ibid.*, (beginning 10 August 1671, 193–4), through XIII; the exchange was initiated by Cassini.

56. e.g. *Correspondence* X, 319 (Cassini to Flamsteed) and 107 (Flamsteed to Cassini).
57. *Ibid.*, 321 note 4 (Flamsteed to Towneley. November 1673 and 369, 3 December 1673.
58. See, *inter alia*, Francis Willmoth, *Sir Jonas Moore* (Woodbridge, 1993), esp. 176–86 and Baily (note 46 above), 36–43.
59. *Correspondence* XI, 429, 28 July 1675, XII, 193, February 1675–6,198, 22 February 1675/6 (from Cassini).
60. Picard's book was first described in some detail by Vernon on 19 January 1669/70, *Correspondence* VI, 422–35 in reply to a query from Oldenburg. For an English summary sent by Vernon at the end of 1671 and the beginning of 1672, see *Correspondence* VIII, 431–7, 478–81, 497–500.
61. For the conveyance of books from Hevelius to Boulliaud and vice versa all via Oldenburg, see esp. *Correspondence* IX, 48–9 *et seqq*, and XI, 126; for the connection with Flamsteed, see VIII, 207 and 209 note 1, and XII, 237–8.
62. Boulliaud's original letter, probably sent on 5 April 1676 is lost, cf. *Correspondence* XII, 235–8; cf. Wallis's letter of 27 May 1676, *ibid.*, 302–4, of which Oldenburg sent a transcript on 8 June, *ibid.*, 330–1.
63. *Correspondence* VIII, 143–5, 8 July 1671.
64. See above, Chapter 5.
65. *Correspondence* VIII, 191–3, 10 August 1671.
66. *ibid.*, 281–4, 10 October 1671.
67. Cf. Oldenburg's reply, *ibid.*, 412–16.
68. Cf. his letters of 3 January 1671/2, *ibid.*, 451–60 and 30 March 1672, 626–8. For his comments on Newton, see *Correspondence* IX, *passim* and Chapter 5 above.
69. *Correspondence* XIII, 421–3, 26 July 1670.
70. *Correspondence* VII, 66–7, 13 July 1670.
71. *Ibid.*, 107–14, 10 August 1670.
72. *Ibid.*, 162–70, 18 September 1670.
73. *Ibid.*, 308–13, 8 December 1670.
74. This was sent to Oldenburg on 1 March 1670/71 and presented to the Royal Society on 23 March 1670/71.
75. It was sent by Leibniz on 21 April 1671 and its reception acknowledged by Oldenburg on 12 June 1671, after it had been shown to the Society, and then both parts were sent to the printer; *Correspondence* VIII, 97–104.
76. On 10 August 1671, *Correspondence* VIII, 192.
77. At the meeting on 11 May 1671.
78. *Correspondence* VII, 562–4, 7 April 1671 and VIII, 103, 12 June 1671.
79. *Correspondence* VIII, *passim*.
80. Cf. *Leibniz A Paris (1672–76)*, t.I, *Les Sciences*, Studia Leibnitiana Supplementa, Wiesbaden, 1978), esp. H.J.M. Bos, 'The Influence of Huygens on the Formulation of Leibniz's Ideas', 59–68 and M.B. Hall, 'Leibniz and the Royal Society 1670–76', *ibid.*, 171–82.
81. As Oldenburg told Leibniz on 30 January 1672/3, *Correspondence* IX, 431–32; the appointment was for the next day.

82. *Ibid.*, 438–47, dated by Oldenburg 3 February, the conversation at Boyle's house having been the day before.
83. *Ibid.*, 525.
84. Leibniz wrote 'a grateful letter' (his words) to Oldenburg soon after his arrival in Paris, *ibid.*, 488–96, 26 February 1672/3, sending also respects to Boyle.
85. After hearing news of his election, Leibniz expressed his gratitude to Oldenburg (*Correspondence* IX, 599, 16 April 1673); in his immediate reply of 8 May Oldenburg in instructing Leibniz to write directly to the Royal Society, also requested him to remember to send his arithmetical machine for inspection by the Society, *ibid.*, 644–5.
86. Cf. his reported remarks at the meeting of 5 February 1672/3 and 2 March 1672/3. What Leibniz made of it is indicated in his letter of 26 February (note 84, above), 493–4.
87. Leibniz's last letter in 1673 is dated 14 May, *Correspondence* IX, 648–52 to which Oldenburg replied on 26 May, *ibid.*, 666–9.
88. *Correspondence* XI 42–6, 5 July 1674.
89. On 15 July according to Oldenburg's meticulous endorsement on Leibniz's letter received three days earlier.
90. Cf. *Correspondence* XI, 141, 8 December 1674 and 165–74, 12 April 1675 with a Latin translation of what Collins wrote for Oldenburg to transmit, 253–62. More of what Leibniz received can be deduced from his letter of 20 March 1674/5, *ibid.*, 236–43.
91. *Ibid.*, 301–6, 10 May 1675.
92. The letter of 20 March 1674/5, note 90, above.
93. Cf. Oldenburg's letter of late May 1675, based on Collins's account of 25 May; *Correspondence* XI, 325–6 and 323–4, respectively. Tschirnhaus replied in early July, *ibid.*, 397–400, shortly before going to Paris where he worked closely with Leibniz. In subsequent correspondence Collins hardly distinguished between them.
94. This began in 1672 (*Correspondence* IX, 210, 13 August) and was to continue to occupy his correspondence, with Vogel begging Oldenburg to communicate his problems to English orientalists—as he did, *Correspondence* IX–XI, *passim*.
95. Edward Pococke (1604–91) was a very learned orientalist who had spent six years in Aleppo in the 1630s,and was from 1648 Professor of Hebrew at Oxford. William Seaman (1606–80) had lived in Constantinople during the 1630s and was a specialist in Turkish. Paul Rycaut (1628–1700) served as Secretary to the ambassador at Constantinople before going to Smyrna as Consul of the Levant Company when he was elected F.R.S. (1666). See *Correspondence* III, 343–4, 16 February 1666/7 and 602–7, 23 November 1667, also XII, 130, 261 (1676).
96. For the Society's interest, *Correspondence* V, 403 note 3; a brief account was published in *Philosophical Transactions* no. 21 (21 January 1666/7), 375; see also *Correspondence* XIII, 223. At this time Boyle was interested in substances phosphorescent through decay. For a general review, primarily concerned with the Royal Society's attitude towards experiment, see J.V. Golinski, 'A Noble Spectacle. Phosphorus and the Public Culture of Science in the early Royal Society', ISIS, 80 (1989), 11–39, largely based on *Correspondence*.

97. For Findekeller, *Correspondence* XII, 273 note; the first mention of Balduin's 'phosphorus' (see below) is 363, 6 July 1676; he first wrote to Oldenburg on 1 September 1676, *Correspondence* XIII, 65–6.
98. For Mentzel, see *Correspondence* V, 289–92, 1 January 1668/9; he was prompted by meeting Dury.
99. *Correspondence* XII, 363.
100. On 1 September, *Correspondence* XIII, 65–6.
101. Hooke's diary, 4 January 1676/7.
102. On Oldenburg's letter of 29 October 1676, *Correspondence* XIII, 126–7; the annotation is transcribed in note 2, 127.
103. *Correspondence* XIII, 131–2, 11 November 1676.
104. *Ibid.*, 9 February 1676/7, 208–9, to Major and 209–10 to Kirchmeyer.
105. *Ibid.*, 214–15, 16 February 1676/7.
106. *Ibid.*, 220–21, 22 February 1676/7.
107. For the rival claims of Kunckel and Kraft, *ibid.*, 222 from Kirchmeyer, ?25 February 1676/7.
108. *Ibid.*, 271, 3 May 1677.
109. *Ibid.*, 233–4, 25 March 1677 and Oldenburg's reply, 261–663, 30 April 1677; for his planned backing of Kirchmeyer with a draft petition to the King, 263 note 1. See also 278–80, 8 May 1677.
110. *Ibid.*, 322, 12 July 1677.
111. Boyle recounted his encounter with Kraft in ' A Short Memorial of some Observations made upon an Artificial Substance, that shines without any precedent *Illustration*' appended to R. Hooke, *Lectures and Collections* (London, 1678) which begins with Hooke's own 'Cometa'.
112. De Graaf had opened the exchange at the encouragement of Paisen who had passed through England and the Netherlands on his way home to Hamburg, *Correspondence* IV, 523–4, 10 July 1668 and 527–8, *c.* 11 July. For the sending news of Swammerdam's work to Italy, see *Correspondence* VI, 467, 7 February 1669/70.
113. *Correspondence* VIII, 396, 10 December 1671 and 397 note.
114. With his first letter (note 113); Oldenburg thanked him on 24 April, *Correspondence* IX, 40–2.
115. Sent with a letter to the Society on 4 June 1672.
116. Cf. Oldenburg's acknowledgement, 19 December 1672, *Correspondence* IX, 367–9.
117. *Correspondence* VIII, 617–18.
118. See below.
119. *Correspondence* IX, 104–6.
120. *Correspondence* XIII, 343–4.
121. *Correspondence* IX, 105.
122. On 2 July, *ibid.*, 137–8.
123. *Correspondence* XI, 346–9, 16 June 1675.
124. *Correspondence* XII, 247–8, 11 April 1676.
125. 18 April 1673, *Correspondence* IX, 602–3, in reply to a letter from Oldenburg of 15 March, of which no trace survived.

126. *Ibid.*, 653–4, 15 May 1673.
127. For a summary, *Correspondence* IX, 604; the letters are to be found in full in *The Collected Letters of Antoni van Leeuwenhoek* (Amsterdam, 1939–); the letters to Oldenburg are in volumes 1 and 2.
128. The dates of Oldenburg's letters to Leeuwenhoek are mostly mentioned in the replies; cf. Leeuwenhoek's letter of 23 May 1674 summarised in *Correspondence* XI, 23.
129. Cf. Constantijn Huygens's letter of 2 February 1673/4, *Correspondence* X, 456–9 (in French, but he knew English well).

Chapter 9

1. See Chapter 4 above, note 88.
2. *Correspondence* II, 78.
3. Birch, *Boyle* VI, 487.
4. *Ibid.*, 495.
5. *Ibid.*, 25 August 1664.
6. *Ibid.*, 499.
7. *Correspondence* II, 262, 20 *October* 1664.
8. *Ibid.*, 383–7.
9. *Ibid.*, 610, 16 November 1665.
10. *Correspondence* III, 9, 8 January 1665/6.
11. *Ibid.*, 32. De Son's watches seem never to have worked; at least nothing was heard of them ten years later when Huygens invented his watch movement.
12. *Ibid.*, 9, 8 January 1665/6.
13. *Ibid.*, 15, 15 January 1665/6.
14. *Ibid.*, 32, 32–3, January 1665/6.
15. Thomas Sprat, *History of the Royal Society* (London, 1667) pp. 173–9, 147.
16. For Hooke's letter of *c.* 20 February 1666/7, see *Correspondence* III, 347–9. Oldenburg's translation was sent on 27 February, *ibid.*, 352–4.
17. *Correspondence* V, 115, 28 October 1668.
18. *Correspondence* IV, 103, 7 January; the meeting had been on 2 January.
19. *Ibid.*, 301, 4 April 1668.
20. *Ibid.*, 318, 14 April 1668.
21. *Correspondence* V, 539, 10 May 1668.
22. *Ibid.*, 557, 19 May 1669.
23. *Ibid.*, 583, 31 May 1669.
24. *Ibid.*, 45 (Huygens, 16 June 1669), 93 (Oldenburg, 5 July), 164 (Huygens, 31 July), 223 (Oldenburg, 6 September), 305 (Oldenburg 1 November 1669).
25. The diary is extremely cryptic for its first few months, nor is there any reason given for its beginning then (if it did) and remains laconic even when more detailed. Hooke sometimes wrote up the entries at the end of a week rather than daily, which explains some discrepancies of detail. In the following pages the references are by date.
26. *Correspondence* X, 432–3, 12 January 1673/4.

27. *Ibid.*, 430, 9 January 1673/4 and 532, 23 March 1673/4.
28. *Correspondence* XI, 3. Hooke was attempting to observe annual parallax.
29. *Correspondence* XI, 25, 25 May 1674.
30. He wrote, mostly briefly, twice in September and once in October. See *ibid.*, 88–89, 90, 105 and 278–9 when he sent several books.
31. Cf. Chapter 7; for differing but complementary and detailed views by modern historians see A. Rupert Hall and Marie Boas Hall, 'Why Blame Oldenburg?', *Isis*, 53, 1962, 482–91 and Rob Iliffe, 'In the Warehouse: Privacy, Property and Priority in the Early Royal Society', *History of Science*, 30, 1992, 29–68.
32. Cf. Chapter 7, note 57.
33. Although there is no mention of it in his Diary, Hooke then drew up a draft for a patent (undated) in the form of an order to the Attorney General and Solicitor General for the preparation of a bill which would grant him a patent.
34. 2 February 1674/5, see Chapter 7, note 53.
35. See Chapter 7, note 63.
36. Unfortunately Hooke was at this time very unmethodical about his diary and there are no entries for the fortnight 15 through 28 April.
37. See *Correspondence* XI, 360 *et seqq* where Hooke's activities are mentioned and 509, 27 September. His Diary describes these things under the date of Sunday, 13th June, some days after the fact.
38. For Hevelius's feelings, see *Correspondence* XI, 468–75 and XII, 216, 259.
39. It was first mentioned in a letter to John Aubrey at the end of 1675, when Hooke believed that the Royal Society was 'too much enslaved to foreigners'. See R.T. Gunther, *Early Science in Oxford*, VII, 435. The progress of his club is detailed in Hooke's diary.
40. Facsimile in Gunther, op. cit., VIII, 'The Cutlerian Lectures'. See Diary entry 3 September, 'Writ against Oldenburg'.
41. For Moray's letters to Huygens, see *Oeuvres Complètes*, vols. 5 and 6.
42. *Correspondence* XII, 40, 8 November 1675.
43. *Ibid.*, 12–13, 11 October 1675 and, for Huygens to Oldenburg and Brouncker, 21 October 1675, 20–1 and 23–4, respectively.
44. *Philosophical Transactions* X, no. 118 (25 October 1675), 4402.
45. See Diary entries for, e.g., 11 November 1675 and 20 January, 1675/6.
46. See Evelyn's diary entry for 29 February 1675/6.
47. Diary entries for 24 June, 29 June, 2, 15, 20 and 31 July–10 August, *et seqq.*
48. Facsimile in Gunther (note 40 above) Tract no. 4.
49. Under the date 12 October, but all the entries for the period 8–14 October were written on Sunday 14 October.
50. Diary entries for 20, 21 October and 2 November 1676.
51. One is printed in *Correspondence* XIII, 148–51.
52. *Ibid.*,147–8, 30 November 1676.
53. Diary entry for 30 November.
54. Diary entry for 2 December 1676.

Chapter 10

1. *Correspondence* IX, 275, and Hooke's Diary, 1672–7.
2. Jonathan Gibson, 'Significant Space in Manuscript Letters', *The Seventeenth Century*, 12, 1997, 1–9.
3. *Correspondence* III, 307.
4. *Correspondence* IV, 80.
5. See *Correspondence* VIII, xxvii.
6. Birch, *History*, meetings on 30 September 1667 and 2 and 11 January 1667/8.
7. *Correspondence* IV, 112–13.
8. *Ibid.*, 116–17, 120–1 and 145.
9. *Ibid.*, 129, 133, 167–8, 397–8, 424.
10. *Ibid.*, 437 and Council Minutes 19, 22 June and 13 July 1668.
11. *Ibid.*, 454–5, Wren to Oldenburg.
12. *Correspondence* III, 593.
13. *Ibid.*, 613.
14. *Correspondence* IV, 58.
15. *Notitiae Anglicae*, quoted from the 6th edition (London, 1673), I, 343.
16. *Correspondence* IV, 235.
17. From the draft in BL MSS. Add. 4441, f.27, printed in *Correspondence* xxiv–xxv.
18. See *Correspondence* IV, 59. Martin's receipt survives as a scrap of paper in the Royal Society archives. It reads, 'Rec. October 15th 1669 from Mr Oldenburgh Eighteen shillings for this Voll: of Transactions by me John Martyn.' I cannot now trace the source of the photocopy in my possession. There are said to be clearer details in a copy of the journal now in private possession but I have not been given any further information about them or it.
19. Collins to Wallis, 17 June 1668, Rigaud II, 515–16.
20. For details see Hall & Hall, *Notes & Records* 18, 94–103 and 23, 33–42. The latter contains the comments by Collins and Pell and detail's of Dora Katherina's family connections. She was baptised in 1657 at the Church of St Antholins, Budge Row, London, as recorded in the Church records. Her mother died seven years later, probably also in London. Before her marriage to Dury she had maintained an intimate correspondence with Lady Ranlelagh, of which her side is preserved in the Hartlib Papers. No trace of any later connection with Lady Ranelagh is now preserved. I have to thank Lynette Hunter for the reference to this correspondence. See also Turnbull, *Hartlib, Dury & Comenius*, esp. pp. 219, 273.
21. *Correspondence* V, 26, whose second clause is in Latin, and 28.
22. Sir John Wroth (d. 1671) of Blendon Hall, Bexley, Kent, first baronet; his wife was a cousin of Dora Katherina's and when he and his son Thomas travelled to France in 1669 Oldenburg provided him with letters of introduction; see *Correspondence* VI, 20–2 and 553.
23. *Correspondence* V, 562.
24. *Correspondence* IV, 222, 29 February 1667/8. It is a rough draft scribbled scribbled on the draft of a letter of the same date to Francis de le Boe Sylvius, Professor of Medicine at Leiden.

25. See *Correspondence* VII, 530–1 and X, 267–9, 312–13.
26. *Correspondence* XI, 22, 22 May 1674.
27. *Correspondence* IX, 310.
28. *Correspondence* V, 141. Colepresse had always called it 'your Transactions' when writing to Oldenburg.
29. See Marie Boas Hall, 'What happened to the Latin edition of Boyle's *History of Cold?' Notes & Records*, 17 (1962), 32–5.
30. Printed in *Correspondence* VIII from the original scrap of paper in Boyle Papers III, f.178. Oldenburg apparently mistakenly reckoned that £5 was owing to him, for the calculation has been corrected in pencil to £4. The poor rates for Pall Mall in 1671 are to be found in P.R.O. F 399, Film 155^3 item 6 p. 141.
31. Boyle Papers XXXVI, f. 88.
32. *Correspondence* XI, 151–2 The translation is unsigned but is ascribed to Oldenburg by the BL, which seems plausible.
33. Martin Lister, *A Journey to Paris in the year* 1698 (3rd ed., London, 1699), p. 81.
34. *Correspondence* IX, 162.
35. Birch, *History* gives a summary of Oldenburg's life under the date 30 November 1677, vol. 3, 355. True, Birch was often inaccurate, but one cannot assume that he was necessarily so here.
36. For what follows, see *Correspondence* IX, xxix, letter of Boyle to Moray mentioning the illness and Moray to Williamson, P.R.O. MS. SP 29/316, no. 184.
37. *Correspondence* IX, 457 and 458–9, both 9 February 1672/3.
38. *Correspondence* IX, 463–4, 16 February 1672/3.
39. *Ibid.*, 574, 8 April 1673; the enclosed translation of one of the letters (the others are only minuted) is printed, *ibid.*, 575–7.
40. *Correspondence* XII, 263.
41. Cf. Vol. 4 (Theology), ff.8–25, Vol. 9 (Philosophy), ff. 173–7, Vol. 14 (Science), pp.87–134, 151–68, 185, and perhaps others.
42. All this is revealed in Wallis's letter of 15 October, *Correspondence* XI, 108–9; the Council had decided to ask him to read on gravity at the meeting on 29 September.
43. *Correspondence* XII, 364, 6 July 1676.
44. *Ibid.*, 327–8, 332, 364, and XIII, 29, all written in the summer of 1676.
45. *Correspondence* XI, 378–80, 1 July 1675 and 381, note 5.
46. *Ibid.*, XI, 413, 22 July 1675.
47. *Ibid.*, 439, Papin to Huygens, 31 July 1675. It is not clear why or when he gave up this post.
48. *Ibid.*, 495.
49. Like Hooke, Papin had devised a new and more efficient airpump, with which he and Boyle pursued pneumatic experiments until 1680; these Papin was allowed to write up for publication in Latin as the second continuation of *New Experiments Physico-Mechanical . . .* ,the Latin version being published in 1680 at London while the English version was published at Oxford in 1682.
50. *Correspondence* XII, *passim*. Most of Oldenburg's letters to Williamson have not survived but their existence can be inferred.

51. The licence is printed in *Correspondence* XII, 265.
52. *Ibid.*, 254 and 255 note 1 to Williamson, 18 April 1676 and 263–4, 29 April 1676.
53. *Correspondence* X, XI, XII and XIII, *passim*.
54. See the elder Huygens's letter to Oldenburg of 2 February 1673–4, *Correspondence* XI, 233–4, 19 March 1674/5, enclosing a letter from his son, dated 7–17 August 1674, *ibid.*, 235–6, the latter in English presumably translated by the father.
55. *Correspondence* XIII, 184–88, 16 January 1676/7.
56. *Correspondence* XII, 57–8. The book (in Dutch) was by Hermannus Busschof (Amsterdam, 1675); the Society was responsible for suggesting its translation as *Two Treatises* (London, 1676). Moxa was to be burned on the affected limb.
57. *Correspondence* XII, 145, XIII, 71–2 (5 September 1676) and 184–8, 16 January 1676/7. It was in widespread use.
58. *Correspondence* XI, 22, 22 May 1674.
59. *Correspondence* XIII, 79.
60. *Ibid.*, 138–9, 17 November 1676 (to Boulliaud) and 140, 17 November (to Huet).
61. As recorded in the printed House of Commons Journal and reproduced by the Huguenot Society of London, 18, 1911, 117–18. There were, curiously, two separate Bills, introduced three days apart and passed six days apart. I have to thank Dr P.D. Buchanan for finding the reprint.
62. *Correspondence* XIII, 334, 9 August 1677.
63. *Ibid.*, 335, 9 August 1677.

Chapter 11

1. Birch, *Boyle*, VI, 531, letter of 1 September 1677, reprinted in R.E.W. Maddison, *The Life of the Honourable Robert Boyle* (London, 1969), p. 136. The letter records the fact that Boyle paid for Oldenburg's funeral.
2. 17/27 September 1677; see Leibniz, *Sämtlichten Schriften, Briefwechsel*, I, ii, no. 269, pp. 293–4.
3. H. Adelmann, ed. *The Correspondence of Marcello Malpighi* (Ithaca, N.Y., 1975) pp. 759–60.
4. To Richard Towneley, 13 February 1679/80, Royal Society MS. 143, no. 45, printed in *The Correspondence of John Flamsteed*, ed. E.G. Forbes, Lesley Murdin and Frances Willmoth (Bristol and Philadelphia, 1995), I, 734.
5. Cf. *Correspondence* II, 575 from Moray, 19 October 1665.
6. Register of Beckley Parish Church now in the Kent County Archives.
7. P.R.O. Prob. 4/11823. I have to thank Dr P.D. Buchanan for finding this, and the P.R.O. for allowing a copy to be made. Unfortunately the surviving document is very faint with age and cannot all be made out clearly.
8. As mentioned in the will of Mrs Margaret Lowden; see below Appendix, note 7.
9. It is not known what happened to them after the death of her daughter. Oldenburg's private library later passed to the Earl of Anglesey, whose books were to be sold by auction, for which the catalogue still exists in the British Library.
10. Burial Register of St Martins in the Fields, in Westminster Archives.

11. Quoted in Hall & Hall, *Notes & Records*, 18, p. 99, from the original in Royal Society Boyle Papers vol. 40, ff. 80–1.
12. Margaret Lowden's appointment as Administratrix is in P.R.O. MS. PROB 6/52 1677, p.105. For more on Mrs Lowden, see below, Appendix.
13. See Hooke's diary for 18 October 1677.
14. The draft receipt dated 1678 is quoted in the article cited in note 11 above from the original in the Royal Society archives.
15. BL MS Sloane 1942, f.1; that received by Huygens is printed in his *Oeuvres Complètes*, VIII, 66.
16. *Correspondence* XIII, 322, to Leibniz, 12 July 1677.

Appendix

1. These include Husting Roll of Land Conveyance 1671, no. 342 (2), which shows that the Lowdens were mortgaging their property to pay Francis Lowden's debts, the 1689 Poll Tax Record (Corporation of London Record Office) which shows Mrs Lowden as living in the Precinct of Barrs still near Goodman Fields, and Mrs Lowden's will, dated 30 November 1710, Prerogative Court Canterbury, PROB 11/529, sig 193. When she was naturalized in 1677 it was stated that she had been born at The Hague, which is confirmed by her statement in a petition to the King, P.R.O. MS. SP29/281A, 107, where she asked that sequestration of her property in Harrow Alley, consequent on new laws relating to aliens, be set aside. For her father's travels, see Calendar of State Papers Domestic. He was related to prominent citizens of Guernsey, for which see Edith F. Carey in *Report Transactions of the Guernsey Society of Natural Science and Local Research*, 1913.
2. Hooke's diary under date 15 January 1677/8.
3. Pell Papers, BL MSS. Add. 4365, f.7, 'From Cassel this 28 of May/7 June 1678'. By 'education' Dury must have meant what would later be called 'upbringing'.
4. *Ibid.*, f.25.
5. See Millicent Rose, *The East End of London* (London,1951).
6. Corporation of London Record Office for the Ward of Portsoken (at the extreme eastern edge of the City and including St Botolphs without Aldgate, where Harrow Alley lay). Mrs Lowden paid 11 shillings.
7. See note 1, above; the will is dated 30 November 1710 Probate 4 October 1712, suggesting that she died in the latter year.
8. In 1717 Robert Lowden Smith, one of Margaret Lowden's grandsons, obtained Letters of Administration for Sophia Oldenburg, empowering him to dispose of her goods and chattels (P.R.O. PROB 6/92); he was one of his grandmother's executors.
9. His indenture is in Corporation of London Records Office, CF1/133178; see also MCFF/228, Marriage Assessment no. 6 p. 8, s.v. 2 August 1698.
10. Apothecary's Company Freedom Book, Guildhall Library.
11. Edward Hasted, *History and Topographical Survey of the County of Kent* (Canterbury, 1778), I, 40.
12. See Mrs Lowden's will; by 1711 he owed her £80, which Mrs Lowden assigned to be given to Sophia when he repaid it (which he probably never did).

13. See John S.Farmer, *Regimental Records of the British Army* (1901) and C. Dalton, *English Army Lists and Commissions Register 1677–1714* (1904), vol. 6, 1707–14.
14. *Compil'd and Collected by Mr. A. Boyer*, xxviii, September 1724, pp. 316–18.

Brief bibliography

Althaus, Friedrich, Munich *Beitrage zur Allgemeinen Zeitung*, 1888–89, esp. vol. 212, 2 August 1889, pp. 1–3, also 1888, nos. 229–33, 1889, nos. 212–14.

Birch, Thomas, *The History of the Royal Society*, 4 vols. (London 1756–7). Oldenburg's activities are chronicled in the first 3 volumes. There are modern reprints.

Birch, Thomas, *The Life and Works of the Honourable Robert Boyle* 6 vols. (London, 1772). There are modern reprints.

Bray, William, ed. *The Diary of John Evelyn*, 2 vols. (London 1907).

Brown, Harcourt, *Scientific Organzations in Seventeenth Century France (1620–50)* (Baltimore, 1934).

Hall, A. Rupert and Hall, Marie Boas, eds. *The Correspondence of Henry Oldenburg*, 13 Vols., (Madison and London etc.). And 'Additions and Corrections to *The Correspondence of Henry Oldenburg*', *Notes and Records of the Royal Society*, 44 (1990), 143–50.

——, 'Some hitherto unknown Facts about Henry Oldenburg', *Notes and Records of the Royal Society*, 18 (1962), 94–103 and 'Further Notes on Henry Oldenburg', *ibid.*, 23 (1968), 33–42.

Hooke, Robert, *The Diary of Robert Hooke M.A.,M.D.,F.R.S.*, ed. H.W. Robinson and Walter Adams (London, 1935).

Hunter, Michael, *Establishing the New Science. The Experience of the Royal Society* (Woodbridge, 1984).

——, *The Royal Society and its Fellows* 1660–1700, British Society for the History of Science Monographs, 1982.

——, ed. *Letters and Papers of Robert Boyle: A Guide to the Manuscripts and Microfilm* (University Publications of America, 1992).

Hutton, Sarah, 'Lady Ranelagh', forthcoming in *The New Dictionary of National Biography*.

Huygens, Christiaan, *Oeuvres Complètes* (The Hague, 1888–1950).

Iliffe, Rob, 'Foreign Bodies: Travel, Empire, and the Early Royal Society of London' *Canadian Journal of History* 24 (1999), 23–50.

——, '"In the Warehouse": Privacy, Property and Priority in the Early Royal Society' *History of Science* 30 (1992), 29–68.

Jones, J.R., *Anglo-Dutch Wars of the Seventeenth Century* (London, 1996).

Kronick, David A., 'Notes on the Printing History of the early Philosophical Transactions', *Libraries and Culture: A Journal of Library History*, 25 (1990) 243–68.

Lillywhite, Bryant, *London Coffee Houses* (London, 1963).

Maddison, R.E.W., *The Life of the Honourable Robert Boyle F.R.S.* (London, 1969).

Mours, Samuel, *Le Protestantisme en France au XVIIe Siècle (1598–1685)* (Paris, 1967).

Oldenburg, Henry, ed. *The Philosophical Transactions* (London and Oxford) 12 vols., 1665–77.

Pepys, Samuel, *The Diary of Samuel Pepys* ed. Robert Latham and William Matthews, 11 vols., (London, 1970–83).

[Rigaud,S.J.], *Correspondence of Scientific Men of the Seventeenth Century . . . in the Collection of . . . the Earl of Macclesfield*, 2 vols. (Oxford, 1841).

Stubbs, Mayling, 'John Beale, Philosophical Gardener of Herefordshire. Part I. Prelude to the Royal Society (1608–1663)' *Annals of Science*, 39 (1982), 463–89 and 'Part II. The Improvement of Agriculture and Trade in the Royal Society (1663–1683', *ibid.*, 46 (1989), 323–63.

Turnbull, G.H., *Hartlib, Dury and Comenius* (Liverpool and London, 1947).

Turnbull, H.W., Scott, J.F., Hall, A.Rupert, Tilling, Laura, *The Correspondence of Isaac Newton* 6 vols., (Cambridge, 1959–77). For correspondence with and about Oldenburg see Vol. I, 1661–75 and II, 1676–87.

Venning, Timothy, *Cromwellian Foreign Policy* (London and New York, 1995).

Wood, Anthony A, *Athenae Oxoniensis*, ed. Philip Bliss, Part II (London, 1820).

Table 1 Number of surviving letters 1663–77

	From Oldenburg	*To Oldenburg*	*Total*
1663	20	31	51
1664	36	26	62
1665	49	66	115
1666	53	63	116
1667	73	75	148
1668	137	189	326
1669	147	150	297
1670	103	129	232
1671	118	152	270
1672	120	145	265
1673	146	151	297
1674	95	81	176
1675	108	123	230
1676	96	132	228
1677	44	54	98
			2911

Note: Oldenburg certainly wrote more letters than this suggests; those listed here represent letters which his recipients kept, those of which he prepared drafts or copies, and those of which he preserved minutes. More letters were presumably received by him than are listed here, but certainly the preponderance are preserved.

Table 2 Number of correspondents 1663–77 per year

	Domestic	*Foreign*	*Total*
1663	8	13	21
1664	7	9	16
1665	13	11	24
1666	15	12	27
1667	19	26	45
1668	39	39	68
1669	31	39	70
1670	23	46	69
1671	22	35	57
1672	20	36	56
1673	28	42	70
1674	16	30	46
1675	18	26	44
1676	22	44	66
1677	15	28	43

Note: 1674 was a year in which the Fellows of the Royal Society were notably inactive; hence presumably the decline in the number of domestic correspondents. In 1676 many Germans became correspondents over the discovery of phosphorus.

Index

Nigel Cross –
Harold Love –
E. Junno